T0245110

THE DEVELOPMENT OF ATMOSPHERIC GENERAL CIRCULATION MODELS

Complexity, Synthesis, and Computation

Over the last fifty years, models that predict the state of the atmosphere have evolved from conceptual frameworks to advanced computational tools for short- and medium-range weather prediction and climate simulation. This book presents a comprehensive discussion of general circulation models of the atmosphere – covering their historical and contemporary development, their societal context, and current efforts to integrate these models into wider Earth system models. Leading researchers provide unique perspectives on the scientific breakthroughs, overarching themes, critical applications, and future prospects for atmospheric general circulation models. Key interdisciplinary links to other subject areas such as chemistry, oceanography, and ecology are also highlighted.

This book is a core reference for academic researchers and professionals involved in atmospheric physics, meteorology, and climate science, and can be used as a resource for graduate-level courses in climate modeling and numerical weather prediction. Given the critical role that atmospheric general circulation models are playing in the intense public discourse on climate change, it is also a valuable resource for policymakers and all those concerned with the scientific basis for the ongoing public-policy debate.

Leo Donner received his Ph.D. in Geophysical Sciences from the University of Chicago in 1983. His research focuses on atmospheric general circulation modeling, especially the treatment of clouds and convective processes. He has served as the science chair of the Global Atmospheric Model Development Team at NOAA's Geophysical Fluid Dynamics Laboratory at Princeton University and a co-chair of the Atmospheric Model Working Group for the Community Atmosphere Model at NCAR in Boulder, Colorado. These models are often regarded as the two leading atmospheric general circulation models for climate studies in the United States. Leo Donner is also a lecturer in the Program in Atmospheric and Oceanic Sciences at Princeton University. He serves on the advisory board for the journal *Tellus* and has been an editor of the *Journal of Climate*.

Wayne Schubert received his Ph.D. in Atmospheric Science from UCLA in 1973, and then went on to join the faculty of the Department of Atmospheric Science at Colorado State University, where he presently teaches graduate-level atmospheric dynamics. His research covers tropical meteorology, atmospheric dynamics, and numerical weather prediction. Professor Schubert is a Fellow of the American Meteorological Society and has served as Co-Chief Editor of the *Journal of the Atmospheric Sciences* and as the AMS Publications Commissioner. He presently serves as an Editor of the *Journal of Advances in Modeling Earth Systems*.

Richard Somerville is Distinguished Professor Emeritus and Research Professor at Scripps Institution of Oceanography, University of California, San Diego. He received his Ph.D. in Meteorology from New York University in 1966 and has been a professor at Scripps since 1979. He is a theoretical meteorologist and an expert on climate change – he is a Coordinating Lead Author of the Fourth Assessment Report of the Intergovernmental Panel on Climate Change. Professor Somerville has received awards from the American Meteorological Society for his research and is a Fellow of the American Association for the Advancement of Science and of the American Meteorological Society.

THE DEVELOPMENT OF ATMOSPHERIC GENERAL CIRCULATION MODELS

Complexity, Synthesis, and Computation

Edited by

LEO DONNER
NOAA, Princeton University, New Jersey, USA

WAYNE SCHUBERT
Colorado State University, USA

RICHARD SOMERVILLE
Scripps Institution of Oceanography, University of California, San Diego, USA

CAMBRIDGE
UNIVERSITY PRESS

University Printing House, Cambridge CB2 8BS, United Kingdom

One Liberty Plaza, 20th Floor, New York, NY 10006, USA

477 Williamstown Road, Port Melbourne, VIC 3207, Australia

4843/24, 2nd Floor, Ansari Road, Daryaganj, Delhi - 110002, India

79 Anson Road, #06-04/06, Singapore 079906

Cambridge University Press is part of the University of Cambridge.

It furthers the University's mission by disseminating knowledge in the pursuit of education, learning and research at the highest international levels of excellence.

www.cambridge.org
Information on this title: www.cambridge.org/9781108445696

First published 2011
First paperback edition 2017

A catalogue record for this publication is available from the British Library

Library of Congress Cataloging in Publication data
The development of atmospheric general circulation models : complexity, synthesis, and computation / edited by Leo Donner, Richard Somerville, Wayne Schubert.
p. cm.
Includes bibliographical references and index.
ISBN 978-0-521-19006-0
1. Atmospheric circulation – Mathematical models. I. Donner, Leo Joseph, 1956–
II. Schubert, Wayne H. III. Somerville, Richard.
QC880.4.A8D48 2011
551.51′7–dc22
2010037289

ISBN 978-0-521-19006-0 Hardback
ISBN 978-1-108-44569-6 Paperback

Contents

Contributors

Alberto Arribas
United Kingdom Meteorological Office, Exeter, UK

Andrew R. Brown
United Kingdom Meteorological Office, Exeter, UK

Kirk Bryan
Princeton University, Princeton, New Jersey, USA

Michael J. P. Cullen
United Kingdom Meteorological Office, Exeter, UK

Robert E. Dickinson
Jackson School of Geosciences, University of Texas, Austin, USA

Leo Donner
Geophysical Fluid Dynamics Laboratory/NOAA, Princeton University Forrestal Campus, Princeton, New Jersey, USA

James Rodger Fleming
Science, Technology, and Society Program, Colby College, Waterville, Maine, USA

Isaac M. Held
Geophysical Fluid Dynamics Laboratory/NOAA, Princeton University, Princeton, New Jersey, USA

Timothy C. Johns
United Kingdom Meteorological Office, Exeter, UK

Akira Kasahara
National Center for Atmospheric Research, Boulder, Colorado, USA

Ngar-Cheung Lau
NOAA/Geophysical Fluid Dynamics Laboratory, Princeton University, Princeton, New Jersey, USA

Peter Lynch
Meteorology & Climate Centre,
School of Mathematical Sciences,
University College Dublin, Ireland

Gillian M. Martin
United Kingdom Meteorological
Office, Exeter, UK

Sean F. Milton
United Kingdom Meteorological
Office, Exeter, UK

David Randall
Department of Atmospheric Science,
Colorado State University, Fort Collins,
Colorado, USA

Wayne Schubert
Department of Atmospheric Science,
Colorado State University, Fort Collins,
Colorado, USA

Catherine A. Senior
United Kingdom Meteorological
Office, Exeter, UK

Richard C. J. Somerville
Scripps Institution of Oceanography
University of California
San Diego, USA

Warren M. Washington
National Center for
Atmospheric Research, Boulder,
Colorado, USA

Stuart Webster
United Kingdom Meteorological
Office, Exeter, UK

Keith D. Williams
United Kingdom Meteorological
Office, Exeter, UK

Foreword

It has become a commonplace to state that we are in the midst of performing a profound, albeit inadvertent, experiment on the Earth's climate. The need for a virtual Earth upon which we can perform experiments to determine how climate responds to emissions of carbon dioxide and other agents of climate change, to inform mitigation and adaptation decisions, is acknowledged as one of the great challenges to science. This collection of essays provides diverse perspectives on the evolution over time of models that have been developed for climate simulation, what they are capable of today, and what some of the challenges are for the future.

Numerical weather prediction provided the starting point for this evolution, with the realization immediately following the invention of computers that weather prediction was a perfect application for this new technology. Atmospheric models evolved rapidly with practical impetus from the needs of weather prediction. Atmospheric general circulation models then evolved from this base as it was realized that one could integrate these systems for arbitrarily long time intervals, gathering statistics of the weather thereby generated, producing simulation of our climate.

From this starting point, focused on the fluid dynamics of the atmosphere, climate models have steadily grown in comprehensiveness and complexity. Ocean models developed later than their atmospheric counterparts, and have presented many distinctive challenges for simulations, most fundamentally due to the sparseness of observations but also because the energy-containing scales of motion are smaller in the oceans, creating an especially challenging computational problem. Ocean models have taken on the added complexity of the cycling of carbon, nitrogen, oxygen, and other elements needed to understand how oceanic biology, and the uptake of carbon by the oceans, will evolve in the future. Land models have more recently increased in realism driven by the need to estimate how the uptake of carbon by land vegetation will evolve and how the hydrological cycle over land responds to climate change. Models of the cryosphere initially tackled the problem of modeling sea ice in the Arctic and Antarctic, and are now focusing on the challenging problem of incorporating the Greenland and Antarctic ice sheets into comprehensive models. Meanwhile atmospheric models have taken on the task of simulating atmospheric chemistry, both in the stratosphere, where a key interest is the distribution of ozone, and the troposphere, with a focus on the ability to simulate the species of relevance to air quality. Modeling particulates in the atmosphere, diverse in their composition, chemistry, and sources, has taken on a special urgency due to the potential for aerosol pollution to mask the effects of increasing greenhouse gases, and for the importance of aerosol/cloud interactions. Simultaneously with this evolution towards greater comprehensiveness, all of these components of our "Earth system models" are evolving towards finer resolution as fast as computational resources allow.

Our climate models have suggested and refined many of the basic tenets on which our understanding of climate change and variability are built. These include: the inability of internal variability to explain the observed twentieth century warming; the existence of a robust and strong water vapor feedback; the increase in precipitation in subpolar latitudes and decrease in the subtropics as the climate warms; and the manner in which the deep oceans take up heat and delay the full equilibration of climate to a perturbation for centuries, even millennia. Ongoing research is making progress on difficult questions such as how changes in the stratosphere affect the tropospheric circulation, how phenomena such as El Niño, Atlantic hurricanes, or the dynamic pole-to-pole circulation in the Atlantic Ocean will react to warming. There are other crucial questions, such as how the cloud cover will respond to warming, how the Greenland and Antarctic ice sheets will respond, how rainfall will change on the small scales of most relevance to agriculture and water resources, and how the uptake of carbon by the land surface will change, for which compelling projections remain elusive. The credibility of climate simulations is steadily increasing. By normal standards, this performance would be considered exemplary, but in our present predicament more is required.

Given the dramatic increase in complexity of climate models, and the pressures that exist to improve simulations rapidly, it should come as no surprise that there is uncertainty within the climate modeling community concerning the best course for the future. I mention a few of these concerns here, which the reader may wish to keep in mind when studying these essays.

The climate modeling enterprise today is dominated by a few large groups with substantial computational resources at their disposal. The rise in complexity has encouraged this trend towards "big science", in that a diverse range of expertise is clearly needed to incorporate many diverse processes into these models.

Additionally, both because of the heterogeneity of the land surface and because atmospheric and oceanic flows are turbulent with a wide range of relevant scales of motion, climate models typically improve as they move to finer spatial resolutions, encouraging the use of the largest computers in existence for these simulations. As a result of these tendencies, is the field of climate modeling becoming too monolithic, thereby stifling innovation? Should there be a greater effort to create frameworks within which a much larger community of researchers can contribute, perhaps patterned after the open-source movement in software development?

It is not uncommon for groups to organize their effort in a modular fashion, with sub-groups around the world contributing individual sub-components. But when one connects different sub-components together, there are invariably issues with the resulting simulation; otherwise the problem would effectively have been solved. Say, for example, that one finds that the simulation of the frequency or structure of El Niño events that emerges from the model is unrealistic. Is the problem in the treatment of small-scale atmospheric moist convection, the stratus cloud formulation, the mixing scheme used in the ocean model, the surface flux computation, the resolution of details of the ocean basin geometry controlling the flow of water from the Pacific to the Indian Ocean, or the way in which the solar flux absorbed in the upper ocean is affected by the transparency of the water as controlled by phytoplankton distributions? (There is literature on each of these processes affecting El Niño simulations.) It is more difficult to distribute model development broadly when it requires a holistic understanding that can only be developed by experimenting with the full system as opposed to individual components.

The increasing complexity of climate models creates problems in communication and education. The world is vitally interested in the results of these simulations, but few understand what a climate model is, or what the strengths and weaknesses of today's models are. Even students desiring to work with and develop climate models have difficulty getting their minds around the full complexity of these models. There is a clear pedagogical need for a hierarchy of models of increasing complexity, and there is much work along these lines. Indeed, it has been argued that such a hierarchy is fundamental to understanding the behavior of these complex models and that this understanding is crucial for creating an efficient model improvement process.

The analogy with biology is useful in this context. Evolution provides the biologist with a natural hierarchy of systems to study, ranging from the simplest bacteria to humans. We do not have a hierarchy of climates of increasing complexity handed to us by nature. Planetary atmospheres are too few and, as a result, too idiosyncratic. Laboratory simulations relevant for most issues in climate research are not possible. Climate scientists must create their own model hierarchies. Does the continuing development of models with more and more interactive components eventually become untenable without the solid foundations provided by the careful analysis of a hierarchy of simpler models?

Should climate models try to build steadily on the unquestioned successes of numerical weather prediction, focusing, for example, on improving seasonal predictions of El Niño and then moving systematically on to prediction of decadal climate variations? Many nations have at least partially integrated their numerical weather prediction and climate modeling efforts. After all, it is the same system that is being simulated in both cases. Some convergence is natural, especially as climate modelers focus on shorter time-scales and prediction efforts move to longer time-scales.

But beyond a couple of decades, we no longer have the luxury of the time needed to accumulate a useful set of forecasts for model validation. Additionally, as time-scales lengthen the prediction problem becomes one of determining the forced response to external perturbations rather than the evolution of internal variability from specific initial conditions. If one is interested in how the El Niño phenomenon will change in the future, is one better off focusing on improving predictions of El Niño or on improving the climatological structure of El Niño variability? If one is close to the final answer, there should be no conflict between these two objectives. If, hypothetically, there is only one uncertain parameter that remains to be determined, with all other aspects of the model perfected, then the same value of this parameter will optimize any metric of interest. But if there is still substantial uncertainty concerning how best to encompass all of the factors that effect the El Niño phenomenon in models, one can easily envision a distinction between model settings that optimize seasonal predictions and settings that optimize long-term El Niño statistics.

The climate-change problem is also distinctive as compared to numerical weather prediction in the importance placed on the attribution of past changes. If one can extract the response to increasing greenhouse gases from observations of the recent past, one can test models of this forced response and, in the simplest case, extrapolate into the future. The attribution problem and the prediction problem then become effectively synonymous. This importance of attribution is not found in weather prediction. Simulations of paleoclimates also play a fundamental role in testing climate models in a way that has no counterpart in standard weather prediction. These are some of the reasons why weather prediction and climate models might diverge and why climate modeling and numerical weather prediction groups might require differing expertise.

Climate researchers face challenges in a host of research areas examining climate-relevant processes in the atmosphere, in the oceans, and on land. Climate modelers also face challenges on more holistic levels in understanding how these processes interact in all their complexity to create our climate and control its sensitivity. And there are challenges with regard to how nations, and the world, should structure their modeling efforts, and on how to communicate results and limitations most effectively with the public and policymakers. These are clearly exciting and challenging times for the climate modeling enterprise.

Isaac M. Held
Geophysical Fluid Dynamics Laboratory/NOAA,
Princeton University, Princeton, New Jersey, USA

1

Introduction

LEO DONNER, WAYNE SCHUBERT, AND RICHARD C. J. SOMERVILLE

The aim of this volume is to describe the development of atmospheric general circulation models. We are motivated to do so by the central and essential role of these models in understanding, simulating, and predicting the atmosphere on a wide range of time scales. While atmospheric general circulation models are an important basis for many societal decisions, from responses to changing weather to deliberations on responding to anthropogenic climate change, the scientific basis for these models, and how they have come about and continue to develop, are not widely known. Our objective in editing this volume is to provide a perspective on these matters.

Atmospheric general circulation models (GCMs) are an important tool in basic scientific research with important applications in numerical weather prediction and climate simulation and projection. These varying purposes to which GCMs are put have dictated varying development strategies. While the applied foundations of GCMs lie in weather prediction, the demands of climate simulation have often led to distinct development paths for these applications. The chapters in this volume illustrate the early relationship of general circulation modeling to numerical weather prediction, the subsequent emphasis on climate in GCM development, and contemporary synergies between the two.

The atmosphere interacts strongly with the physical and biological systems which bound it: the oceans, ice sheets, land, and vegetation. These interactions are important for the behavior of both the atmosphere and the bounding systems. The behavior of the atmosphere and the development of atmospheric GCMs are as consequences intimately related to the behavior of these systems and the coupling of models for them to atmospheric GCMs. We aim also to describe major aspects of ocean and land models, and their coupling to atmospheric GCMs, in this volume.

The pervasive role of the atmosphere in human activities has resulted in a close link between the goals of GCM development and the societal context in which this development has taken place from the inception of the endeavor. Concern about anthropogenic climate change became widespread in the last decades of the twentieth century, but a scientific awareness of humanity's potential to alter the general circulation extended back decades earlier. This awareness was strong among some of the scientific leaders guiding early GCM development in the middle of the century. Human impacts were viewed rather differently than in contemporary times, during which they have been most prominently considered by the Intergovernmental Panel on Climate Change (IPCC). This volume will recount the issues outside of the GCM development community, which have both influenced and been influenced by GCMs, over the historical course of the research.

This volume is a project of the Center for Multi-Scale Modeling of Atmospheric Processes (CMMAP), a science and technology center funded by the U.S. National Science Foundation, directed by Professor David Randall and based at Colorado State University. The center was established to address problems in climate modeling and numerical weather prediction arising from the broad spectrum of time and space scales that determine atmospheric behavior. An early objective of the center was to place its activities in the larger context of the historical and contemporary development of atmospheric GCMs by publishing this monograph. In doing so, the editors, in consultation with other participants in CMMAP, chose to elicit reviews of important themes in GCM development from scientists who themselves have played key roles in this development. The result is a volume in which the individual perspectives and styles of these scientists shine through. While we hope our readers will appreciate the personality which is thereby imparted to the volume, we also realize it poses some demands on the reader that an approach leavened by a single author's unified construction would not. We are hopeful ours will not be the last treatment of this important topic.

Acknowledgments

The editors are most appreciative of the authors who have contributed their time and perspectives to this volume. Less evident, but also essential, have been our colleagues who provided, mostly anonymously, external reviews of early chapter drafts. We also thank Rodger Ames (Center for Multi-Scale Modeling of Atmospheric Processes) for editorial and technical support and Jeff Varanyak (Geophysical Fluid Dynamics Laboratory) for assistance with figure preparation.

2

From Richardson to early numerical weather prediction

PETER LYNCH

The development of computer models for numerical simulation of the atmosphere and oceans is one of the great scientific triumphs of the past fifty years. These models have added enormously to our understanding of the complex processes in the atmosphere and oceans. The consequences for humankind of ongoing climate change will be far-reaching. Earth system models are the best means we have of predicting the future of our climate.

The basic ideas of numerical forecasting and climate modeling were developed about a century ago, long before the first electronic computer was constructed. However, advances on several fronts were necessary before numerical prediction could be put into practice. A fuller understanding of atmospheric dynamics allowed the development of simplified systems of equations; regular observations of the free atmosphere provided the initial conditions; stable finite difference schemes were developed; and powerful electronic computers provided a practical means of carrying out the calculations required to predict the changes in the weather.

In this chapter, we trace the history of computer forecasting from Richardson's prodigious manual computation, through the ENIAC (Electronic Numerical Integrator and Computer) integrations to the early days of operational numerical weather prediction and climate modeling. The useful range of deterministic prediction is increasing by about one day each decade. We set the scene for the story of the remarkable progress in weather forecasting and in climate modeling over the past fifty years, which will be treated in subsequent chapters.

2.1 Pioneers of scientific forecasting

The fundamental idea of computing changes in the weather by numerical means was formulated around the turn of the twentieth century, before electronic computers were available. The great American meteorologist Cleveland Abbe recognized that meteorology is essentially the application of hydrodynamics and thermodynamics

L. F. Richardson, 1931

Figure 2.1 Lewis Fry Richardson (1881–1953), signed and dated 1931, when Richardson was aged 50 (photograph by Walter Stoneman; copy courtesy of Oliver Ashford).

to the atmosphere (Abbe, 1901), and he identified the system of mathematical equations that govern the evolution of the atmosphere. Abbe's work was reviewed recently by Willis and Hooke (2006). The Norwegian scientist Vilhelm Bjerknes undertook a more explicit analysis of the weather prediction problem from a scientific perspective (Bjerknes, 1904). His stated goal was to make meteorology an exact science, a true physics of the atmosphere.

Later, Lewis Fry Richardson (Figure 2.1) attempted a direct solution of the equations of motion, and presented the results in his book *Weather Prediction by Numerical Process* (Richardson, 1922). The book opened with a discussion of then-current practice. Richardson described the use of an index of weather maps, constructed by classifying old synoptic charts into categories. The index assisted the forecaster to find previous maps resembling the current one and thus deduce the likely development by studying the evolution of these earlier cases. This "analog approach" was at the heart of operational forecasting until the modern era, and forecast skill remained rather static. Indeed, Sverre Petterssen (2001) described the advances prior to the computer era as occurring in "homeopathic doses".

Table 2.1 *Six-hour changes in pressure (units: hPa/6 h). [LFR: Richardson; MOD: Model; DFI: Filtered.]*

Level (km)	LFR	MOD	DFI
1 (11.8)	48.3	48.5	−0.2
2 (7.2)	77.0	76.7	−2.6
3 (4.2)	103.2	102.1	−3.0
4 (2.0)	126.5	124.5	−3.1
Surface	**145.1**	**145.4**	**−0.9**

The full story of Richardson's work, the reason for his catastrophic results, and a complete reconstruction of the forecast are described in a recent book (Lynch, 2006). Richardson calculated a change in surface pressure over a six-hour period of 145 hPa, a totally unrealistic value. His extrapolation over six hours exacerbated the problem, but was not the root cause of it. Lynch showed that, when the analyzed data are balanced through the process of initialization, a realistic value of pressure change is obtained. In Table 2.1 we show the six-hour changes in pressure at each level of the numerical model he used. The column marked LFR (Lewis Fry Richardson) has the values obtained by Richardson (1922). The column marked MOD (model) has the values reconstructed using a computer model based directly on Richardson's method: they are very close to Richardson's values. The column marked DFI (digital filter initialization) is for a forecast from data initialized using a digital filter: the initial tendency of surface pressure is reduced from the unrealistic 145 hPa/6 h to a reasonable value of less than 1 hPa/6 h (bottom row, Table 2.1). These results indicate clearly that Richardson's unrealistic prediction was due to imbalance in the initial data that he used. Complete details of the forecast reconstruction may be found in Lynch (2006).

Richardson's forecasting scheme involved a phenomenal volume of numerical computation and was quite impractical in the pre-computer era. But he was undaunted, speculating that

> "some day in the dim future it will be possible to advance the computations faster than the weather advances".

The work of Max Margules, published almost twenty years before Richardson's forecast, pointed to serious problems with Richardson's methodology. Margules examined the relationship between the continuity equation (which expresses conservation of mass) and changes in surface pressure (Margules, 1904). He considered the possibility of predicting pressure changes by direct use of the mass conservation principle. He showed that, due to strong cancellation between terms, the calculation is very error-prone, and may give ridiculous results. Therefore, it is not possible, using the continuity equation alone, to derive a reliable estimate of synoptic-scale changes in pressure. Margules concluded that any attempt to forecast the weather was *immoral and damaging to the character of a meteorologist* (Fortak, 2001).

To make his forecast of the change in pressure, Richardson used the continuity equation, employing precisely the method that Margules had shown to be seriously problematical. The resulting prediction of pressure change was completely unrealistic. The question of what influence, if any, Margules' results had on Richardson's approach to forecasting was considered by Lynch (2003). At a later stage, Richardson did come to a realization that his original method was unfeasible. In a note contained in the Revision File, inserted in the manuscript version of his book, he wrote,

> "Perhaps the most important change to be made in the second edition is that the equation of continuity of mass must be eliminated".
>
> *(Richardson's underlining)*

He went on to speculate that the vertical component of vorticity or rotation rate of the fluid might be a suitable prognostic variable. This was indeed a visionary adumbration of the use of the vorticity equation for the first successful numerical integration in 1950.

Of course, we now know that Margules was unduly pessimistic. The continuity equation is an essential component of primitive equation models which are used in the majority of current computer weather prediction systems. Primitive equation models use the exact equations of motion except for the hydrostatic approximation: they support gravity wave solutions and, when changes in pressure are computed using the continuity equation, large tendencies can arise if the atmospheric conditions are far from balance. However, spuriously large tendencies are avoided in practice by an adjustment of the initial data to reduce gravity wave components to realistic amplitudes, by the process of initialization, as indicated in Table 2.1.

2.2 Pre-computer forecasting

Weather forecasts are now so reliable, accurate, and readily available that it is easy to forget how things were only a few decades ago. Before the computer era, forecasting was imprecise and undependable. Analysis of the global atmospheric state was severely hampered by lack of observations, and the principles of theoretical physics played little role in practical forecasting. Although much of the underlying physics was known, its application to the prediction of atmospheric conditions was impractical. Observations were sparse and irregular, especially for the upper air and over the oceans.

Forecasting was a haphazard process, very imprecise and unreliable. The forecaster used crude techniques of extrapolation, knowledge of local climatology, and guesswork based on intuition; forecasting was more an art than a science. The observations of pressure and other variables were plotted in symbolic form on a weather map and lines were drawn through points with equal pressure to reveal the pattern of weather systems – depressions, anticyclones, troughs, and ridges. The forecaster used experience, memory, and a variety of empirical rules to produce a forecast map. The primary physical process attended to by the forecaster was *advection*, the transport of fluid characteristics and properties by the movement of the fluid itself. But the crucial quality of advection is that it is *nonlinear*; the human forecaster may extrapolate trends using an assumption of constant wind, but is quite incapable of intuiting the subtleties of complex advective processes.

The technique of "weather typing" was used with limited success. This was the method underlying the index of weather maps, mentioned by Richardson. Current meteorological conditions were compared with the historical record. If a close match was found, it was assumed that the evolution of the flow for the following days would be similar to that observed on the previous occasion. However, the atmosphere shows little tendency to repeat itself. Richardson was not optimistic about this method. He wrote in his Preface that

"The forecast is based on the supposition that what the atmosphere did then, it will do again now . . . The past history of the atmosphere is used, so to speak, as a full-scale working model of its present self".

(Richardson, 1922)

Bjerknes had contrasted the precision of astronomical prediction with the "radically inexact" methods of weather forecasting. Richardson returned to this theme, pointing out that the *Nautical Almanac*,

"that marvel of accurate forecasting"

is not based on the principle that astronomical history repeats itself. Given the complexity of the atmosphere, why should we expect a present weather map to be exactly represented in a catalog of past weather?

In Europe, meteorology was studied in many universities, and researchers applied physical principles to atmospheric problems. Bjerknes had dreamed of mathematical forecasting but, finding it impractical, had marshaled his team in Bergen to develop more feasible methods. They developed mechanistic models of extratropical weather systems and described the life-cycles of mid-latitude depressions and their associated warm and cold fronts. Although the models of polar fronts and the life-cycles of extratropical depressions are conceptual in nature, they are founded on sound scientific principles. Frontal and air-mass theory gradually came to have a profound influence on operational practice on both sides of the Atlantic. Bjerknes visited America to promote the new ideas and many of his Bergen students played major roles in the advancement of American meteorology, both in government agencies and universities.

2.3 Key developments, 1920–1950

Richardson was several decades ahead of his time in what he attempted to do. At the time of the First World War, computational weather forecasting was impractical for at least four reasons.

First, observations of the three-dimensional structure of the atmosphere were available only on a very occasional basis, with inadequate coverage and never in real time. The registering balloons had to be recovered and the recordings analyzed to recover the data, a process that took days or even weeks. Second, the numerical algorithms for solving the atmospheric equations were subject to instabilities that were not understood. Thus, the numerical solution might bear little or no resemblance to the solution of the continuous equations. Third, the balanced nature of atmospheric flow was inadequately understood, and the imbalances arising from observational and analysis errors confounded Richardson's forecast. Fourth, the massive volume of computation required to advance the numerical solution could not be done, even by a huge team of human computers. Indeed, Richardson's estimate of 64,000 computers, the number of people needed to do the calculations for a useful forecast in real time, was a serious under-estimate. It has been reckoned that one million people would have been required for the task (Lynch, 1993). Thus, what Richardson devised was a "method without a means".

A number of key developments in the ensuing decades set the scene for progress. Developments in the theory of meteorology provided crucial understanding of atmospheric dynamics, in particular the balance of the atmosphere and the means of eliminating spurious high-frequency gravity waves. Advances in numerical analysis led to the design of algorithms which were stable and faithfully replicated the true solution. Timely observations of the atmosphere in three dimensions were becoming available following the invention of the radiosonde, which provided measurements of pressure, temperature, humidity, and winds through a vertical column of the atmosphere. Finally, the development of digital computers provided a way of attacking the enormous computational task involved in weather forecasting.

In addition to the technical developments, the socio-political framework of the mid century provided a crucial impetus. Progress in meteorology has often followed from natural or human-made catastrophes. The Second World War was a spectacular example. Military operations on land, air, and sea all depend heavily on accurate weather forecasts. The role of weather in Operation Overlord – the D-day invasion of Normandy – was recounted by Pettersen (2001). In the United States an intensive training program for meteorologists was organized under the inspiration of Carl-Gustav Rossby. As a result, the professional meteorological community grew by a factor of fifteen during the war, from 400 before to 6000 afterwards. Many of the new entrants were highly skilled in mathematics and physics and wished to develop rigorous methods of forecasting that were based on scientific principles.

2.4 The ENIAC integrations

The first general-purpose electronic computer, ENIAC (Electronic Numerical Integrator and Computer) was commissioned by the U.S. Army for use in calculating the dynamics of projectiles. The principal designers of ENIAC were John Mauchly and Presper Eckert. It is noteworthy that Mauchly's interest in computers arose from his desire to forecast the weather by calculation. The computer was originally called the Electronic Numerical Integrator. A U.S. Army colonel suggested adding the words "and Computer" to give the catchy acronym ENIAC (McCartney, 1999). The ENIAC, which had been completed in 1945, was the first general-purpose electronic digital computer ever built. It was a gigantic machine, with 18,000 thermionic valves, filling a large room and consuming 140 kW of power. Input and output were by means of punch-cards. McCartney (1999) provides an absorbing account of the origins, design, development, and destiny of ENIAC.

John von Neumann recognized weather forecasting, a problem of both great practical significance and intrinsic scientific interest, as ideal for an automatic computer. He was in close contact with Rossby, who was the person best placed to understand the challenges that would have to be addressed to achieve success in this venture. Von Neumann established a Meteorology Project at the Institute for Advanced Study in Princeton and recruited Jule Charney to lead it. Arrangements were made to compute a solution of a simple equation, the barotropic vorticity equation (BVE), on the only computer available, the ENIAC. Barotropic models treat the atmosphere as a single layer, averaging out variations in the vertical. The resulting numerical predictions were truly ground-breaking. Four 24-hour forecasts were made, and the results clearly indicated that the large-scale features of the mid-tropospheric flow could be forecast numerically with a reasonable resemblance to reality.

The ENIAC forecasts were described in a seminal paper by Jule Charney, Ragnar Fjørtoft, and John von Neumann (Charney, *et al.*, 1950, referenced below as CFvN). The story of this work was recounted by George Platzman in his Victor P. Starr Memorial Lecture (Platzman, 1979). The atmosphere was treated as a single layer, represented by conditions at the 500 hPa level, modeled by the BVE. This equation, expressing the conservation of absolute vorticity following the flow, gives the rate of change of the Laplacian of the height of the 500 hPa surface in terms of the

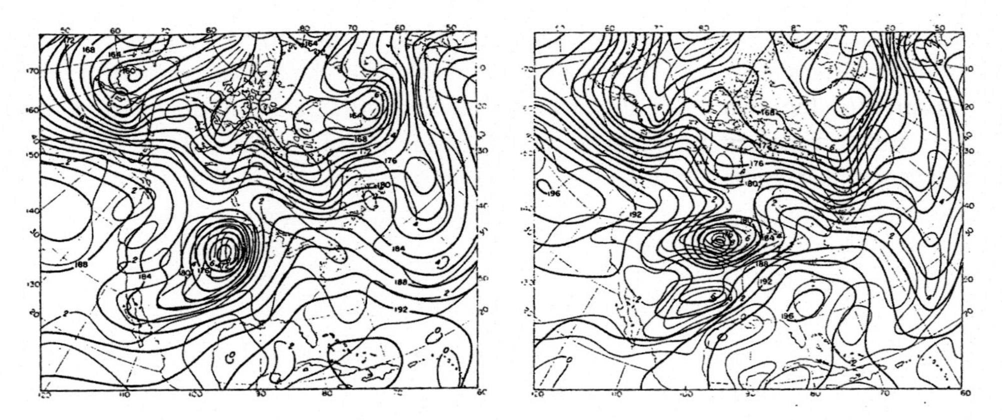

Figure 2.2 The ENIAC forecast starting at 0300 UTC, January 5, 1949. Left panel: analysis of 500 hPa height (thick lines) and absolute vorticity (thin lines). Right panel: forecast height and vorticity (from Charney, *et al*., 1950). Height units are hundreds of feet, contour interval is 200 ft. Vorticity units and contour interval are 10^{-5}s^{-1}.

advection. The tendency of the height field is obtained by solving a Poisson equation with homogeneous boundary conditions. The height field may then be advanced to the next time level. With a one-hour timestep, this cycle is repeated 24 times for a one-day forecast.

The initial data for the forecasts were prepared manually from standard operational 500 hPa analysis charts of the U.S. Weather Bureau, discretized to a grid of 19 by 16 points, with grid interval of 736 km. Centered spatial finite differences and a leapfrog time-scheme were used. The boundary values of height were held constant throughout each 24-hour integration. The forecast starting at 0300 UTC, January 5, 1949 is shown in Figure 2.2 (from CFvN). The left panel is the analysis of 500 hPa height and absolute vorticity. The forecast height and vorticity are shown in the right panel. The feature of primary interest was an intense depression over the United States. This deepened, moving NE to the 90°W meridian in 24 hours. A discussion of this forecast, which under-estimated the development of the depression, may be found in CFvN and in Lynch (2008).

The success of the ENIAC forecasts had an electrifying effect on the world meteorological community. Several baroclinic (multi-level) models were developed in the following years. They were all based on the filtered or quasi-geostrophic system of equations, an approximate system derived using geostrophic balance between the pressure and winds. Later, models using the more accurate primitive equations were introduced. Charney had anticipated this as a necessary development, and indeed André Robert later identified it as the key development in numerical weather prediction (see Lin *et al.* 1997).

The Princeton team studied the severe storm of Thanksgiving Day, 1950 using two- and three-level quasi-geostrophic models. After some tuning, they found that the cyclogenesis could be reasonably well simulated. Thus, it appeared that the central problem of operational forecasting had been cracked. However, it transpired that the success of the Thanksgiving forecast had been something of a fluke: early multi-level models were consistently worse than the simple barotropic equation; and it was the single-level model that was used when regular operations commenced in 1958. A fuller discussion can be found in Lynch (2006). The trials and triumphs of the Joint Numerical Weather Prediction Unit, which will appear again in Chapter 3, and the establishment of operational computer forecasting are described comprehensively in a recent book (Harper, 2008).

2.5 Advancing computer technology

Advances in computer technology over the past half-century have been spectacular. The increase in computing power is encapsulated in an empirical rule called Moore's Law, which implies that computing speed doubles about every 18 months. Thus, a modern micro-processor has far greater power than the ENIAC had. Recently, Lynch and Lynch (2008) decided to repeat the ENIAC integrations using a programmable cell-phone, which was called the Portable Hand-Operated Numerical Integrator and Computer, or PHONIAC. This technology has great potential for generation and display of operational weather forecast products.

The oft-cited paper in *Tellus* (CFvN) gives a complete account of the computational algorithm and discusses four forecast cases. Lynch (2008) presented the results of repeating the ENIAC forecasts using a MATLAB program `eniac.m`, run on a laptop computer (a Sony Vaio, model VGN-TX2XP). The main loop of the 24-hour forecast ran in about 30 ms. Given that the original ENIAC integrations each took about one day, this time ratio – about three million to one – indicates the dramatic increase in computing power over the past half-century. The program `eniac.m` was converted from MATLAB to a Java application, `phoniac.jar`, for implementation on a cell-phone. The program was tested on a PC using emulators for three different mobile phones. A basic graphics routine was also written in Java. When working correctly, the program was downloaded onto a Nokia 6300 cell-phone for execution.

Charney *et al.* (1950) provided a full description of the solution algorithm for the BVE. The programs `eniac.m` and `phoniac.jar` were constructed following the original algorithm precisely, including the specification of the boundary conditions and the Fourier transform solution method for the Poisson equation. Hence, given initial data identical to that used in CFvN, the recreated forecasts should be

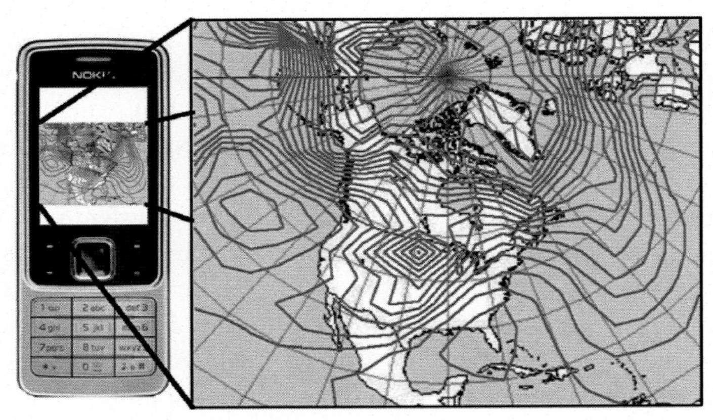

Figure 2.3 The Nokia 6300, dubbed PHONIAC (left) and the forecast for 0300 UTC, January 6, 1949 (right) made with the program `phoniac.jar`. The contour interval is 50 m (from Lynch and Lynch, 2008). See also color plate.

identical to those made in 1950. Of course, the reanalyzed fields are not identical to those originally used, and the verification analyses are also different. Nevertheless, the original and new results are very similar.

The initial fields for the four ENIAC forecasts were valid for dates in January and February, 1949. A retrospective global analysis of the atmosphere, covering more than fifty years, has been undertaken by the National Centers for Environmental Prediction (NCEP) and the National Center for Atmospheric Research (NCAR) (Kistler *et al.*, 2001). This reanalysis extends back to 1948, including the period chosen for the ENIAC integrations. The reanalyzed data are available on a 2.5° by 2.5° grid. The GRIB fields of the 500 hPa analyses were downloaded from the NCEP/NCAR reanalysis website and interpolated to the ENIAC grid.

In Figure 2.3 we show PHONIAC and the forecast for 0300 UTC, January 6, 1949 made with the program `phoniac.jar`. The main features of the forecast (right panel) are in broad agreement with the originals (right panel, Figure 2.2). In CFvN it is noted that the computation time for a 24-hour forecast was about 24 hours, that is, the team could just keep pace with the weather provided the ENIAC did not fail. This time included off-line operations: reading, punching, and interfiling of punch-cards. PHONIAC executed the main loop of the 24-hour forecast in less than one second. The main steps in the solution algorithm are presented in Lynch (2008, Appendix B). For the benefit of students, the MATLAB and Java codes are available on a website (http://maths.ucd.ie/~plynch/eniac/). Maps of the four original and recreated forecasts are also available there, along with miscellaneous supplementary material relating to the ENIAC integrations.

2.6 Climate modeling

We can trace the beginnings of climate modeling back to 1956, when Norman Phillips carried out the first long-range simulation of the general circulation of the atmosphere. He used a two-level quasi-geostrophic model on a beta-plane channel, ignoring the effects of sphericity except for variations of the Coriolis force with latitude. The computation used a spatial grid of 16 × 17 points, and the simulation was for a period of about one month. Starting from a zonal flow with small random perturbations, a wave disturbance with a wavelength of 6000 km developed. It had the characteristic westward tilt with the height of a developing baroclinic wave, and moved eastward at about 20 m s^{-1}. Figure 2.4 shows the configuration of the flow after twenty days' simulation. Phillips examined the energy exchanges of the developing wave and found good qualitative agreement with observations of baroclinic systems in the atmosphere. He also examined the mean meridional flow, the average circulation in a vertical cross-section along a meridian, and found circulations corresponding to the Hadley, Ferrel, and Polar cells, large-scale

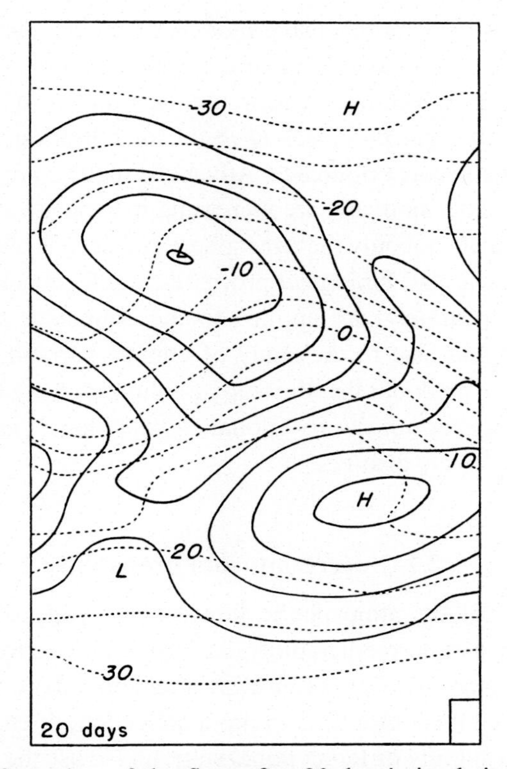

Figure 2.4 Configuration of the flow after 20-days' simulation with a simple, two-level filtered model. Solid lines: 1000 hPa heights at 200-foot intervals. Dashed lines: 500 hPa temperatures at 5 °C intervals (Phillips, 1956).

circulations with ascending motion at one latitude coupled to descending motion at another through meridional flow:

> "We see the appearance of a definite three-celled circulation, with an indirect [or reverse] cell in middle latitudes and two somewhat weaker cells to the north and south. This is a characteristic feature of . . . unstable baroclinic waves".
>
> *(Phillips, 1956, p. 144)*

Von Neumann was hugely impressed by Phillips' work, and arranged a conference at Princeton University in October 1955, *Application of Numerical Integration Techniques to the Problem of the General Circulation*, to consider its implications. The work had a galvanizing effect on the meteorological community. Within ten years, there were several major research groups modeling the general circulation of the atmosphere, the leading ones being at the Geophysical Fluid Dynamics Laboratory, the National Center for Atmospheric Research, the University of California, Los Angeles, the Lawrence Livermore National Laboratory, and the United Kingdom Meteorological Office These research efforts will appear again in Chapters 3 and 9.

The development of comprehensive models of the atmosphere is undoubtedly one of the finest achievements of meteorology in the twentieth century. Following Phillips' seminal work, several general circulation models (GCMs) were developed, including various physical processes such as solar heating, terrestrial radiation, convection, and small-scale turbulence. Advanced models are undergoing continuing refinement and extension, and are increasing in sophistication and comprehensiveness. They simulate not only the atmosphere and oceans but also a wide range of geophysical, chemical, and biological processes and feedbacks. The models, now called *Earth system models*, are applied to the eminently practical problem of weather prediction and also to the study of climate variability and humankind's impact on it. There is no doubt that the study of climate change and its impacts is of enormous importance for our future. Global climate models are the best means we have of anticipating likely changes.

2.7 Uncertainty and probability

The chaotic nature of the atmospheric flow, most clearly elucidated by Lorenz (1963), imposes a limit on predictability, as unavoidable errors in the initial state grow rapidly and render the forecast useless after some days. The most successful means of confronting this obstacle is to run a collection, or ensemble, of forecasts, each starting from a slightly different initial state, and to use the combined outputs to deduce probabilistic information about future changes in the atmosphere. Probability forecasts for a wide range of weather events are generated for use in the operational centers. These have become the key guidance for medium-range prediction.

Computer prediction models are now the primary input for operational forecasters and are vital for a wide range of applications. Perhaps the most important application is to provide timely warning of weather extremes. The ensemble approach provides valuable quantitative guidance on the probability and likely severity of extreme events. The warnings which result from computer guidance enable great saving of life and property.

Transportation, energy consumption, construction, tourism, and agriculture are all sensitive to weather conditions. There are expectations from all these sectors of increasing accuracy and detail of forecasts, as decisions with heavy financial implications must continually be made. Numerical weather prediction models are used to generate special guidance for the marine community. Trajectories for modeling pollution drift, for nuclear fallout, and smoke from forest fires are easily derived. Aviation benefits significantly from computer guidance, which provides warnings of hazards. Automatic generation of terminal aerodrome forecasts enables servicing of a large number of airports from a central forecasting facility.

Interaction between the atmosphere and ocean becomes a dominant factor at longer forecast ranges. For seasonal forecasting, coupled atmosphere–ocean models are essential. Once again, the ensemble approach is an effective means of addressing uncertainty in the predictions. Good progress has been made in seasonal forecasting for the tropics. Considerable effort is being made to produce useful long-range forecasts for temperate regions, but many challenges remain. The ensemble approach has also become a central aspect of climate change prediction.

2.8 Dreams fulfilled

Developments in atmospheric dynamics, instrumentation, observing practice, and digital computing have made the dreams of Abbe, Bjerknes, and Richardson an everyday reality. Numerical weather prediction models are now at the center of operational forecasting. It is no exaggeration to describe the advances made over the past half-century as revolutionary. Progress in weather forecasting and in climate modeling has been dramatic. We can now predict the weather for several days in advance with a high degree of confidence, and the useful range of deterministic prediction is increasing by about one day each decade. Using Earth system models, we are gaining great insight into the factors causing changes in our climate, and their likely timing and severity.

Meteorology is now firmly established as a quantitative science, and its value and validity are demonstrated daily by the acid test of any science, its ability to predict the future. The development of comprehensive models of the atmosphere is undoubtedly one of the finest achievements of meteorology in the twentieth century. The story of how the models have developed over the past fifty years, and the current state of numerical weather prediction and climate modeling, is told in the following chapters of this volume.

References

Abbe, C. (1901). The physical basis of long-range weather forecasts. *Monthly Weather Review*, 29, 551–61.

Bjerknes, V. (1904). Das Problem der Wettervorhersage, betrachtet vom Standpunkte der Mechanik und der Physik. *Meteorologische Zeitschrift*, 21, 1–7.

Charney, J. G., Fjörtoft, R., and von Neumann, J. (1950). Numerical integration of the barotropic vorticity equation. *Tellus*, 2, 237–254.

Fortak, H. (2001). Felix Maria Exner und die österreichische Schule der Meteorologie. Pp. 354–386, in Hammerl, Christa, Wolfgang Lenhardt, Reinhold Steinacker, and Peter Steinhauser, 2001: *Die Zentralanstalt für Meteorologie und Geodynamik 1851–2001. 150 Jahre Meteorologie und Geophysik in Österreich*. Leykam Buchverlags GmbH, Graz.

Harper, K. J. (2008). *Weather by the Numbers: the Genesis of Modern Meteorology*. MIT Press, 308pp.

Kistler, R., Kalnay, E., Collins, W., Saha, S., White, G., Woollen, J., Chelliah, M., Ebisuzaki, W., Kanamitsu, M., Kousky, V., van den Dool, H., Jenne, R., and Fiorino, M. (2001). The NCEP-NCAR 50-year reanalysis: Monthly means CD-ROM and documentation. *Bulletin of the American Meteorological Society*, 82, 247–267.

Lin, C. A., Laprise, R., and Richie, H. ed. (1997). *Numerical Methods in Atmospheric and Oceanic Modelling*. The André J. Robert Memorial Volume. Canadian Meteorological and Oceanographic Society.

Lorenz, E. N. (1963) Deterministic nonperiodic flow. *Journal of the Atmospheric Sciences*, 20, 130–141.

Lynch, P. (1993). Richardson's forecast factory: the $64,000 question. *Meteorological Magazine*, 122, 69–70.

Lynch, P. (2003). Margules' tendency equation and Richardson's forecast. *Weather*, 58, 186–193.

Lynch, P. (2006). *The Emergence of Numerical Weather Prediction: Richardson's Dream*. Cambridge University Press, 279 pp.

Lynch, P. (2008). The ENIAC forecasts: a recreation. *Bulletin of the American Meteorological Society*, 89, 45–55.

Lynch, P. and Lynch, O. (2008). Forecasts by PHONIAC. *Weather*, 63, 324–326.

McCartney, S. (1999). *ENIAC: The Triumphs and Tragedies of the World's First Computer*. Berkley Books, New York, 262pp.

Margules, M., (1904). Über die Beziehung zwischen Barometerschwankungen und Kontinuitätsgleichung [On the relationship between barometric variations and the continuity equation]. *Ludwig Boltzmann Festschrift*. Leipzig, Barth, J. A., 930 pp.

Petterssen, S. (2001). *Weathering the Storm. Sverre Petterssen, the D-Day Forecast, and the Rise of Modern Meteorology*. Fleming, J. R. ed. American Meteorological Society.

Phillips, N. A. (1956). The general circulation of the atmosphere : A numerical experiment. *Quarterly Journal of the Royal Meteorological Society*, 82, 123–164.

Platzman, G. W. (1979). The ENIAC computations of 1950 – Gateway to numerical weather prediction. *Bulletin of the American Meteorological Society*, 60, 302–312.
Richardson, L. F. (1922). *Weather Prediction by Numerical Process*. Cambridge University Press. 2nd Edn. with Foreward by Peter Lynch (2007).
Willis, E. P. and Hooke, W. H. (2006). Cleveland Abbe and American meteorology, 1871–1901. *Bulletin of the American Meteorological Society*, 87, 315–26.

3

The evolution and future research goals for general circulation models

WARREN M. WASHINGTON AND AKIRA KASAHARA

3.1 Introduction

Two notable events happened in 1955 that led to a definition of research goals for computer modeling of the Earth's general circulation. One was a document entitled *Dynamics of the General Circulation* which was written by John von Neumann on July 29, 1955 and edited with handwritten notes by Harry Wexler, Chief of Scientific Services Division, U.S. Weather Bureau (USWB); see Figure 3.1. (Note: Harry Wexler was one of the most influential meteorologists of the first half of the twentieth century as described by James Fleming in Chapter 4 of this volume.) The purpose of this historic document was to propose the initiation of a research project to understand the dynamics of the general circulation using a high-speed electronic computer. As discussed in Chapter 2, in 1947 von Neumann had already started the Electronic Computer Project (ECP) at the Institute for Advanced Study in Princeton, New Jersey, with a particular objective of developing dynamical methods for weather forecasting (Thompson, 1983) using ENIAC. Therefore, this document was another bold attempt by von Neumann to contribute to understanding how the atmospheric general circulation works.

The first part of this historic document reviews the accomplishments of the ECP, led by Jule Charney, including the first successful barotropic forecasts in 1950 by J. Charney, R. Fjørtoft, and J. von Neumann (Platzman, 1979) and other encouraging quasi-geostrophic baroclinic forecasts (Charney and Phillips, 1953). On the basis of these results the U.S. Weather Bureau (later the National Weather Service, NWS), the Air Weather Service of the U.S. Air Force, and the Naval Weather Service established in 1954 the Joint Numerical Weather Prediction Unit (JNWPU) at Suitland, Maryland, with George Cressman as its first Director, and began issuing numerical predictions after installation of an IBM 701 in March 1955 (Shuman, 1989).

Then, the document describes von Neumann's vision on what he called "longer-range forecasts", namely understanding the general circulation of the atmosphere, which was encouraged by another significant event which took place at the ECP,

Draft
July 29, 1955

DYNAMICS OF THE GENERAL CIRCULATION

1. In 1947, a project was started in Princeton by the U.S. Navy and U.S. Air Force, for theoretical and computational investigations in meteorology, with particular regard to the development of methods of numerical weather forecasting. After a few years of experimenting, the project concentrated on exploring the validity and the use of the differential equation methods developed by Dr. J. Charney for numerical forecasting. For this purpose, the U.S. Army Ordnance Corps ENIAC computing machine was used in 1950 and in 1951, and the Institute for Advanced Study's own computing machine from 1952 onward. Subsequently, use was also made of the IBM 701 machine in New York City. With the help of these computing tools, it was found that forecasts over periods like 24 (and up to 48) hours are possible, and give significant improvements over the normal, subjective method of forecasting. Certain experiments demonstrated that even phenomena of cyclogenesis could be predicted. A considerable number of sample forecasts were made, which permitted the above mentioned evaluation of the validity of the method. A large number of variants were also explored, particularly with respect to eliminating successively the major mathematical approximations that the original method contained. It must be noted, however, that the method, and also all its variants which exist at the present, are still affected with considerable simplifications of a physical nature. Thus, the effects of radiation have only been taken into consideration in exceptional cases, the same is true for the effects of geography and topography, while humidity and precipitation have not been considered at all. That significant results could, nevertheless, be obtained, is due to the relatively short time-span of the forecasts. Indeed, over 24 or 48 hours the above mentioned effects do not yet come into play decisively.

Figure 3.1 This is the first page of a four-page letter (provided by Geophysical Fluid Dynamics Laboratory/NOAA, Princeton, New Jersey, USA) prepared by John von Neumann and the edited handwritten comments from Harry Wexler. The document establishes the framework and goals written by the first group on the general circulation modeling effort beyond numerical weather prediction.

i.e. the pioneering work of Norman Phillips (1956) on a simulation of the general circulation with a two-layer quasi-geostrophic baroclinic model (see Figure 2.4 and related discussion in Chapter 2). The document further discussed the need to make "mathematical and geometrical" improvements, changing from a limited area to the entire hemisphere, adding the seasonal and diurnal solar forcing, and improving both the horizontal and vertical resolution of the model. It was also explicitly stated that geography and topography should be included, along with atmospheric humidity and evaporation from the Earth's surface. Note that the reference to humidity really implies the atmospheric water cycle, including precipitation. The document goes on to say,

"It also requires the introduction of, presently reasonably understood, semi-empirical rules regarding the dependence of the rate of evaporation on the local atmospheric and oceanic temperatures, atmospheric stability, and wind velocity. Some, as yet necessarily very imperfect empirical rules about the delay-relationship of over-saturation, cloud formulation, and precipitation. Also, some empirical rules about the decrease of solar irradiation by clouds . . .".

The document clearly recognizes that treatment of "the effect of atmospheric humidity on the solar irradiation" is extremely difficult and requires "a great deal of experimental work." As is well known in the atmospheric and climate research field, we are still grappling with these fundamental problems more than a half-century later. In our view one should not be critical of the researchers of the 1950s because no one at the time understood the true complex dynamics and thermodynamics of the atmosphere and the rest of the climate system.

The other notable event in 1955 was a climate conference in Princeton (discussed in Chapter 2) at which research goals for general circulation modeling were articulated. The proceedings were not actually published until five years later by the editor, Richard L. Pfeffer (1960). The book's title is *Dynamics of Climate* with a subtitle *Proceedings of a Conference on the Application of Numerical Integration Techniques to the Problem of the General Circulation held October 26–28, 1955.* Note this conference was held just a few months after the von Neumann and Wexler proposal was written.

In the welcoming address at the conference, J. Robert Oppenheimer, the famous scientific director of the Manhattan Project drew a parallel between the present

---

Caption for Figure 3.1 (cont.)
It mentions Norman Phillips' pioneering effort. It is interesting to note that the planners thought that it could be a project accomplished in just a few years. Of course they did not realize the true complexity of the climate system or the difficulty of computer modeling for the climate system. Toward the end of the memo, the handwritten note by Wexler indicates that Jule Charney was consulted in the preparation of this document. The complete document is in the appendix for this chapter.

conference dealing with problems of the general circulation of the Earth's atmosphere and the conference held at Los Alamos, New Mexico, in preparation for work on the atomic bomb. He recalled that the two weeks which were spent talking and arguing about various points at the Los Alamos meeting served as an important foundation for the future work on the atomic bomb. He pointed out that the problem which faces the participants of the present conference,

> "a problem dealing with the complicated dynamics of the atmospheric motions . . . is a much more difficult one."

The summary of the welcoming addresses was prepared by J. Robert Oppenheimer and Milton Greenberg (Pfeffer 1960, pp. 3–4).

In these proceedings, von Neumann presented his view on the approach to forecasting weather and climate fluctuations using the general circulation model (GCM). We rephrase his remarks in light of current practice in the prediction of weather and simulation of climate. The prediction problem is divided into three different categories depending on the time-scale of the prediction. The first category is a short- to medium-range weather prediction for a few days to two weeks which is mostly governed by the initial condition. The second category is a climate simulation which is practically independent of the initial condition, but essentially governed by the physical processes in the atmosphere involving both external radiative forcing and internal rearrangement of various forms of thermodynamic energy by the motions. Between the two extreme situations, there is the third category of medium- to seasonal-range forecasting of weeks to years. For this time-scale, both the initial condition and details of physical processes must be taken into account. The current and future research goals for general circulation modeling will be aimed at the need of improving the third category prediction.

Robert M. White, protégé of Victor P. Starr of MIT, who served at the Geophysics Research Directorate in the late 1950s, later became the Chief of USWB, and former President of the National Academy of Engineering, defined in his talk that the purpose of the conference was to consider the problems of studying the general circulation of the Earth's atmosphere by the use of numerical methods. He went on to say

> "It seems clear that the application of numerical techniques to the general circulation can lead to a better understanding of the fluctuations of the general circulation, with accompanying improvements in long-range weather forecasting and increased appreciation of the factors governing climatic changes."
>
> *(See Pfeffer 1960, pp. 5–6.)*

It is indeed true that presently this goal for understanding the first-order principles of the general circulation has been substantially met. However, there is a great need for important and challenging future modeling improvements for the third category of prediction, and also for extending our current general circulation modeling efforts to build Earth system models.

Over the next few decades the research goals for various groups engaged in general circulation modeling evolved from a variety of objectives. For example, the objectives of the Meteorology Project at Princeton were to advance the science of weather forecasting in the first category and to understand some basic aspects on the mechanism of the atmospheric general circulation in the second category. These goals were met by a successful short-range prediction of the large-scale flow patterns and Phillips' simulation of the atmospheric general circulation, together with a study of the energy transformations which were similar to those in the observed atmosphere (Lewis 1998). It is interesting that Phillips (1956) noted

> "Thus, one is almost forced to the conclusion that at least the gross features of the general circulation of the atmosphere can be predicted without having to specify the heating and cooling in great detail."
>
> *(See Pfeffer 1960, pp. 18–25)*

Phillips demonstrated this by specifying only the zonal mean (averaged over longitude and thus a function of height and latitude) heating/cooling in his pioneering numerical model experiment. Much of the present improvements in climate simulations are due to more accurate treatment of heating/cooling terms whose formulations have benefitted from having observational measurements that are used to validate the model performances.

One of the GCM subjects that we have not touched on so far is performance verification of the GCMs. Nowadays, the performances of GCMs are judged not only from comparison of model outputs with long-term climate statistics which are obtained from the multiple sources of satellite and field observations together with reanalysis (post-analysis of historical observational data with improved models in contrast to real-time analysis) data, but also from the skill of medium-range weather forecasts starting from global initial conditions. The situation was totally different during the 1950–60s when the earlier GCMs were developed. Except for the statistics of ground observations, very few synoptic upper-air statistics were available. Therefore, each GCM development team had to assign a specialist to check the outputs of GCM against observations to keep the modelers honest. In the case of UCLA, Yale Mintz himself was an expert in diagnostic studies, having collaborated with Professor Jacob Bjerknes in the climatic analysis of upper-air data (Johnson and Arakawa, 1996). At GFDL Smagorinsky asked Abraham Oort (1964) to join the lab; Oort had engaged in the observational analysis of atmospheric energetics under Professor Victor Starr of MIT. At NCAR, David Baumhefner, who was educated at UCLA in synoptic meteorology, served as our critic and even performed real-data forecasts using subjectively analyzed global IGY (International Geophysical Year) data sets as the first coherent input to a global GCM (GCM steering committee, 1975).

Before we leave this section it should be pointed out that there are other approaches to understanding the climate system than using GCMs. These approaches involve using a simpler form of the basic equations and not explicitly computing the motions in the general circulation. Schneider and Dickinson (1974) describe in detail the use of energy balance models for studies of climate change and C. Leith (1975) describes the use of statistical models. Both techniques have been useful tools for understanding climate and climate change; however they cannot provide the full range of information that GCMs can, nor can they be compared in detail with observations.

3.2 Evolution of general circulation modeling efforts

Within a few years from the aforementioned conference, various groups initiated general circulation modeling activities of the atmosphere and oceans as reviewed recently by Edwards (2000). This chapter will describe some of their stated organizational goals for building general circulation models (GCMs) and how they evolved with time. However, in the earlier efforts of GCM development, it was not always clear that there were detailed strategic planning documents that laid out step-by-step goals for building GCMs. In addition to historical perspective, we will describe in the later part of this chapter some of the future goals of general circulation modeling for projecting changes in the climate of the entire Earth system.

In addition to Chapter 2, more information on the history of the earliest efforts in numerical modeling of the atmosphere is presented in two notable references. One is an article by George W. Platzman (1967) on a retrospective view of the book on numerical weather prediction by Lewis Fry Richardson. Platzman describes in great detail not only the state of meteorological research and weather forecasting at the time when the book was written in 1922, but also a review on Richardson's step-by-step numerical procedure in the light of present-day practice. Another is an informative book by Peter Lynch (2006) which documents the difficulties and solutions to L. F. Richardson's dream to predict the weather by numerical process. Lynch states in his book that

> "the development of comprehensive models of the atmosphere is undoubtedly one of the finest achievements of meteorology in the twentieth century."

Without question there has been remarkable progress from Richardson's early attempt to present-day numerical weather prediction practice.

3.2.1 GFDL models

Receiving the proposal of von Neumann and Wexler mentioned in the Introduction, Francis Reichelderfer, then Chief of the U.S. Weather Bureau, set up a group in 1955 called the General Circulation Research Section, and appointed Joseph

Smagorinsky to lead the section, which is now called the NOAA Geophysical Fluid Dynamics Laboratory (GFDL). This group is the earliest organized team of scientists to tackle the problem of numerical simulation of the general circulation. John Lewis (2008) has written an insightful account of Smagorinsky's early influence by Jule Charney and Harry Wexler (whose role is taken up in more detail in Chapter 4), as well as his early interactions with the ECP at the Institute for Advanced Study in Princeton. According to Smagorinsky (1983),

"it seemed the next logical step beyond Phillips' model was to allow nongeostrophic models (Atmospheric models without the assumption of flow being geostrophic) which could be of great significance in how the tropics operated in, and interacted with, the general circulation."

The Phillips experiment had no realistic tropical simulation, so that GFDL's goal was to actually include tropical circulation as an integral part of the general circulation. For this purpose it was necessary to consider at least the hemispheric domain and hopefully extend the domain to the entire globe to deal with the interactions between the tropics and the extratropics. Also the team's goal was to include moisture and precipitation processes which were not explicitly included in the Phillips experiment.

In the late 1950s it became clear that the quasi-geostrophic approximation is too severe for dealing with hemispherical and global flows, so GFDL decided to adopt the primitive-equation (PE) formulation from the beginning. However, running the PE model stably with the explicit prediction of water vapor and the inclusion of condensation heating was a difficult task in those days. Even in the Smagorinsky *et al.* (1965) paper, the moist processes were not included explicitly in the model. Instead, the stabilizing effect of moist convection was emulated by adjusting the temperature lapse rate when it exceeded the moist adiabatic rate (the rate of temperature decrease with height when a rising air parcel is saturated by moisture). One notable solution was proposed by Manabe *et al.* (1965). When air is saturated, the temperature lapse rate is adjusted without violating the conservation of moist entropy. This procedure is now referred to as moist convective adjustment. As we will discuss later, the fact that a primitive-equation model with moist physics could be run stably became a turning point in the development of GCMs.

Thus, the goals of the GFDL group were to attack the following problem areas: condensation processes including clouds, radiation modeling, dynamics of moist convection, parametric internal diffusion, numerical schemes of finite differences, and the modeling of ocean circulation. It should be noted that the name and the location of the group had changed several times before the present name and location became established. The development of an ocean circulation model at GFDL, which is described by Kirk Bryan in Chapter 7, was clearly a necessary component of general circulation modeling for understanding the mechanisms of atmospheric circulations on climate time-scale. Suki Manabe has made significant contributions in climate modeling working with K. Bryan and others. As Lewis' article (2008) states, Smagorinsky achieved his mission by assembling an excellent team of atmospheric and ocean scientists along with programming experts in a close-knit group somewhat separated from the operational numerical prediction group of the U.S. Weather Bureau, which is now part of the National Centers for Environmental Prediction of NOAA.

Over the years, the GFDL team of scientists developed a hierarchy of models that were applied to investigate questions of atmospheric predictability and sensitivity of the climate system. (See GFDL (1980) for "A Review of Twenty-Five Years of Research at GFDL"). It is interesting to note that a 1962 U.S. Weather Bureau press release discussed some of the research goals in connection with the installation of a new IBM STRETCH computer. They referred to the computer as the "Global Weather Simulator" and went on to say

> "This simulator provides a means for controlled experimental studies of weather problems which cannot be tackled in other types of laboratories."

The model at that time had only 15,000 instructions.

Of course, the GFDL team continues to refine their GCM (GFDL Global Atmospheric Model Development Team, 2004; Donner *et al.*, 2010); their goals have evolved and broadened. Presently, GFDL's goals are stated as:

> "The Geophysical Fluid Dynamics Laboratory is NOAA's center for climate change research through state-of-the art climate modeling methodology and advanced study of ocean and atmospheric dynamics. The goal of GFDL's research is to understand and predict the Earth's climate and weather, including the impact of human activities. GFDL conducts leading edge research on many topics including weather and hurricane forecasts, El Nino prediction, stratospheric ozone depletion, and climate change."

One of the merits of the "large center" approach that GFDL had in the early days of modeling is that they had adequate personpower and computer resources to explore several aspects of GCM research ranging from weather prediction to simulation of past, present, and future climates and exploration of planetary circulation. Clearly their modeling focused mostly on one approach. There is no single unique method of solving the basic atmospheric and ocean equations involved in GCM research. Other groups tried different approaches and that spirit of modeling innovation has remained healthy for the science.

3.2.2 UCLA Mintz–Arakawa model

Yale Mintz, Professor of Meteorology of UCLA, was another pioneer of GCM development. His modeling work with Akio Arakawa has been used both for educational purposes as well as research. As well documented in an article by Donald Johnson and Akio Arakawa (1996), reviewing the scientific contributions of Yale Mintz, it was clear that Mintz was profoundly influenced by the general circulation model experiment of Norman Phillips. Mintz had already established himself as an expert on the diagnostic analysis of the Earth's general circulation, particularly the angular momentum and energy balance. He had written an earlier paper (Mintz, 1958) that described in some detail how to go about building an atmospheric model. We repeat the quote from that paper given in Johnson and Arakawa (1996), which set a goal that

"A rational, physical method for long-period forecasts of the transient general circulation – that is, for long-range seasonal forecasts is an ambition or dream for the future. But we will never arrive at this goal (or even know whether we can arrive) unless we begin with the first step."

Not being a theoretician in the usual sense, Mintz was looking for a collaborator who had a strong background in numerical modeling. When Mintz was attending the first international conference on numerical weather prediction held in Tokyo in 1960, the organizer of the conference, Sigekata Syono, arranged for him to meet with Akio Arakawa, who agreed to join Mintz's project (Lewis, 1993). At UCLA during 1961–63, Arakawa (1966) developed a unique quadratic-conserving finite-difference scheme, which effectively dealt with the problem of nonlinear computational instability. By mid 1963, they had put together a two-level prototype atmospheric model based on the primitive equations with terrain-following sigma coordinates (normalized pressure coordinates in the vertical divided by the surface pressure). They did not have an explicit water vapor prognostic variable, but included a convective adjustment scheme that simulates the effects of shallow and deep convection by adjusting the temperature to follow a moist lapse rate and increasing the mean temperature (Mintz, 1965). More details on the later UCLA GCM were described by Arakawa and Lamb (1977) which includes a sophisticated Arakawa–Schubert cumulus parameterization (a scheme to take into account the effect of cumulus convection whose scale is smaller than the computational mesh) for handling moist convective heating.

Many scientists took the basic formulation of the UCLA model and developed it into a family of models to suit their research. Larry Gates and his collaborators developed versions of the model at the Rand Corporation and Oregon State University (Gates *et al.*, 1984). Somerville *et al.* (1974) developed a nine-level global model at the Goddard Institute for Space Studies (GISS) based on the UCLA formulation. Later, Hansen *et al.* (1983) reformulated the GISS model with their own physical parameterizations, but kept Arakawa's original dynamical core. They aimed for a model design that had "computational efficiency which allows long-range climate experiments". David Randall of Colorado State University founded a university-based GCM research and education center by extending further the UCLA model. For more detailed historical perspectives on GCM development, see Randall (2000) from the symposium proceedings in honor of Akio Arakawa. The symposium highlighted the significant influence that the Mintz–Arakawa modeling effort has had on many aspects of general circulation modeling research, including novel innovation of new numerical methods, cumulus convection, and developing successive generations of climate modeling scientists.

3.2.3 The Livermore atmospheric model (LAM)

Chuck Leith of Lawrence Livermore National Laboratory (LLNL) is another trailblazer who built a general circulation model single-handedly in the early 1960s. Leith had received a bachelor's degree in mathematics at the University of California, Berkeley in 1943 and worked on the Berkeley and Oak Ridge parts of the Manhattan Project. After the Second World War he returned to graduate school to complete a Ph.D. in mathematics while at the same time working on high energy particle physics at Livermore. In the late 1950s, he decided to work on the numerical simulation of the atmosphere. The interview of Leith by George Michael gives some insights about the motivation and goals of Leith's effort to build a GCM (Michael, 1996).

Leith wanted to use his experience of writing large computer codes for nuclear and thermonuclear devices to explore computer simulation of the atmosphere as one of the complicated physical systems which require an electronic computer to solve. He was familiar with von Neumann's activity on numerical weather prediction and talked with the staff of the ECP at Princeton, who encouraged him to get into atmospheric modeling. In 1960, LLNL was getting a UNIVAC computer that was named the Livermore Atomic Research Computer (LARC), which was the first transistor-based computer at the laboratory. To learn about atmospheric dynamics, Leith took leave from LLNL and spent a few months at the International Institute of Meteorology in Stockholm to develop a computer model for the LARC computer, which was dubbed by some as "the Leith Atmospheric Research Calculator" because he was such a big early user of the machine.

The Livermore atmospheric model (LAM) is a five-layer global model with the hydrostatic primitive equations (Leith 1965). What is unique about the Leith model, which ran successfully in 1961, is that water vapor is explicitly predicted and condensation heating is one of the energy sources, together with the effects of solar and terrestrial radiation. Recall that in the early days of running primitive-equation moist models, a difficulty was encountered in incorporating condensation heating. Its solution had to wait several years before the idea of convective adjustment was developed at GFDL and UCLA. Leith realized that a grid-point condensation heating must be reduced in proportion to the scale of moist convection in a coarse computational grid area and devised essentially a convective parameterization. This contribution was especially important and allowed for his modeling success. Another of Leith's unique contributions in the early 1960s was the visualization of computer output using a cathode ray tube (CRT), which made possible the production of color movies of flow patterns. The trick was to take three successive black and white frames of 35 mm film, which later in the printing process were printed through filters and superposed into a single color frame. Following Leith's advice, the NCAR Computing Center installed a similar CRT visualization device and developed a graphics package which enormously helped the evaluation of computer output from GCMs. Simply stated, the goal of Leith's effort was to take advantage of early computers and demonstrate that a workable primitive-equation moist model of the atmosphere could be built even by just one person. Although Leith did not continue his modeling effort, he has since made significant contributions to atmospheric dynamics and the statistical theory of atmospheric variability.

3.2.4 NCAR models

The earliest document that explains the basis for the development of a GCM at NCAR is a publication called the *NCAR Quarterly*, which in its January 1963 issue contained an article entitled "The General Circulation: A Testing Ground." In particular, it stated

> "Any examination of the total behavior of the atmosphere – its current state, its future, and what influences might be brought to bear to change its course – must therefore be hemispheric or global in scale, and take into account the rich and often bewildering complexity of interacting processes that combine to produce great variations in climate and shorter-range weather patterns."

The article goes on to say

> "The study that ties the whole together, the general circulation problem, is naturally a major focus of the atmospheric sciences . . . Progress towards a fundamental theory of the general circulation – that is, of comprehensive physical description in precise mathematical terms – is the necessary first step toward useful understanding of climate change . . ."

The author of this document is unknown. However, it is very likely that Philip Thompson, who was Associate Director of NCAR and Director of the Atmospheric Science Laboratory at that time, had provided the materials to the editor of *NCAR Quarterly* to set the goals of research to use the GCM as an integrator and testing ground for atmospheric sciences.

Many researchers joined NCAR in the early 1960s, including the two authors of this chapter. It was characteristic of Phil Thompson that he had never mentioned on what problem we should work, despite his wish to develop a GCM. One day, we asked Thompson at a lunch meeting if he would be supportive of us constructing a GCM. He was delighted by our proposal and raised his eyebrows a couple of times as his sign of approval. That was it and we finished the lunch. In those days there was no need to prepare any detailed strategic plan of how we would do it. We both had some experience building computer models and that was all we needed to get started in this emerging field. Since GCM modeling was already taking place at GFDL, UCLA, and Livermore, we thought that we should formulate our model somewhat differently from others so that we could still contribute to our understanding of how atmospheric modeling works.

We decided to build our model following closely the dynamical formulation of Richardson (1922), which used geometric height for the vertical coordinate. Incidentally, Richardson apparently thought about using the pressure as the vertical coordinate under hydrostatic equilibrium. He wrote down the well-known transformation formula between height and pressure coordinates and said

> "This (pressure) system readily yields elegant approximations".
>
> *(Richardson, 1922, p. 17)*

Since other groups had adopted the pressure-related coordinates, we decided to experiment with the height coordinate. The first- and second-generation NCAR models adopted a mountain blocking procedure in the computational mesh as described in Kasahara and Washington (1971). Later, for computational efficiency the terrain-following height coordinate was adopted to handle mountains as described in the third-generation NCAR GCM by Washington and Williamson (1977).

When development of a GCM at NCAR began in 1964, it was intended to make the model widely available to the atmospheric science community as a facility for numerical experimentation, as well as to fulfill a need at NCAR for research on weather and climate. When Francis Bretherton was appointed as President of UCAR and Director of NCAR in 1974, he decided to undertake a thorough 10-year review on the development and use of NCAR GCMs. Since the development of a GCM consumed large resources of personpower and computer time, Bretherton himself served as Chairman of the GCM steering committee, which made major decisions concerning the model development, the allocation of computer time, and scientific reviews of proposals related to the use of GCMs by university scientists as well as the NCAR staff. Details of the activities of the committee were published as a report of the GCM steering committee (1975).

It was clear that the development of a GCM at NCAR was intended to make global modeling available to the university community right from the beginning, which is to be expected by a university-managed organization. The concept of the NCAR community climate model (CCM) gradually developed over the years, but a broadening of the CCM concept took place in 1981 when the NCAR management decided to adopt a spectral model developed by W. Bourke, B. McAvaney, K. Puri, and R. Thurling in Australia as the basic computational framework (Washington, 1982). In mid 1980, the first version of the NCAR CCM, called CCM1 became available to the community (Williamson *et al.*, 1987). Later, by adding various component models of land processes, oceans, and sea ice, the model evolved to the Community Climate System Model (CCSM). Over the years, the NCAR CCSM effort has developed with support and encouragement from the National Science Foundation and in more recent years with substantial additional financial and computational support from the Department of Energy, such that each funding agency jointly manages the research effort. The CCSM is now one of the few models in the world that are freely downloadable from the Internet and available to the general scientific community for use.

Simply stated, the goal of the CCSM was to provide a fully coupled, global climate model that was state of the art and capable of performing computer simulations of the Earth's past, present, and future climate states, including understanding the causes of observed variability and change.

3.2.5 World-wide proliferation of GCM activities

The excitement among meteorologists after learning of the successful attempts at numerical weather prediction (NWP) by the Princeton Project in the 1950s and initiation of operational numerical weather forecasts by the Joint Numerical Weather Prediction (JNWP) Unit at Suitland, Maryland had spread world-wide like wild fire. During the following two decades many countries in the world started to engage in academic research on NWP and to promote formation of operational groups at weather services. Anders Persson (2005a,b,c) wrote

"The story of numerical weather prediction is a great story".

He published a detailed three-part historical account of the early development of NWP outside of the USA. Part 1 covers NWP activity in Sweden, where the leadership of Carl Gustaf Rossby (1959) revolutionized modern meteorology. It is no exaggeration to say that without the meteorological insight of Rossby and the numerical and computational foresight of von Neumann, the birth of NWP would have been considerably delayed. Part 2 describes NWP activities in many other countries of the world. In Part 3 Perrson singles out the case of British NWP activity to elucidate the reasons why the entry to operational NWP at the U.K. Met Office

(UKMO) was rather late compared with the practice of weather services in some of the other countries. The UKMO went into operational NWP in November 1965.

Chapter 5 discusses in detail synergies between NWP and GCM activities. Here, we discuss some key differences and mutual interactions between them. Both started at approximately the same time, yet took somewhat different paths even though the basic mathematical and physical principles are the same. In terms of von Neumann's terminology mentioned in the introduction, NWP is a prediction problem of the first category in which the initial condition plays a critical role, while climate modeling is a prediction problem of the second category, which is essentially governed by the balance of external energy sources and sinks in statistical equilibrium states and is independent of variations in the initial conditions. It is understandable that the immediate needs of weather services in the world required engagement in the practice of NWP rather than development of GCMs, which consume a great deal of computing resources for extended runs. Outside the USA, earlier GCM efforts at the U.K. Met Office by Gilchrist *et al.* (1973), Corby *et al.* (1977), and others in Bracknell, UK can be noted. Their specific goal at that time was to use the model for investigating the impact of sea surface temperature anomalies on atmospheric climate systems.

International NWP activities convinced the weather services of many countries about the need for increased observations on a global scale to more accurately specify initial conditions. In fact, the first weather satellite, TIROS I, was launched in 1960 and revolutionized the way we observe the atmosphere. Prior to the satellite era, atmospheric observations were limited and there were large gaps in our view of global circulation systems. In 1967 the International Council of Scientific Unions and the World Meteorological Organization proposed a Global Atmospheric Research Program (GARP) to advance the state of atmospheric science by conducting a global observation and analysis experiment which utilized every observational platform, including airplanes, buoys, ships, high-flying balloons, and satellites, as well as traditional surface and upper-air observational networks. In order to justify the huge expenditure for this international endeavor, which also had to be supported by increased research efforts, the meteorological community wanted to demonstrate the great benefits of extending the forecasting range of NWP due to more accurate initial conditions.

One of the fundamental questions in NWP concerns the predictability of the atmosphere. If we had a perfect model, how far could we forecast the atmosphere? Since the forecast models are nonlinear, small differences (errors) in the initial conditions grow with time and eventually reach the level of day-to-day differences. At that time it is said that the predictability is lost. At the suggestion of the Committee on Atmospheric Sciences of the National Academy of Sciences, a group of scientists led by J. Charney conducted numerical experiments using the early versions of GCMs developed at UCLA and GFDL and found an error-doubling time of 5 to 7 days (Charney *et al.*, 1966). With this error-doubling time the deterministic predictability limit becomes 2 to 4 weeks, depending on observational errors in the initial conditions. In contrast, theoretical estimates of error-doubling time based on nonlinearity of fluid motions were obtained by Thompson (1957) and Lorenz (1965), who were the most outstanding investigators in the 1960s involved in the theoretical limits of the predictability of weather. They gave the values of error-doubling time of 2 to 3 days, which are considerably shorter than those given by the numerical models. The discrepancy between these estimates came from the fact that the level of transient eddy activity in early GCMs was rather low due to the coarseness of computational grids and many shortcomings in the physical processes in the models. Nevertheless, these numerical experiments with GCMs met one of the goals of GCMs, i.e. to test a hypothesis regarding the behavior of the atmosphere. We will discuss more on this objective in connection with our anticipation of climate changes. Incidentally, the estimate of error-doubling time with current NWP models shows values of 1.5 to 1.7 days, which gives a deterministic predictability limit of about 10 days (Simmons *et al.*, 1995). This implies that the prediction problem beyond 10 days must be considered in ensemble mode, an idea which was not anticipated in the early days of NWP.

In anticipation of a major improvement in medium-range weather forecasts with the use of expanded observations collected during the Global Weather Watch in 1978–79, a group of European countries decided to set up the European Centre for Medium-range Weather Forecasts (ECMWF) in Reading, UK, in 1975, with Aksel Wiin-Nielsen as the first Director. As noted by Lennart Bengtsson (2000), at the beginning ECMWF adopted an extended forecasting model developed by Miyakoda *et al.* (1974) at GFDL, which was on Kurihara's global grid instead of a hemispheric model. It is clear that extended-range NWP requires a model comparable to a GCM with full physics. The development of various physical parameterizations for GCMs contributed to the advance in this third category of prediction problems. One important aspect of the third category is the analysis of observations for the input data to NWP. It turned out that the initialization for the primitive equation (PE) model was a considerable challenge, similar to the problem that L. F. Richardson had experienced. Because the PE model allows motions of inertia-gravity (IG) (wave oscillations due to the restitutive forces of the Earth's gravity and rotation) type as well as Rossby type (planetary scale waves due to the latitudinal variation of the Earth's rotation which were first observationally confirmed by C.-G. Rossby and collaborators), small-scale errors in the initialization are projected onto the IG modes, then reappear during the time integration and mask the large-scale quasi-geostrophic motions, unless the magnitudes of errors are reduced in the process of initialization. Therefore, the data initialization requires an adjustment to the observed data to suppress unrealistically large IG motions. After a great deal of research and experimentation, the NWP researchers were finally able to solve this challenging problem during the 1970s and 1980s. The history of meeting this challenge is another great scientific story.

During the 1980s the likelihood of changing the atmospheric and oceanic climates through human activities became a particular concern for our society. Actually, even during the 1950s some meteorologists were worried about the climatic consequences of burning fossil fuel and the need to study the global transport of air pollution, as evident from the aforementioned article by Rossby (1959). Rossby suggested to Paul Crutzen, who was his programmer and who later, in 1995, became a Nobel laureate in Chemistry, that he tackle the problem of mathematically modeling the chemical reactions to be eventually included in GCMs, while most of us were struggling with NWP (personal communication, 1973). In 1988 the World Meteorological Organization (WMO) and the United Nations Environmental Programme (UNEP) established the Intergovernmental Panel on Climate Change (IPCC)

> "to provide the decision-makers and others interested in climate change an objective source of information about change".
>
> *(http://www.ipcc.ch/about/index.htm)*

Chapter 10 describes the important relationships that have developed between the IPCC and GCM development.

The governments of many countries in the world responded to the request by the IPCC to provide scientific findings on the possibility of climate changes by organizing climate research centers in their weather services or elsewhere and by further accelerating climate research using GCMs, which had been carried out since the 1970s. For example, the U.K. Met Office established the Hadley Centre for climate prediction research in 1990 and scientists started to develop a series of climate models by combining a GCM with ocean and land-surface models of various complexities in numerics and physical parameterizations, depending on the availability of computer resources at the time of model development. Similarly, in 1993 the Canadian Center for Climate Modelling and Analysis (CCCma), which was originally set up in the early 1970s, was upgraded. The Canadians are particularly interested in rising temperatures in the Arctic and have been developing a series of climate models in a traditionally strong background of NWP development. Likewise, the development of a series of climate models is currently underway at the Max Planck Institute for Meteorology (MPI-M) in Hamburg, Germany jointly with ECMWF, but the objective of MPI-M is much broader than just modeling. They are also aiming to tackle chemical and biological problems connected with global and regional climate change. Of course, all industrial countries of the world have research organizations in their own governments and/or in universities which develop climate system models of various complexities. However, their goals are all more or less similar, i.e. to contribute to understanding the past, current, and future variations of global and regional climate due to natural and anthropogenic causes. As far as a projection of future climate change is concerned, the use of climate prediction models is the only possibility. For a reliable projection, it is crucial to have a climate model in which all the relevant physical, chemical, and biological processes are represented as accurately as possible. This is a tall order, but it is the modeler's "dream".

Since its inception in 1988, the IPCC has conducted scientific assessments of climate change and published their findings in the form of periodic reports. The fourth and most recent IPCC assessment report (AR4) was published as Climate Change 2007. Working Group I of the report (IPCC, 2007) presented the physical science basis of findings from the climate model results conducted by eighteen modeling groups around the world, following a standard set of specific scenarios involving changes in environmental conditions resulting from human activities such as the increased amount of CO_2 due to consumption of fossil fuels. The reliability of climate models is evaluated based on analyses of model results of the twentieth-century runs with comparison against observed data. We will not go into details of specifications of the climate models used by various modeling groups, since the IPCC report (IPCC, 2007, pp. 596–598) gives the references to their component models of the atmosphere, ocean, sea ice, and land-surface processes. Rather we should emphasize that various groups adopt different computational resolutions together with various formulations of the physical processes in each model component. Another delicate issue unique to climate modeling is how to combine these various model components into a single climate model, such as coupling the atmosphere with the oceans. Altogether there are huge disparities in computational and physical specifications among different models. Nevertheless, the goals of each modeling group are identical, i.e. to simulate the long-term variations of the Earth's climate in the most faithful way.

We should point out that this goal of modeling is only a part of the overall goals of GCM research, which include not only a faithful reproduction of nature, but also aim at understanding how the atmosphere and oceans behave. For the purpose of understanding we must perform numerical experiments with and without a particular effect under consideration, or focus on one particular process by eliminating all other complications not directly relevant to that particular process. As an example, to understand the moist dynamics in the atmosphere, a numerical experiment is perhaps the only way, unlike the case of dry dynamics for which an analytical approach is also available. One such example is the use of the aquaplanet configuration in GCMs in which zonally uniform sea surface temperatures are specified by stripping down all complicated processes coming from a realistic topography over the land surface. Since the Madden–Julian oscillation (MJO; Madden and Julian, 1994) is a typical planetary-scale disturbance in the tropics with time-scales of 30 to 50 days, the simulation of the MJO has been used as a benchmark for successful GCMs, but so far none of the current GCMs are able to reproduce the phenomenon to the extent that we can claim we understand its mechanism (Slingo *et al.*, 2005).

Because the MJO appears over the oceans and is generated by moist convection, which is dominated by small-scale motions, the use of an aquaplanet configuration is ideal for investigation of the nature of the MJO by gaining the computational accuracy of high-resolution calculations but retaining simplification of its environment. In fact, Nasuno *et al.* (2007) have done just that in their very-high-resolution modeling study of the MJO by analyzing aquaplanet runs with a horizontal grid size ranging from 3.5 to 7 km in a 40-day period using explicit cloud physics without cumulus parameterization. They showed the structure of very detailed moist convective multi-scale systems which are embedded in the planetary-scale equatorial waves and move in different directions. Whether the nature of MJO-like phenomena in the aquaplanet runs is identical to what would be expected in a full-blown GCM of comparable grid resolution can be debated, but their results look very realistic based on our knowledge of cloud systems obtained from satellite observations. If their findings on the need for high-resolution grids with explicit cloud physics to realistically simulate tropical convective systems in a GCM are correct, it is not likely that we can design a cumulus parameterization which can simulate the MJO in present-day GCMs with relatively coarse computational grids any time soon.

3.2.6 Timeline of climate model development

The development of climate system models mostly started with atmospheric models in the 1960s, along with an almost singular effort on ocean modeling by Kirk Bryan at GFDL, which is discussed in Chapter 7. Figure 3.2 shows a schematic of the time evolution of a generic climate system model by adding various model components to handle details of physical, chemical, and biological processes from the 1960s to 2010. It should be noted that each of the components has undergone continuous improvements through the addition of more realistic physical processes. In the early development of GCMs there was emphasis on keeping the computational costs to a minimum. However, that is of less concern nowadays with the use of multiprocessor supercomputer systems. As models become more complex, more computer processors can be added easily to the model such that the wall clock time for execution will not be greatly impacted.

It should be pointed out that there was no master plan for GCM model development. Scientists from the various modeling groups frequently interact at both national and international scientific meetings, where they present their model development plans and research results. There remains a general consensus in the community that we should continue to build future versions of the models such that

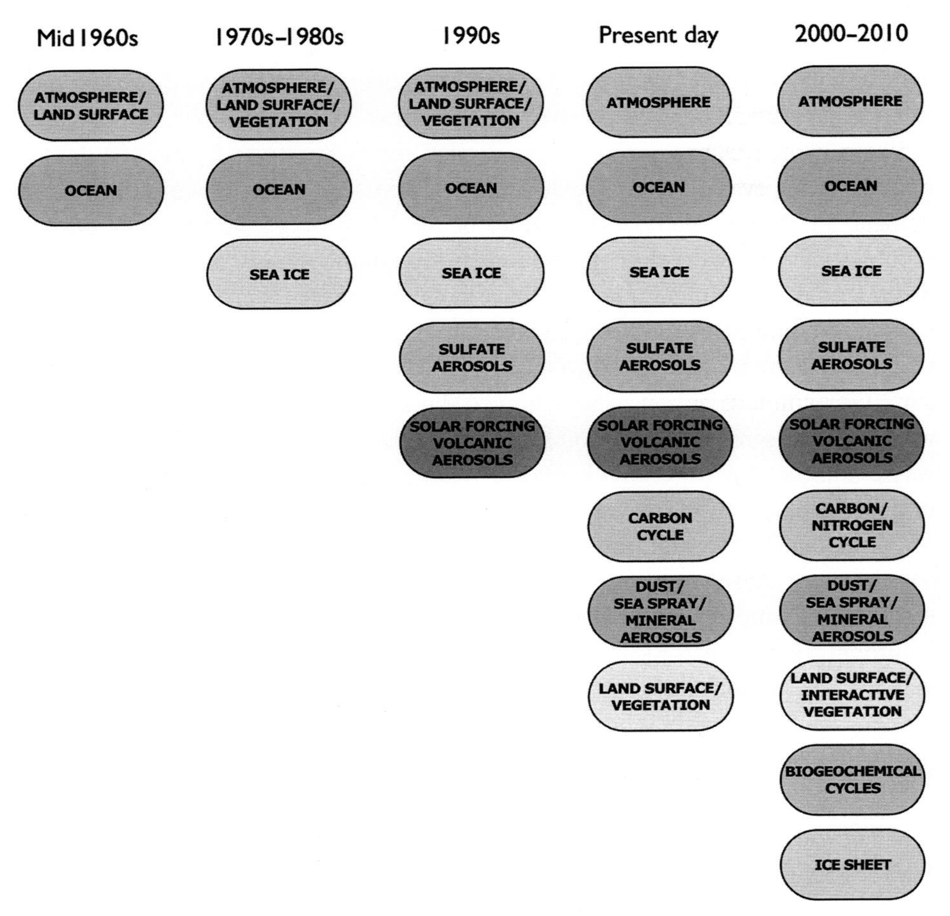

Figure 3.2 The development of climate models as a function of time. The major climate and Earth system modeling groups are developing a goal for increasingly more complex models over time that include biogeochemical cycles such as the carbon cycle. The newest additional features still in the process of being added are detailed ice sheet and integrative assessment components.

they reflect the true complexity of the Earth's climate and biogeochemical systems. The science is still in a relatively early stage of development compared to other fields of science such that consolidation of efforts into "one or a few centers" would be counterproductive for the continued development of the science.

3.3 Future goals for climate and Earth system modeling

3.3.1 Evolution of dynamical core for climate and Earth system modeling

Ever since Phillips' GCM experiment, which used a quasi-geostrophic baroclinic model, the dynamical cores of GCMs have all been based on the hydrostatic primitive equations of motion, which adopt two major simplifications – "shallowness" and "traditional" approximations in addition to the minor approximation that the geopotential field of the Earth is spherically symmetric. Looking ahead at building Earth system models that include the whole atmosphere and deal with multi-scale weather events, it is crucial to use the most accurate form of the dynamical equations. Hence, we should question the accuracy of these approximations.

The *shallowness approximation* is that any undifferentiated radial distance r appearing in the equations of motion in spherical coordinates is replaced by a constant radius of the Earth, a, and the differentiated form of radial distance r is replaced by dz, where z denotes the altitude from the surface of the sphere. When this approximation is introduced into the equations of motion directly to simplify the equations, the resulting system does not retain the principle of angular momentum conservation. This means that, as Phillips (1966) pointed out, direct introduction of the shallowness rule to the equations of motion leads to a physically inconsistent set. To have a physically consistent set under the shallowness approximation, he argues that the Coriolis terms coming from the horizontal component of the Earth's rotation and a few other minor terms related to the curvature of the coordinate system must be omitted. The resulting set becomes the nonhydrostatic shallow (NHS) model, following the nomenclature of White *et al.* (2005). In the NHS model the vertical acceleration is still present. If this term is ignored, the vertical equation of motion reduces to hydrostatic equilibrium. Then, the resulting system becomes the hydrostatic primitive equation (HPE) model, or simply the "primitive equation (PE)" model.

The *traditional approximation*, which is abbreviated here as the TA, was coined by Eckart (1960), and is essentially a simplification to the equations of motion of a rotating fluid system by neglecting the horizontal component of the Coriolis vector. It is a powerful simplification used traditionally in atmospheric and ocean dynamics. However, Eckart wondered about the physical significance of the TA and initiated a study on the missing physics. He concluded with the conjecture that

> "the effects of the horizontal component of the Coriolis vector may be "very marked" for (inertia-gravity wave) frequencies in the neighborhood of the inertial frequency".

The inertial frequency is the Coriolis parameter, defined by $2\Omega\sin\phi$, where Ω denotes the Earth's angular velocity of rotation and ϕ is the latitude. Although the full meaning of "very marked" is not yet evident, research on understanding the role of the Coriolis terms due to $2\Omega\cos\phi$, referred to here as non-traditional effects, has been rather sporadic in the past.

There has been a renewed interest in understanding non-TA effects over the last 10 years or so, associated with the desire to improve prediction models for multi-scale weather and oceanic phenomena. It is well known that there are two important wave modes in the atmosphere and ocean, beside the acoustic modes which do not play a major role in the general circulation. One is the planetary or Rossby mode, which arises from the latitudinal variation of the vertical component of the Coriolis vector, i.e. the β- effect. The Rossby mode is essentially rotational and therefore quasi-geostrophic. The other is the inertia-gravity mode, which results from the restoring forces of buoyancy and rotation and is essentially irrotational and therefore non-geostrophic. They are essential to describe medium- and small-scale flows such as frontal motions, severe storms, and tropical cyclones. The importance of inertia-gravity motions, even in large-scale flows, has begun to be recognized particularly in the tropics (Žagar *et al.*, 2009). Without going into details of recent findings, such as from a review article by Gerkema *et al.* (2008) which provides a good source of references, we can assert that the non-traditional Coriolis effects permit a more accurate description of the class of motions belonging to the inertia-gravity wave spectrum. In other words, the accuracy of inertia-gravity motions is compromised, sometimes spectacularly, by the TA.

Now, the question is how much model accuracy is sacrificed by the adaption of these approximations. The answer depends on how we plan to use the next-generation GCMs. If we intend to include the whole atmosphere, even the shallowness approximation can lead to simulation errors of more than O (1%) (O: order of) in flow patterns. Moreover, the assumption of constant gravity brings another error of similar magnitude. Can we accept simulation errors of O (1%)? The usual answer is that such errors can be tolerated, because the uncertainty in physical parameterizations may yield much larger errors in flow simulation. Perhaps their largest uncertainty may come from the calculation of various hydrological phenomena involving the phase transition of water vapor, such as in clouds and precipitation. They in turn interact with solar and terrestrial radiation. The following example may indicate the magnitude of uncertainty in hydrological parameterizations.

In 2006, version 3 of the Community Climate System Model, called CCSM3 (Collins *et al.*, 2006), was released to replace CCSM2. To indicate the magnitude of changes in the physical packages, we quote a following example.

"The atmosphere in CCSM3 absorbs 7.1 Wm^{-2}, more shortwave radiation under clear-sky conditions than CCSM2. The new aerosols increase the absorption by 2.8 Wm^{-2} ... The new treatment of near-infrared extinction by H_2O increases the global mean clear-sky and all-sky atmospheric absorption by 4.0 and 3.1 Wm^{-2}, respectively".

Since a mean flux of incoming solar energy at the top of the model is about 340 Wm^{-2}, the magnitudes of change in the physical processes from CCSM2 to CCSM3 are of O (1%). Even though the uncertainty of each component in the physics package is thought to be much higher than O (1%), the degree of adjustable amount in tuning a physical process remains on O (1%) to maintain the total energy balance required to run the model in equilibrium.

Now, we should evaluate the magnitude of errors due to the traditional approximation, namely the inaccuracy of motions belonging to the class of inertia-gravity modes by neglecting the cosine Coriolis terms. The magnitude of atmospheric horizontal flows is O (10 ms^{-1}), equivalent to that of the quasi-geostrophic or rotational part of the motions. The non-geostrophic or irrotational part of atmospheric motion is on O (1 ms^{-1}) which is not observable and must be inferred mostly from the dynamics of inertia-gravity modes. Therefore, non-traditional effects improve calculation of the irrotational part of motions, but if the amount of improvement in the calculations is only a fraction of 1 ms^{-1}, the gain may look marginal. It turns out, however, that the non-traditional effects can significantly enhance the strength of inertia-gravity motions when suitable forcing excites the motions of near-inertial periods due to resonance of singular-type solutions unique to the presence of the cosine Coriolis terms. A discussion of such unusual behavior of inertia-gravity modes without the TA and its implications for understanding near-inertial oscillations in the oceans can be found in Kasahara (2010). It is very likely that the non-traditional effects are similarly present in the atmosphere due to the fact that the role of large-scale diurnally recurring motions such as manifested by thermal tides are more prevalent in the middle to upper atmosphere. Moreover, the non-traditional effects give rise to active vertical dynamic coupling. Under the TA, the vertical dynamical coupling is only passive through vertical mass continuity responding to the action of thermodynamics. Therefore, all together the non-traditional effects provide a more sophisticated description of the motions under rotation than expected from the PE model.

In summary, a dynamical core based on the PE formulation for current GCMs has model errors which are at least O (1%) and become substantially more under resonant conditions. This magnitude likely exceeds what one encounters while tuning various physical parameterizations. Because the influence of smaller-scale inertia-gravity motions to large-scale planetary motions through nonlinear effects takes a time of O (10 days), the benefits of non-TA effects will appear more on climate time-scales than the time-scales for short- to medium-range NWP. For the design of the next-generation Earth system model, it seems imperative to eliminate both the shallowness and traditional approximations and adopt the dynamical core based on the nonhydrostatic deep (NHD) formulation using the radial coordinate r from the center of sphere. Global circulation models of the atmosphere based on the NHD formulation have been developed at the U.K. Met Office (Staniforth and Wood, 2008) and the Frontier Research Center for Global Change (FRCGC) in Japan (Satoh *et al.*, 2008). For very-high-resolution numerical models, inclusion of the vertical acceleration is most beneficial for motions of cloud scale as well as the non-traditional Coriolis effects.

3.3.2 Goals for improving physical processes in climate and Earth system models

There have been a number of groups that have stated future goals for modeling the climate system. We have chosen two recent examples of planning goals, one the United States Department of Energy workshop on Scientific Grand Challenges in Climate Change Science and the Role of Computing at the Extreme Scale, held in Washington, DC on November 5–6, 2008 and chaired by one of us (WMW). The part of the workshop that discussed goals was co-chaired by Dave Bader of the Lawrence Livermore National Laboratory and William Collins of the Lawrence Berkeley National Laboratory (DOE report, Bader and Collins, 2009). The other example is the World Modeling Summit (WMS) for Climate Prediction held in May 2008 in Reading, United Kingdom. These two examples show how the community of scientists has striven to produce very detailed goals, which is in sharp contrast to earlier pioneering efforts. The reason for this difference is that we now understand the climate system and its complex interactions much better, thus we can be more specific in our scientific and societal goals. Clearly the last few decades of research have answered fundamental scientific questions; however, the research has also generated additional questions that need to be answered.

Over the last four decades there has been substantial improvement in the treatment of physical processes used in climate and Earth system models. Generally, present models have the following basic physical process components: atmosphere, ocean, land/vegetation, sea ice, cloud macrophysics and microphysics, carbon cycle, other biogeochemical cycles, and most recently ice sheet dynamics and thermodynamics. As expected in each of the existing components there have been constant refinements and improvements. Also, we now have fewer "tunable" parameters because the range of adjustment must be within the bounds of the observable range of certain parameters such as snow and sea ice surface albedo. Over time these bounds will become tighter and tighter.

The near-term goals are expected to be increased evaluation of the shortcomings and biases of present-generation models. As we gain more knowledge from *in situ* field studies and satellite measurements, we will continue to objectively compare our model simulations with observed data. From such detailed comparisons, we can improve the treatment of the physical processes. Without such observed data it would be nearly impossible to judge whether we had more accurate process modeling in the GCMs. A possible list of goals is shown below, not necessarily in priority order:

1. Interactive or dynamic vegetation that changes with climate including the statistical effect of forest fires
2. Changes in anthropogenic land cover

3. Interactive carbon, methane, nitrogen, and sulfur cycles
4. Interactive treatment of carbon, sulfate, dust aerosols
5. Interactive tropospheric and stratospheric chemistry
6. Volcanic eruption impacts on climate
7. Cloud microphysics interactions
8. Provide information on various "climate tipping points" and instabilities
9. Develop global and regional climate change models with emphasis on impacts that can be used for future climate change mitigation and adaptation studies
10. Develop high-resolution models that can scale well on present and next generation high performance supercomputer systems. This will likely involve including a non-hydrostatic deep formulation in future high-resolution GCMs
11. Investigate changes in extreme events such as precipitation changes
12. Examine the possibility of abrupt climate change

The climate modeling community is presently testing the coupling of integrative assessment models (IAMs) to climate and Earth system models. Modern integrative assessment models deal with changes in global and regional economics, demographics, energy and associated greenhouse gas emissions, atmospheric composition, climate, land cover, and their impacts on society. The scientific issues concerning mitigation and adaptation to climate change will require new sets of future climate simulations that will be very useful to policymakers. Clearly such projections will have sizeable uncertainties; however, they can provide some guidance on policy with regard to the impacts of future climate change. So far the climate part of integrative assessment models is rather rudimentary. In order to improve this limitation it is expected that several international groups will strive for coupling Earth system and climate models to IAMs. Because it is expected that sea level and temperature will rise substantially in the future, and there will be significant changes in precipitation, depending on location, we will need to couple our models to detailed high-resolution hydrological models such that we can predict surface water changes more accurately. We already know that regional climate models (RCMs) that use boundary forcing from global GCMs can provide much more detailed prediction of precipitation estimates, especially in mountainous and coastal areas. Ultimately, as a goal, we would like to have global high-resolution GCMs of a few km in the horizontal, but until then we will have to rely on coupling of the global and regional models.

A major issue for future climate models is to use effectively modern parallel supercomputing systems. This requires spreading the computations required to solve climate model equations over very large numbers of interconnected computer processor systems. Washington *et al.* (2009) describe various methods for solving the equations used by most modeling groups. This involves having a coupler or driver that is responsible for merging fields between various model components, mapping fields with different resolutions, coordinating communications, and sequencing the component models for execution. It is also interesting to see new computation methods being developed, such as the use of finite element and other numerical techniques as described in the DOE report referred to earlier (Bader and Collins, 2009). A specific example of this can be seen in Taylor *et al.* (2007).

One of the most important goals that is not often mentioned in a review like this is the need to constantly replenish the scientific talent. Each generation of GCM-related scientist has an obligation to leave a legacy to the next generation of scientists. They can do this by serving as mentors and providing knowledge on the successes and failures in our field. From time to time there are calls for consolidation of the many scientific efforts into one big group effort. We would caution against such a direction for the progress of science. Every aspect of GCMs in dynamics, numerics, physics, and applications is still evolving and requires new and innovative approaches as nothing is "completely settled" science. On the other hand, building of a state-of-the-art GCM is beyond the capability of an individual or even a small group. The path forward is assembling a cadre of experts to work jointly with other groups to continue the development of more accurate GCMs.

Acknowledgments

The authors of this chapter thank Leo Donner for his significant contribution to atmospheric sciences and for leading this effort to document the historical development of general circulation models. We also thank an anonymous reviewer and John Lewis who read our manuscript and offered useful suggestions for improvement of the text.

References

Arakawa, A. (1966). Computational design for long-term numerical integration of the equations of fluid motion: Two-dimensional incompressible flow. Part I. *Journal of Computational Physics*, 1, 119–143.

Arakawa, A. and Lamb, V. R. (1977). Computational design of the basic dynamical processes of the UCLA general circulation model. In *Methods in Computational Physics*, 17, General Circulation Models of the Atmosphere, ed. Change J., Academic Press, pp. 173–265.

Bader, J. and Collins, W. M., Co-leads (2009). Model Development and Integrative Assessment, DOE report on: Scientific grand challenges: challenges in climate change science and the role of computing at the extreme scale, Chair: Warren Washington. http://www.sc.doe.gov/ober/ClimateReport.pdf

Bengtsson, L. (2000). The development of medium range forecasts. In *50th Anniversary of Numerical Weather Prediction Commemorative Symposium*. Deutsch Meteorologische Gesellschaft e. V., 119–138.

Charney, J. G. and Phillips, N. A. (1953). Numerical integration of the quasi-geostrophic equations for barotropic and simple baroclinic flows. *Journal of Meteorology,* 10, 71–99.

Charney, J. G. and members of Committee of Atmospheric Sciences (1966). The feasibility of a global observation and analysis experiments. *Bulletin of the American Meteorological Society*, 47, 200–220.

Collins, W. D., Bitz, C. M., Blackmon, M. L., *et al.* (2006). The Community Climate System Model version 3 (CCSM3). *Journal of Climate*, 19, 2122–2143.

Corby, G. A., Gilchrist, A., and Rowntree, P. R. (1977). United Kingdom Meteorological Office five-level general circulation model. In *Methods in Computational Physics*, 17, General Circulation Models of the Atmosphere, ed. Change, J., Academic Press, pp. 67–110.

Donner, L. J., Wyman, B. L., Hemler, R. S., Horowitz, L. W., Ming, Y., Zhao, M., and 20 co-authors (2010). The dynamical core, physical parameterizations, and basic simulation characteristics of the atmospheric component of the GFDL Global Coupled Model CM3. J. Climate. Article submitted.

Eckart, C. (1960) *Hydrodynamics of Oceans and Atmospheres*. Pergamon Press.

Edwards, P. (2000). Chapter 2: A Brief History of Atmospheric General Circulation Modelling. In *General Circulation Model Development – Past, Present, and Future*, ed. Randall, D. A., International Geophysics Series; Vol. 70, San Diego, CA, Academic Press, pp. 67–90.

Gates, W. L., Han, Y.-J., and Schlesinger, M. E. (1984). *The Global Climate Simulated by a Coupled Atmosphere-Ocean General Circulation Model*: Preliminary results. Report No. 57, Climatic Research Institute, Oregon State University, Corvallis, OR, 31 pp.

GCM steering committee (1975). *Development and Use of the NCAR GCM*: A report of the GCM steering committee. NCAR Tech. Note. NCAR-TN/STR-101. 177 pp.

Gerkema, T., Zimmerman, J. T. F., Maas, L. R. M., and van Haren, H. (2008). Geophysical and astrophysical fluid dynamics beyond the traditional approximation. *Reviews of Geophysics*, 46, RG2004.

GFDL (1980). A Review of Twenty-Five Years of Research at GFDL, See GFDL Activities-FY 80 Plans-FY81 Report, pp.1–200.

GFDL Global Atmospheric Model Development Team (2004). The new GFDL global atmosphere and land model AM2-LM2: Evaluation with prescribed SST simulations. *Journal of Climate*, 17, 4641–4673.

Gilchrist, A., Corby, G. A., and Newson, R. L. (1973). A numerical experiment using a general circulation model of the atmosphere. *Quarterly Journal of the Royal Meteorological Society*, 99, 2–34.

Hansen, J., Russell, G., Rind, D., Stone, P., Lacis, A., Lebedeff, S., Ruedy, R., and Travis, L. (1983). Efficient three-dimensional global models for climate studies: Models I and II. *Monthly Weather Review*, 111, 609–662.

IPCC (2007). Climate Change 2007: The Physical Science Basis. Contribution of Working Group I to the Fourth Assessment Report of the Intergovernmental Panel on Climate Change. In *The IPCC Fourth Assessment Report, or AR4*, eds. Solomon, S., Qin, D., Manning, M., Chen, Z., Marquis, M., Averyt, K. B., M. Tignor, and Miller, H. L., Cambridge, United Kingdom and New York, NY, USA, Cambridge University Press.

Johnson, D. and Arakawa, A. (1996). On the scientific contributions and insight of Professor Yale Mintz. *Journal of Climate*, 9, 3211–3223.

Kasahara, A. (2010). A mechanism of deep-ocean mixing due to near-inertial waves generated by flow over bottom topography. *Dynamics of Atmospheres and Oceans*, 49, 124–140.

Kasahara, A. and Washington, W. M. (1971) General circulation experiments with a six-layer NCAR model, including orography, cloudiness, and surface temperature calculations. *Journal of the Atmospheric Sciences*, 28, 657–701.

Leith, C. (1965). Numerical simulation of the earth's atmosphere. In *Methods in Computational Physics*, 4 (Applications in Hydrodynamics). ed. Alder, B. *et al.* Academic Press, 1–28.

Leith, C. (1975). The Design of a Statistical-Dynamical Climate Model and Statistical Constraints on the Predictability of Climate, In *The Physical Basis of Climate and Climate Modeling*, Joint Organizing Committee, Global Atmospheric Research Programme Publications Series, No. 16, 137–141.

Lewis, J. M. (1993). Meteorologists from the University of Tokyo: Their exodus to the United States following World War II. *Bulletin of the American Meteorological Society*, 74, 1351–1360.

Lewis, J. M. (1998). Clarifying the dynamics of the general circulation: Phillips's 1956 experiment. *Bulletin of the American Meteorological Society*, 79, 39–60.

Lewis, J. M. (2008). Smagorinsky's GFDL: Building the Team. *Bulletin of the American Meteorological Society*, 89, 1339–1353.

Lorenz, E. N. (1965). A study of the predictability of a 28-variable atmospheric model. *Tellus*, 17, 130–141.

Lynch, P. (2006). *The Emergence of Numerical Weather Prediction, Richardson's Dream*. Cambridge Univ. Press, 279 pp.

Madden, R. A. and Julian, P. R. (1994). Observations of the 40–50 day tropical oscillation – a review. *Monthly Weather Review*, 122, 814–837.

Manabe, S., Smagorinsky, S, J., and Stickler, R. F. (1965). Simulated climatology of a general circulation model with a hydrological cycle. *Monthly Weather Review*, 93, 769–798.

Michael, G. (1996). An Interview with Chuck Leith, http://www.computer-history.info/Page1.dir/pages/Leith.html

Mintz, Y. (1958). Design of some numerical general circulation experiments, *Bulletin of the Research Council of Israel*, 76, 67–114.

Mintz, Y. (1965). Very long-term global integration of the primitive equations of atmospheric motion. In *WMO-IUGG Symposium on Research and Development Aspects of Long-range Forecasting*. Tech. Note No. 66, WMO-No.162.TP.79, World Meteor. Org., pp. 141–167.

Miyakoda, K., Sadler, J. C., and Hembree, G. D. (1974). An experimental prediction of the tropical atmosphere for the case of March 1965. *Monthly Weather Review*, 102, 571–591.

Nasuno, T., Tomita, H., Iga, S., Miura, H., and Satoh, M. (2007). Multiscale organization of convection simulated with explicit cloud processes on an aquaplanet. *Journal of the Atmospheric Sciences*, 64, 1902–1921.

Oort, A. (1964). On estimates of the atmospheric energy cycle. *Monthly Weather Review*, 92, 483–493.

Persson, A. (2005a). Early operational numerical weather prediction outside the USA: an historical introduction. Part I: Internationalism and engineering NWP in Sweden, 1952–69. *Meteorological Applications*, 12, 135–159.

Persson, A. (2005b). Early operational numerical weather prediction outside the USA: an historical introduction. Part II: Twenty countries around the world. *Meteorological Applications*, 12, 269–289.

Persson, A. (2005c). Early operational numerical weather prediction outside the USA: an historical introduction – Part III: Endurance and mathematics – British NWP, 1948–1965. *Meteorological Applications*, 12, 381–413.

Pfeffer, R. L. (Ed) (1960). *Dynamics of Climate: The Proceedings of a Conference on the Application of Numerical Integration Techniques to the Problem of the General Circulation held October 26–28, 1955*. Pergamon Press.

Phillips, N. A. (1956). The general circulation of the atmosphere: a numerical experiment. *Quarterly Journal of the Royal Meteorological Society*, 82, 123–164.

Phillips, A. (1966). The equations of motion for a shallow rotating atmosphere and the "traditional approximation". *Journal of the Atmospheric Sciences*, 23, 626–628.

Platzman, G. W. (1967). A retrospective view of Richardson's book on weather prediction. *Bulletin of the American Meteorological Society*, 48, 514–550.

Platzman, G. W. (1979). The ENIAC computations of 1950 – Gateway to numerical weather prediction. *Bulletin of the American Meteorological Society*, 60, 302–312.

Randall, A. (Ed.) (2000). *General Circulation Model Development*. International Geophysics Series; Vol. 70, San Diego, CA, Academic Press.

Richardson, L. F. (1922). *Weather Prediction by Numerical Process*. Cambridge University Press.

Rossby, C.-G. (1959). Current problems in meteorology. In *The Atmosphere and the Sea in Motion*. ed. Bolin B., Rockefeller Inst. Press, pp. 9–50. [This great article of Rossby's was actually written in Swedish in 1957 and translated into English to be included in this memorial volume. A part of this article was presented by Rossby as the special after-dinner speaker at an annual meeting of the American Meteorological Society (AMS) in January 1957 in New York at the occasion of his receiving the Award for Outstanding Services to the AMS (See Vol. 3, 1957, Bulletin of AMS, pp. 180–181.) He died in August 1957 at the age of 59.]

Satoh, M., Matsuno, T., Tomita, H., Miura, H., Nasuno, T., and Iga, S. (2008). Nonhydrostatic icosahedral atmospheric model (NICAM) for global cloud resolving simulations. *Journal of Computational Physics*, 227, 3486–3514.

Schneider, S. H. and Dickinson, R. E. (1974). Climate Modeling, *Reviews of Geophysics, Space Physics*, 12, 447–493.

Shuman, F. G. (1989). History of numerical weather prediction at the National Meteorological Center. *Weather and Forecast.*, 4, 286–296.

Simmons A. J., Mureau, A. R., and Petroliagis, T. (1995). Error growth and estimates of predictability from the ECMWF forecasting. *Quarterly Journal of the Royal Meteorological Society*, 121, 1739–1771.

Slingo, J. M., Inness, P. M., and Sperber, K. R. (2005). Modeling. In *Intraseasonal Variability in the Atmosphere-Ocean Climate System*. eds. Lau, K. M. and Waliser, D. E. Praxis. Springer, pp. 361–388.

Smagorinsky, J. (1983). The beginnings of numerical weather prediction and general circulation modeling: Early recollections, *Advances in Geophysics*, 25, 3–37.

Smagorinsky, J., Manabe, S., and Holloway, J. L. (1965). Numerical results from a nine-level general circulation model. *Monthly Weather Review*, 93, 727–768.

Somerville, R. C., Stone, P. H., Halem, M. *et al.* (1974). The GISS model of the global atmosphere. *Journal of the Atmospheric Sciences*, 31, 84–117.

Staniforth, A. and Wood, N. (2008). Aspects of the dynamical core of a nonhydrostatic, deep-atmosphere, unified weather and climate-prediction model. *Journal of Computational Physics*, 227, 3445–3464.

Taylor, M. A., Edwards, J., Thomas, S., and Nair, R. (2007). A mass and energy conserving spectral element atmospheric dynamical core on the cubed-sphere grid. *Journal of Physics*: Conference Series 78, 012074, 5pp.

Thompson, P. D. (1957). Uncertainty of initial state as a factor in the predictability of large scale atmospheric flow pattern. *Tellus*, 9, 275–295.

Thompson, P. D. (1983). A history of numerical weather prediction in the United States. *Bulletin of the American Meteorological Society*, 64, 755–769.

Washington, W. M. (1982). *Documentation for the Community Climate Model (CCM), Version 0*. NCAR report, Boulder, Colorado, NTIS No. PB82 194192.

Washington, W. M. and Williamson, D. L. (1977). A description of the NCAR global circulation models. In *Methods in Computational Physics*, 17, General Circulation Models of the Atmosphere, ed. Change J., Academic Press, pp. 111–172.

Washington, W. M., L. Buja, and A. Craig (2009). The computational future for climate and earth system models: On the path to petaflop and beyond. *Phil. Trans. of the R. Soc. A*, The environmental eScience revolution, Royal Society Publishing, London, UK, 367, 833–846.

White, A. A., Hoskins, B. J., Roulstone, I., and Staniforth, A. (2005). Consistent approximate models of the global atmosphere: shallow, deep, hydrostatic, quasi-geostrophic and non-hydrostatic. *Quarterly Journal of the Royal Meteorological Society*, 131, 2081–2107.

Williamson, D. L., Kiehl, J. T., Ramanathan, V., Dickinson, R. E., and Hack, J. J. (1987). *Description of NCAR Community Climate Model (CCM1)*. NCAR Tech. Report, NCAR/TN-285+STR, 112 pp.

Žagar, N., Tribbia, J., Anderson, J. L., and Raeder, K. (2009). Uncertainties of estimates of inertia-gravity energy in the atmosphere. Part I: Intercomparison of four analysis systems. *Monthly Weather Review*, 137, 3837–3857.

Appendix

Letter from John von Neumann, with handwritten comments from Harry Wexler in square brackets (provided by Geophysical Fluid Dynamics Laboratory/NOAA, Princeton, New Jersey, USA).

Draft July 29, 1955

DYNAMICS OF THE GENERAL CIRCULATION

1. In 1947, a project was started in Princeton by the U.S. Navy and U.S. Air Force, for [theoretical and] computational investigations in meteorology, with particular regard to the development of methods of numerical weather forecasting. After a few years of experimenting, the project [concentrated on] exploring the validity and the use of the differential equation methods [developed by] Dr. J. Charney for numerical forecasting. For this purpose, the U.S. Army Ordnance Corps ENIAC computing machine was used in 1950 [and in 1951], and the Institute for Advanced Study's own computing machine from 1952 onward. Subsequently, use was also made of the IBM 701 machine in New York City. With the help of these computing tools, it was found that forecasts over periods like 24 (and up to [48]) hours are possible, and give significant improvements over the normal, subjective method of forecasting. Certain experiments demonstrated that even phenomena of [cyclogenesis] could be predicted. A considerable number of

simple forecasts were made, which permitted the above mentioned evaluation of the [validity] of the method. A large number of varian[ts] were also explored, particularly with respect to [eliminating successively] the major mathematical approximations that the original method contained. It must be noted, however, that the method and also all its varian[ts] which exist at the present, are still affected with considerable simplifications of a physical nature. [T]hus, the effects of radiation have only been taken into consideration in exceptional cases, the same is true for the effects of [geography and] topography, while humidity and precipitation have not been considered at all. [That significant] results could, nevertheless, be obtained, is due to the relatively short [time] span of the forecasts. Indeed, over 24 or 48 hours the above mentioned effects do not yet come into play decisively.

On the basis of the results [cited], it was determined [by the sponsoring agency] that a routine 24-hours numerical forecasting service has become possible, and should be set up on a permanent basis. This was done by [a] joint organization of U.S. Army, U.S. Navy, and Weather Bureau [(NWP = Numerical Weather Forecasting Project)], which is being operated by the U.S. Weather Bureau at Suitland, Md. It has now been making daily forecasts for over 3 months, and with very good success.

2. The logical next step after this is to pass to longer range forecasts and, more generally speaking, to a determination of the ordinary general circulation of the terrestrial atmosphere. Indeed, determining the ordinary circulation pattern may be viewed as a forecast over an infinite period of time, since it predicts what atmospheric conditions will generally prevail when they have become, due to [the lapse] of very long time [intervals, causally] and statistically independent of whatever initial conditions may have existed.

   There is reason to believe that the above mentioned "infinite" forecast i.e., deriving the general circulation, is less difficult than intermediate length forecasts, say, to 30 or 90 days. This is just a reflection of the fact that extreme cases are usually easier to treat than intermediate ones, since in extreme cases only a part [of] factors plays a role, [dominating] all others, while in intermediate cases, all factors become of comparable importance. It should be added that both the "infinite" and the "intermediate" forecasts have to be performed for the entire earth, or at least for an entire hemisphere. Indeed, the spread of [meteorological effects] is such that, already after 2 to 3 weeks, every part of the terrestrial atmosphere will have interacted with every other – except for the relative weakness of the interaction between the Northern and Southern hemispheres. Thus, in both cases, a hemispheric forecast is the minimum that can be envisaged.

   In view of the above, it seems logical to investigate now the "infinite" forecast, i.e., the general circulation. It is hoped that this will subsequently lead to a better understanding of the factors involved in the "intermediate" forecasts, (compare above). Thus, the "intermediate" forecasts should enter into the program at a somewhat later stage.

3. With regard to calculating the general circulation in the Northern hemisphere, quite significant progress [was] made in Princeton. Several calculations were made in which the Northern hemisphere – or rather a [quadrant] of it – was treated in a highly simplified way. The simplifications were as follows: The [quadrant] of the hemisphere was treated as a "flat" area, thus distorting the geometry, [primarily in] the arctic, considerably. Instead [of] using a coriolis parameter with its proper meridional variability, the "Rossby plane" was used, i.e. the coriolis parameter was given its mean value, and treated as a constant, however, in all places where the exact theory [makes] reference to the

meridional deriv[a]tive of the coriolis parameter, the (positive) mean value of [that] quantity was used.

The solar radiation impinging [upon] the earth was considered without its seasonal or [diurnal] variations. Indeed, it was treated as a heat source with a linear meridional variation.

This model was treated on a horizontal 16 × 16 lattice, with two vertical strata. Starting with an atmosphere at rest, the integration was carried out over 30 days. The [effects of] humidity, and of geography and topography [were] disregarded. The calculations on this model were started with an atmosphere at rest, and [at a uniform] temperature. The developing motions and adjustments were calculated over a period of 30 "real" days. The circulation pattern which developed was first the one that one usually obtains by verbal discussion: Northward flow of [heated] air aloft, and southward flow of [cooled] air below, with easterly winds on the lower, and westerly winds on the upper level. This [(not real!)] flow was observed to pass its [turbulent] stability limit after 5 "real" days. At this point, its breakdown was inducted by adding (computational) "noise" to the motion. Hereupon, in the course of the next 25 days a cyclone and an anticyclone, of familiar type, developed with westerlies in the middle, and easterlies in the high and low latitudes on the lower level, and westerlies aloft above the lower level westerlies. The [mean] westerly velocity on the lower level was about 30 miles per hour, and on the high level the maximum westerly velocity reached 200 miles per hour. The temperature difference between the tropics and the arctic was, as it should be, about twice what it is in reality. This doubling [should] correspond to the fact, that in reality half the heat transported north is latent heat of humidity, hence, when this [contribution] is neglected, the temperature increment that [is needed to] take care of [all] the requirements, will be double of what it is in reality.

Thus, even this very primitive model disclosed the main features of the general circulation, in a [rather detailed] way, which no verbal, or less elaborate computational analysis have ever been able to do. Several calculations of this type were made that gave concordant results, and also disclosed the limitations of the method used. The above described calculation (repeated for checking) required 30 computing hours on the Princeton machine.

4. It seems clear that these general circulation calculations should now be expanded and improved. Even [applying only] the obvious mathematical and geometric[al] improvements will greatly increase the size of each calculation. As a minimum program, the entire northern hemisphere should be considered; its curvature and the meridioinal variation of coriolis parameter should be properly treated; [and] the meridional variation of the solar energy input, with or without its seasonal [or diurnal] variations, should be introduced into the calculation. In addition to this, we know that the optimum [grid] size is about twice as fine (in linear dimension) than what was used, and that one should properly consider 3 or 4 vertical levels (rather than 2, compare above). All of this, with various secondary complications that it induces, is likely to increase the size of the calculation by at least a factor ten. This would mean a problem time of about 300 hours on the Princeton machine, or if the problem is checked in a less time-consuming way than by repetition, 150 hours. Comparing the Princeton machine with the IBM 701, it appears likely that the latter will be about 5 times faster on this problem. (The intrinsic speed of the IBM 701 is only twice [that] of the Princeton machine, but various memory

limitations of that machine probably increase [this] factor to something line [like] 5.)[1] Thus, on the IBM 701, presumably about 30 hours would be needed per problem, allowing for the above indicated refinements. This means that the time on the IBM 704 would probably be about 15 hours, and on the (Naval Ordnance Research Calculator) NORC, perhaps 7 hours.

Since a research [program] of this type requires large scale experimentation, with computing methods, with variations of parameter, and physical approximations of various kinds, there is no doubt that in any [rational] program, a large number of such problems will have to be solved. Therefore, even the best time mentioned above (7 hours on the NORC) would not be too fast. I.e., even under these conditions, computing would probably take more time than analyzing and planning. This is increasingly true for the IBM and [the] IBM 701. Consequently the use of the IBM 701, or, if [feasible] of one of the faster machines would be, in principle amply justified.

5. It is, therefore, proposed to set up a project which has available to it at least a machine of the IBM 701 type. Since the first improvements and refinements on the problem are sufficiently understood today, to be put immediately into the phase of [mathematical] planning and [coding], it would be important to [think in terms of] a machine which can be made available soon. The only machine of this speed class which is immediately available, is the IBM 701. [W]while this machine exists today in about 20 copies, only a few of them could be acceded to [easily]. At this moment, neither Princeton nor New York offer such a possibility, whereas one exists in Washington, at the Suitland establishment of the U.S. Weather Bureau (the NWP Project referred to earlier). It would, therefore, be very profitable to initiate measures immediately which make it possible to use this machine for the calculations mentioned above.

   The obvious vehicle for this work would be [a] project organized around the Suitland machine, and with the advice [and collaboration] of those who directed the Princeton project, and the above circulation calculations – J. Charney, N. Phillips, and J. von Neumann – readily available.

   It is proposed that such a project be set up as the U.S. Weather Bureau to be located at Suitland, with adequate personnel, physical space and facilities, and with about one shift of the Suitland IBM 701 machine available. It is proposed that within the Weather Bureau organization, Dr. H. Wexler, who has considerable familiarity with this work, be made the project officer. It is contemplated that in scientific and policy matters he would be guided by the decisions of a committee to [consist] of J. Charney, J. von Neumann, and himself.

6. The progress of this project can now be mapped out for about two years. During the first year, the general organization of personnel and facilities should take place, the setting up of computing methods in the sense of "minimum improved" general circulation problem, as outlined above, and the carrying out of [a sufficient] efficient sample of calculations on this basis. In the second year, the obvious physical improvements should be gradually introduced into the treatment. As such, one would consider in order of increasing difficulty the introduction of the following factors:

[1] "Line" would appear to be a typographical error by the original authors; the text makes sense with "like."

a) Purely [cinematic] effects of geography and topography;
b) [A]cquisition of humidity in the atmosphere by evaporation. This necessitates the (geographical) consideration of position of the oceans. It also requires the introduction of, presently reasonably well understood, semi-[empirical] rules regarding the dependen[ce] of the rate of evaporation on [the] local atmospheric and oceanic temperatures, atmospheric stability, and wind velocity.
c) Some, as yet, necessarily very imperfect [empirical] rules about the [delay-] relationships of over-saturation, cloud formation, and precipitation. Also, some [empirical] rules about the decrease of solar irradiation by clouds.
d) The very difficult problem of the effects of atmospheric humidity on the solar irradiation of the earth.

I repeat that (d) is [an] extremely difficult problem, which will probably only be reached at the end of the two year period, and on which progress will only be made at still later stages, and then only in combination with a great deal of experimental work.

c) is, in principle, even more difficult, but in this case, acceptable practi[cal] approximations can probably be made. (a) is quite simple; (b), while not very simple, is nevertheless based on things that we understand reasonably well at present.

Personnel and budget for the project are envisaged as follows:

[(see the notes we made at lunch today.)]

It should be noted that the above figures apply to the first year [only]. They should be reconsidered at the end of [that] year, and the budget of the second year determined on the basis of the experiences gained in the first year. It is expected that [the latter] will not differ very significantly from [the budget] of the first year, but [that it] will probably be somewhat higher.

At the end of the first year we [may] also find that a faster machine than the IBM 701 [is becoming] available.

In addition to the above, the consultations with J. Charney, N. Phillips, and J. von Neumann (without compensation) will be needed.

4

Beyond prediction to climate modeling and climate control: New perspectives from the papers of Harry Wexler, 1945–1962

JAMES RODGER FLEMING

The career of Harry Wexler (1911–1962), who served as Head of Research at the U.S. Weather Bureau, provides new perspectives on the formative era of numerical weather and climate modeling, from the birth of digital computing in 1945 to the first serious analysis of climate engineering presented by Wexler in 1962. This history illuminates technical, scientific, and social agendas for climate modeling by adding important new dimensions to a somewhat familiar story, and by linking the histories of prediction and simulation with the checkered history of purposeful intervention and climate control. Wexler made critical contributions to the development of NWP and GCMs; he also provided early, serious, and thoughtful critiques of the dangers of geoengineering. His prescient work on this subject clearly reminds us that we are not the first generation to be involved with or concerned about geoengineering, and places the current debate in the context of at least a half-century of continuous and usable history (Fleming, 2010a).

4.1 On the possibilities of climate control

"The subject of weather and climate control is now becoming respectable to talk about."

This was Harry Wexler's opening line in his speech "On the Possibilities of Climate Control," given to technical audiences in early 1962 in Boston, Hartford, and Los Angeles (Wexler, 1962a). He based his remarks on newly available technical capacities in climate modeling and satellite remote sensing, new scientific insights into the Earth's heat budget and stratospheric ozone layer, and new diplomatic initiatives, notably President John F. Kennedy's speech at the United Nations proposing,

"cooperative efforts between all nations in weather prediction and eventually in weather control".
(Kennedy, 1961)

Soviet Premier Nikita Khrushchev, flush with the success of two space spectaculars carrying Russian cosmonauts into orbit, had also mentioned weather control in his report to the Supreme Soviet in July 1961. Wexler noted that the subject had recently received serious attention from the President's Scientific Advisory Committee, the National Academy of Sciences Committee on Atmospheric Sciences, and the State Department. The Academy had recommended increased funding and the creation of a National Center for Atmospheric Research. The United Nations, with Wexler's scientific input through the State Department, had recommended measures to advance the state of atmospheric science and technology in outer space

> "to provide greater knowledge of basic physical forces affecting climate and the possibility of large-scale weather modification".
>
> *(United Nations 1961)*

Prior to this, statements about controlling the atmosphere were typically provided by non-meteorologists – generals and admirals, futurists, computer professionals, and cloud-seeding enthusiasts.

Harry Wexler was none of the above. He was one of the most influential meteorologists of the first half of the twentieth century. He was born in 1911 in Fall River, Massachusetts and died suddenly of a heart attack in August 1962 at age 51 during a working vacation in Woods Hole, just after giving these speeches. A 1932 graduate of Harvard University in mathematics, Wexler received his advanced degrees in meteorology at MIT where he was mentored by the noted meteorologists Carl-Gustaf Rossby, Hurd C. Willet, and Bernhard Haurwitz. He was particularly close to Rossby. Wexler held research and teaching positions at MIT, the U.S. Weather Bureau, the University of Chicago, and the U.S. Air Force. Following his honorable discharge from the military in January 1946 with the rank of Lt. Colonel, Wexler returned to the U.S. Weather Bureau, becoming the chief of the Special Scientific Services division. In this capacity, Wexler encouraged the development of new technologies including airborne observations of hurricanes, sounding rockets, weather radar, the use of electronic computers for numerical weather prediction and general circulation modeling, and satellite meteorology. He also served on numerous federal and international panels and committees, including the Pentagon's Research and Development Board, the Subcommittee on Meteorological Problems of the National Advisory Committee for Aeronautics, and was the lead architect, along with Soviet academician V. A. Bugaev, of the World Weather Watch, a program that continues today. In other words, Wexler was a meteorological heavyweight (Yalda, 2007; Fleming, 2010b).

4.2 Computers and meteorology, 1945–1946

Wexler was introduced to digital computing in 1945 by University of Pennsylvania engineer John W. Mauchly and to the possible role of computers in weather and climate control by RCA Associate Research Director, Vladimir K. Zworykin, whose inventions included television transmitting and receiving devices. Wexler served as the official weather bureau liaison to the Institute for Advanced Study computer project, where, with the help of Rossby, he assisted John von Neumann, the multi-talented mathematician extraordinaire who started the meteorology project at the Institute for Advanced Study (IAS) in Princeton, New Jersey, in identifying and recruiting meteorological and mathematical talent. Later, Wexler took steps to institutionalize and operationalize numerical weather prediction and general circulation modeling.

At the birth of modern digital computing, in January 1945, Mauchly visited Washington, DC to explain to various research groups,

> "the general nature and purpose of the machines under development by the Moore School"

at the University of Pennsylvania. Mauchly was seeking examples of difficult sorting and computing applications that would guide the development of machines of maximum usefulness and flexibility. At the weather bureau he met with C. F. Sarle, Assistant Director for Scientific Research. Sarle focused on the need for machine sorting of observational data on punched cards for later climatological analysis, the possibility of "extrapolating the weather map," and speeding up statistical correlations. Mauchly remarked in his report,

> "Although I mentioned several times that EDVAC would be capable of solving partial differential equations, the use of the EDVAC in handling such equations when they arise in meteorology was not touched upon until I asked about such possibilities."

Sarle did offer one good piece of advice; he directed Mauchly to get in touch with Major Harry Wexler at the Pentagon, a permanent weather bureau employee on loan to the Army Air Force Weather Service for the duration. The result was completely different. According to Mauchly, Wexler

> "displayed a great deal of enthusiasm concerning the possibilities of the EDVAC in meteorological research" and suggested immediately that such a machine be employed in "integrating hydrodynamic equations occurring in meteorological work".
>
> *(Mauchly, 1945)*

Interest was growing about the possible application of electronic computers to meteorological problems. Several months after Mauchly's visit, Wexler received a pamphlet on "Modern Computing Devices" from Zworykin at RCA. Later, in September 1945, weather bureau chief F. W. Reichelderfer, visited the RCA Labs in Princeton, New Jersey with other representatives of the Department of Commerce for a preliminary discussion about the use of modern electronic devices in meteorological analysis. In October 1945 Zworykin wrote his influential, but now all but forgotten mimeographed "Outline of Weather Proposal" (Zworykin, 1945).

He began by discussing the importance to meteorology of accurate prediction, which he thought was entering a new era. Modern communication systems were beginning to allow the systematic compilation of scattered and remote observations, and new computing equipment was becoming available that could either solve the equations of atmospheric motion, or at least search quickly for statistical regularities and past analogue weather conditions. He imagined "an automatic plotting board" that would quickly digest and display all this information.

Zworykin suggested that "exact scientific weather knowledge" might allow for effective weather control. If a perfectly accurate machine could be developed that could predict the immediate future state of the atmosphere and identify the precise time and location of leverage points or locations sensitive to rapid storm development, then intervention might be possible. A paramilitary rapid deployment force might then be deployed to intervene in the weather as it happens – literally to pour oil on troubled ocean waters or use physical barriers, giant flame throwers, or even atomic bombs to disrupt storms before they formed, deflect them from populated areas, and otherwise control the weather. Zworykin suggested a study of the origins and tracks of hurricanes, with a view to their prediction, prevention, and even diversion. Even long-term climatic changes could be engineered by large-scale geographical modification projects involving such areas as deserts, glaciers, and mountainous regions. In effect numerical experimentation using computer models would guide field experiments and interventions in both weather and climate. According to Zworykin:

> The eventual goal to be attained is the international organization of means to study weather phenomena as global phenomena and to channel the world's weather, as far as possible, in such a way as to minimize the damage from catastrophic disturbances, and otherwise to benefit the world to the greatest extent by improved climatic conditions where possible. *Such an international organization may contribute to world peace by integrating the world interest in a common problem and turning scientific energy to peaceful pursuits. It is conceivable that eventual far-reaching beneficial effects on the world economy may contribute to the cause of peace.*
>
> *(Zworykin 1945, original emphasis)*

Zworykin's striking claims for global peace and prosperity through climate engineering have been echoed by enthusiasts ever since, but there are also prominent voices warning of the chaos and conflict that might ensue from such attempted control.

John von Neumann, who had been involved in the development of both ENIAC and EDVAC, and who hoped to design a computer of his own, formally endorsed Zworykin's view in a letter enclosed with the proposal dated 24 October 1945. Von Neumann wrote,

> "I agree with you completely . . . This would provide a basis for scientific approach[es] to influencing the weather."

Using computer-generated predictions, von Neumann envisioned that weather and climate systems

> "could be controlled, or at least directed, by the release of perfectly practical amounts of energy"

or by

> "altering the absorption and reflection properties of the ground or the sea or the atmosphere."

It was a project that neatly fit von Neumann's overall agenda and philosophy:

> "All stable processes we shall predict. All unstable processes we shall control".
>
> *(Zworykin, 1945)*

Zworykin's proposal also contained a long endorsement by the noted oceanographer Athelstan Spilhaus, then a U.S. Army major, who ended his letter of 6 November 1945, with these words:

> "In weather control, meteorology has a new goal worthy of its greatest efforts".
>
> *(Zworykin, 1945)*

Reichelderfer, who had read the proposal, extended an invitation to Zworykin, von Neumann, Spilhaus, and others to gather in Washington, DC on 9 January 1946 with E.U. Condon, Director of the National Bureau of Standards, and representatives of the army, navy, and weather bureau including notably, the weather bureau's new head of research, Harry Wexler. The meeting centered around Zworykin's proposal and served to introduce meteorologists to digital computing and non-meteorologists to current methods of weather analysis and forecasting (Reichelderfer, 1945; Harper, 2008). Two days later an article in the *New York Times* both surprised and embarassed the participants by describing the supposedly confidential meeting and revealing plans for a revolutionary, multi-million-dollar super calculator – "12 feet high" – that might take some of the mysteries out of weather forecasting and even weather control (Shalett, 1946; Reichelderfer, 1946a).

The excitement among meteorologists was real. On January 14th they held a second conference in Washington, DC with electrical engineers Mauchly (who had visited a year earlier) and J. Presper Eckert of the University of Pennsylvania, designers of a computing device for the U.S. Army Ordnance Department. To cement relations further, two weeks later Wexler and his colleague and life-long friend, the meteorologist Jerome Namias, met up with Mauchly who was attending the annual meeting of the American Meteorological Society in New York City. After attending the formal dedication ceremony of ENIAC in Philadelphia in February, weather bureau chief Reichelderfer reported to the Secretary of Commerce,

"The importance of this development, if it turns out to be possible, can scarcely be over-estimated, and we are going ahead with our investigation of the possibilities as rapidly as possible".

(Reichelderfer, 1946b)

In Washington, Wexler distributed fifteen more copies of Zworykin's proposal and suggested that the interagency Air Coordinating Committee formulate a definite program to utilize high-speed electronic computers in weather research, forecasting, and possibly control (Wexler, 1946a).

Beginning in February 1946, Wexler made regular trips to Princeton as the official liaison of the weather bureau, reporting that he, Namias, and Captain Gilbert Hunt of the air weather service had discussed with Zworykin and von Neumann

"a purely kinematic problem of extrapolating the isobaric and frontal pattern on the sea-level weather map"

and

"a dynamic problem of computing from the equations of motion the future (24–36 hr.) flow patterns at say the 700 mb. surface."

He also noted that von Neumann would get in touch with Rossby concerning advanced theoretical problems involving mathematics and physics. Reichelderfer noted on his report,

"Good – the first step in a project which I hope will bring important developments".

(Wexler and Namias, 1946)

The close relationship between Wexler and his mentor Rossby cannot be over-emphasized. Notes taken during a telephone conversation on 12 April 1946 indicate that the two men discussed the electronic computing issue; the initial makeup of the meteorological working group (Haurwitz, Walter Elsaesser, Hans Panofsky, and Victor Starr); von Neumann's time commitment to the project (somewhere between 5 and 25 percent); the need to convene a conference on theoretical issues; strategies to connect von Neumann with Commander Daniel Rex, head of the navy's Office of Research and Invention (ORI); and decisions to emphasize first the theory of the general circulation rather than the more applied kinematic problem of weather forecasting (Wexler, 1946b). Rossby reinforced these decisions in a follow-up letter to Reichelderfer four days later, indicating that the ENIAC in Philadelphia

"could be used to solve numerically the problem of the general circulation of the atmosphere."

This first project . . . may have more practical significance than is immediately apparent. If satisfactory working models can be developed, it should perhaps be possible to see to what extent various arbitrarily introduced changes in the amount and character of incoming solar radiation would affect the general circulation pattern; this would certainly be an important first step towards long-range forecasting of climatic fluctuations, but I don't believe much headway can be made with that problem unless such complex factors as cloudiness, polar ice distribution and reflective characteristics of the surface of the earth can be included in the computing scheme.

To accomplish this, Rossby recommended choosing

"a man with a thorough synoptic knowledge of such thermodynamic factors as ice, clouds and radiation"

to be added to a working group that might otherwise be

"far too top-heavy on the mathematical side".

(Rossby, 1946a)

In a detailed letter to von Neumann dated 23 April 1946 [page two is dated a day earlier], Rossby outlined and orchestrated the entire project, including suggesting that IAS submit a proposal to ORI

"to examine the foundations of our ideas concerning the general circulation of the atmosphere,"

a project that

"might throw some light on the nature of climatic fluctuations."

Rossby also provided von Neumann with budgetary guidelines, suggestions for personnel, and nominated his former student, Harry Wexler, to lead the project as a

"thoroughly competent senior investigator" at an annual salary of $7,000.

(Rossby, 1946b)

Wexler was in Chicago meeting with Rossby and discussing these very points with his mentor when this letter was sent (Wexler, 1946c).

Wexler and Rex visited Princeton on 7 May, to meet with von Neumann. This was just one day before the IAS submitted its proposal to ORI for

"an investigation of the theory of dynamic meteorology in order to make it accessible to high speed, electronic, digital, automatic computing."

Not surprisingly, the proposal closely followed Rossby's outline and contained an allusion to Zworykin's original agenda on weather and climate control by claiming

"the first step towards influencing the weather by rational, human intervention will have been made – since the effects of any hypothetical intervention will have become calculable."

As a benchmark for later climate modeling, Wexler thought an important by-product of the effort might be

"additional physical measurements such as the absorption spectrum of water vapor and carbon dioxide under natural atmospheric conditions."

The proposal also nominated nuclear weapons scientist Edward Teller as a consultant to the project (Wexler, 1946d; Aydelotte, 1946; Thompson, 1983).

Working relations between Wexler and von Neumann were cordial that summer, with Wexler promising at least a half-time commitment to the project and looking into possibly relocating to Princeton. He visited von Neumann again on 23 May to discuss the availability of personnel and their salary levels; later he sent a meteorological bibliography of textbooks and periodicals, offering weather bureau library services in support of the project, and followed this up with phone conversations. Von Neumann even asked Wexler to edit the announcement of the meteorology project in the 1946–47 IAS *Bulletin* (Wexler, 1946e, f, g; von Neumann, 1946).

On 19 July Wexler learned from Herman Goldstein, von Neumann's assistant, that IAS had received a contract for the meteorology project and was making arrangements to acquire war surplus housing. Wexler immediately wrote to the meteorologists and selected others listed on the proposal to call a meeting in Princeton in August to discuss both the objectives of the project and practical arrangements. He asked all involved to

"come with some definite ideas regarding objectives, problems, and working procedures… to examine the foundations of meteorology, to solve the basic problems of the general circulation, and to improve our understanding of atmospheric processes".

(Wexler, 1946h)

The meeting, convened by von Neumann at the Princeton Inn, 29–31 August, had its agenda mainly set by Rossby and Wexler. After a morning session by von Neumann, the first three meteorological talks were by Rossby, Haurwitz, and Willett, all of whom had been Wexler's professors; a fourth talk was given by Namias (Wexler, 1946i). Also of note at this meeting were two visiting scholars from Chicago invited by Rossby, one famous and one relatively obscure: Jule Charney, with a newly minted Ph.D. from UCLA, eventually directed the IAS meteorology project to produce a successful machine forecast. His primary focus was to eliminate all equations that were not meteorologically important and to find a numerical solution for those remaining (Nebeker, 1995; Harper, 2008). Jeou-jang Jaw, visiting Chicago at the invitation of the U.S. State Department from the Institute of Meteorology of the Academia Sinica, Nanking, was interested in atmospheric waves, instability criteria, and developing perturbation equations in a baroclinic atmosphere. He is probably the most distinguished meteorologist you have never heard of (Wexler, 1946i; Jaw, 1946)!

Wexler traveled to Princeton about every two weeks in the late fall of 1946. On 14–15 October, he was again concerned with personnel and facilities, but also looked into the possibility that the equations of the general circulation Captain Hunt was developing, were

"in such shape they can be put on the ENIAC in the near future."

He and Hunt dined with Zworykin at his home, where again weather control–perhaps by igniting oil on the sea surface to prevent hurricanes–was

"thoroughly discussed over the vodka."

Annotations on Wexler's memo by weather bureau employees indicate that Zworykin's views

"were not shared by most tropical meteorologists".

(Wexler, 1946j)

Two weeks later Wexler made a special trip to allay the concerns of von Neumann who was worried that a stable research group had not yet been formed and who was contemplating giving up on the project. There they discussed adding Philip D. Thompson, a

"young, enthusiastic, intelligent, and hard-working Army Air Force weather officer"

who was working on numerical forecast computations at UCLA. Von Neumann was appeased when Wexler reiterated his personal commitment to the project and recommended regular meetings with a large group of dynamic meteorologists, since the problems they were facing were likely to be outside the range of expertise of any small, handpicked team that could be assembled on site. This, more than anything, set the tone for the future of the project (Wexler, 1946k). The meeting on 13 November included oceanographers Walter Munk of Scripps and Henry Stommel of Woods Hole who wished to see problems of joint interest to meteorologists run as trials on the ENIAC (Wexler, 1946l), while on 21 November Wexler suggested an objective method of determining the field of divergence in the atmosphere by tracking constant level balloons by radio (Wexler, 1946m). Both of these projects came to fruition decades later. The meeting on 7 December included a discussion of Wexler's interest in polar meteorology, with von Neumann suggesting additional mathematical analysis of Arctic circulation systems (Wexler, 1946n; Fleming, 2010b).

Wexler's final trip of the year was to Chicago to attend the conference on research problems in meteorology and to visit the experimental labs there, topics that were of direct relevance to his job as head of research at the weather bureau and his role as liaison to the Institute meteorology project. There, in the midst of a dazzling array of domestic and international talent assembled by Rossby, he was able to connect again with Thompson and Charney, both recently out of UCLA. On the final day Rossby asked Wexler to present a summary of the conference and

> "to prepare a list of lines of approach in meteorological research that should be pursued."

Wexler's list included observational needs such as measuring out-going longwave radiation at various heights, in and around cloud decks and in the presence of aerosols; numerical forecasting of a smarter variety than the "brutal assault" method; a more detailed study of atmospheric waves; defining and analyzing instability criteria and vortex motion; and a concerted attempt to model the general circulation, with improved observational evidence to define initial conditions. Rossby added the need to pay greater attention to interactions between the hemispheres (Chicago Conference, 1946).

4.3 Fantasies of control

Complicating the picture at the time were suggestions about the use of atomic weapons for climate control and announcements of new discoveries in cloud seeding. The prominent scientist–humanist Julian Huxley, then head of UNESCO, spoke to an audience of 20,000 at Madison Square Garden at an arms control conference about the possibilities of using nuclear weapons as "atomic dynamite" for "landscaping the Earth" or perhaps using them to change the climate by dissolving the Polar ice cap (Kaempffert, 1945). Major Eddie Rickenbacker, First World War flying ace, Medal of Honor recipient, and public persona extrodinaire, was on record as an advocate of using atomic bombs for "cracking the Antarctic icebox" to gain access to its known mineral deposits (*New York Times,* 1946a). "Sarnoff Predicts Weather Control" read the headline on the front page of the *New York Times* on 1 October, 1946. The previous evening, at his testimonial dinner at the Waldorf Astoria, RCA president Brigadier General David Sarnoff had speculated on peaceful projects worthy of the post-war era. Among them were

> "transformations of deserts into gardens through diversion of ocean currents,"

a technique that could also be reversed in time of war to turn fertile lands into deserts, and ordering

> "rain or sunshine by pressing radio buttons,"

an accomplishment that Sarnoff declared, would require a World Weather Bureau in charge of global forecasting and control (*New York Times,* 1946b).

A commentator in the *New Yorker* intuited the problems with such control immediately. "Who," in this Civil Service outfit, he asked, "would decide whether a day was to be sunny, rainy, overcast ... or enriched by a stimulating blizzard?" It would be "some befuddled functionary,"

probably bedeviled by special interests such as the raincoat and galoshes manufacturers, the beachwear and sunburn lotion industries, and resort owners and farmers. Or if a storm was to be diverted,

"Detour it where? Out to sea, to hit some ship with no influence in Washington?".

(New Yorker, *1946)*

On a more serious note, on 13 November the news bureau of the General Electric Corporation announced that laboratory cold-box experiments conducted by Vincent Schaefer had succeeded in making snowflakes and that he would soon try an outdoor experiment to see if he could exercise "some human control over snow clouds." The story reported speculation by Schaefer that it may be possible someday to prevent snow from falling on cities by directing it to fall elsewhere or to control fog over airports and harbors (General Electric, 1946; *New York Times* 1946c).

That very day Schaefer conducted a cloud-seeding test out of doors by dropping six pounds of dry ice pellets into a cold cloud over Mount Greylock in the nearby Berkshires, creating ice crystals and streaks of snow along a three-mile path (Schaefer, 1946; Byers, 1974), and when Bernard Vonnegut, also at GE, and also in November 1946, identified silver iodide as an effective cloud-seeding agent, his colleague, Nobel Laureate Irving Langmuir began to make a series of unsubstantiated claims that the chemical might eliminate severe aircraft icing, suppress hail storms, and perhaps seed the entire atmosphere of the United States, resulting in climatic or large-scale weather changes (General Electric, 1947).

In January 1947 both von Neumann and Zworykin spoke in New York at a joint session of the American Meteorological Society and the Institute of the Aeronautical Sciences, chaired by MIT meteorologist and incoming AMS president Henry G. Houghton. Von Neumann's talk, "Future uses of high speed computing in meteorology," was followed by Zworykin's much more controversial, "Discussion of the possibility of weather control." As covered in the *New York Times*:

Hurricanes may be dispersed, Dr. Zworykin said, and rain may be made, first through the speed which an electronic computer now approaching completion can synthesize all elements in weather problems, and second, through application of energy in small doses from spreads of blazing oil to heat critical portions of the atmosphere or blackened-over areas to cool them.

Zworykin focused on "trigger" mechanisms, such as artificial fogs or even cloud seeding, as examples of adding small amounts of energy to cause enormous effects, claiming the missing ingredient is not the techniques, but how to "make the most of our weather information mathematically." The news account ends editorially with the comment,

"If Dr. Zworykin is right the weather-makers of the future are the inventors of calculating machines".

(BAMS, *1947;* New York Times, *1947a)*

Most scientists thought this premature. The distinguished oceanographer H. U. Sverdrup at Scripps Institution of Oceanography was not convinced by Zworykin that

"the underlying general physical principles governing weather behavior are mostly well understood"

and found his claims premature. Regarding weather control, he wrote,

"it seems that only in rare cases can we expect to know the initial conditions in sufficient detail to predict the consequences of a 'trigger action'".

(Sverdrup, 1946)

Yet talk of triggers was something the military understood. In a 1947 fundraising speech before the annual alumni dinner at MIT, General George C. Kenney, commander of the Strategic Air Command, speaking of future weapons systems asserted,

"if rain could be kept from falling where it has been falling for ages," it is conceivable that "the nation which first learns to plot the paths of air masses accurately and learns to control the time and place of precipitation will dominate the globe".

(New York Times, *1947b)*

4.4 Computers and meteorology, 1947–1948

Just before the AMS meeting in January 1947 Wexler made his seventh liaison trip to Princeton to discuss with von Neumann and others, notably Philip Thompson, problems related to mathematical forecast techniques. These discussions continued in New York where, as mentioned, von Neumann was speaking (Wexler, 1947a). A follow-up trip a week later focused on von Neumann's concerns that the equations of motion for atmospheric disturbances were so general that they could be applied to either the speed of sound or to the trade winds.

"Something has to be done to specify the type of motion we wish to investigate by means of the equations"

– a problem confronted directly by von Neumann, Rossby, and Charney (Wexler, 1947b), and also discussed in relation to the pioneering work of Lewis Fry Richardson (discussed in Chapter 2) and the ongoing work of Caltech meteorologist Robert D. Elliott (Richardson, 1922; Wexler, 1947c). Things were not going well, however. By the end of the year, young Lieutenant Thompson was the "*sole full-time worker"* on the meteorological project [emphasis added by Reichelderfer], but von Neumann and Wexler had discussed adding, by the spring of 1948, Jule Charney, Arnt Eliasson, and Eric Eady (Wexler 1947d).

The rest of the early history of numerical weather prediction and general circulation modeling, at least up to the first successful tests and the operationalizing of the procedures, has been covered in the previous chapters and elsewhere, often by the principals (Charney, 1951; Phillips, 1956; Smagorinsky 1983; Thompson, 1983; Nebeker, 1995; Harper, 2008). Suffice it to say that the Wexler Papers, dated 1948–1952, contain personal letters received from von Neumann, Rossby, Charney, Thompson, and others; numerous liaison reports – twenty-four of them; practical arrangements for use of the ENIAC computer; and accounts of how Wexler recruited and nurtured Joseph Smagorinsky, a veteran of the IAS project with a newly minted Ph.D. from NYU, to lead "an advanced forerunner of actual NWP operations" (Wexler, 1952b). Wexler also saw it institutionalized in 1954 as the U.S. Joint Numerical Weather Prediction Unit in Suitland, Maryland. This was a partnership of the weather bureau with the air force and navy

> "to produce prognostic weather charts on an operational basis using numerical techniques".
>
> *(USJNWPU, 1955)*

A year later, based on a successful numerical experiment by Norman Phillips, in which he was able to simulate realistic features of the general circulation of the atmosphere, von Neumann and Wexler argued for the creation of a General Circulation Research Section (later Laboratory) to be located nearby. This is really the beginning of computer climate modeling (Washington, 2006). In the words of Joseph Smagorinsky:

> The enabling innovation by Phillips was to construct an energetically complete and self-sufficient two-level quasi-geostrophic model that could sustain a stable integration for the order of a month of simulated time. Despite the simplicity of the formulation of energy sources and sinks, the results were remarkable in their ability to reproduce the salient features of the general circulation. A new era had been opened".
>
> *(Smagorinsky, 1983)*

In this new era of computer modeling, Harry Wexler was working as well on many other fronts to advance the scientific study of weather and climate (Rigby and Keehn, 1963).

4.5 Wexler's key contributions in other areas

Wexler joined the nuclear age on 16 July 1945 when he analyzed the pressure waves generated by the Trinity atomic bomb test. He studied the weather's effects on reactor safety and developed techniques for following radioactive tracers downwind and around the globe following atmospheric nuclear tests. He was a member of the Advisory Committee on Reactor Safeguards for the U.S. Atomic Energy Committee, chaired the National Academy study group on Meteorological Aspects of the Effects of Atomic Radiation, and served as a U.S. delegate to the "Atoms-for-Peace" Conference in Geneva, Switzerland in 1955. The weather bureau's publication *Meteorology and Atomic Energy* was prepared under his supervision. Wexler often responded to public concerns linking nuclear tests with adverse weather events, pointing out that the energy released by atmospheric events was far greater than that of even the largest bomb and that, while the immense heat and towering mushroom cloud produced by a nuclear blast had dramatic short-term and local effects, he had found no evidence of any long-term effects on the weather and climate. He kept his eye on radioactive fallout, however, both as a potential health concern and as a global tracer, even in the snowfields of the South Pole (Wexler , 1956a).

Radar meteorology was an important by-product of wartime electronics research. Harry's brother Ray, who worked on weather radar with the U.S. Army Signal Corps Engineering Laboratories in New Jersey, detected and analyzed a frontal storm in 1946 in the Atlantic and published useful photographs of the event. A year later, Harry used these techniques in an early article on the structure of hurricanes as determined by radar (Wexler, 1947e). By 1960 radar images were being combined with photographs from the TIROS satellites to reveal the spirally banded structures of weather systems on a variety of spatial scales, including mesoscale convection on the order of 10 miles in diameter, hurricanes spanning 250 miles, and the cloud structure of mid-latitude cyclones up to 1000 miles across.

Air pollution, broadly defined, was also one of Wexler's interests. Early in his career he had studied the turbidity of the air and the trajectories of the dust bowl clouds of the 1930s as they made their way from the American heartland to the cities of the east coast. He also prepared a meteorological analysis of the killer smog episode of October 1948 in Donora, Pennsylvania and examined the effects of the Canadian forest fires of September 1950 that generated the "great smoke pall" over the Northeast. The latter study is still cited in the literature on nuclear winter since it generated a widespread regional cooling effect (Wexler *et al.*, 1949; Wexler, 1950, 1951).

Since volcanic eruptions could also cool the entire planet by blanketing it with dust and sulfate aerosols, Wexler took up the suggestion of William Jackson Humphreys that a series of such events could have caused the Quaternary ice ages. Assuming a diminution of available incoming radiation caused by volcanism of up to twenty percent, Wexler derived a possible climatic scenario with increased meridional circulation in which an upper cold trough dominated the weather over central North America. While this ice-age pattern could not be sustained by the typical volcanic eruption frequency of four per century, researchers could not rule out a ten-fold increase at least once every 10,000 years. Such evidence might be revealed in deep ice cores (Wexler, 1956b).

Rockets and satellites also came under Wexler's purview as scientific probes and observational platforms to investigate the atmosphere. He served as chairman of several influential committees on this subject, including the Upper-Atmosphere Committee of the American Geophysical Union, the NACA Special Committee for the Upper Atmosphere, and the National Research Council's Space Science Board. In his 1954 lecture at the Hayden Planetarium symposium on space travel, Wexler pointed out that a V-2 flight in 1947 had photographed clouds from an altitude of 100 miles, and an Aerobee rocket launch in 1954 had identified a previously unknown tropical storm in the Gulf of Mexico. Wexler displayed an artist's impression of the Earth from space showing clouds, land, and ocean and depicting weather features such as a family of three cyclonic storms along the polar front, a small hurricane embedded in the trade winds, evidence of jet stream winds, and fog off the coasts (Fleming, 2007a). Encouraged by the novelist and futurist Arthur C. Clarke, Wexler published versions of his remarks in the *Journal of the British Interplanetary Society* and in the *Journal of Astronautics,* where he made strong claims for the utility of the meteorological satellite, not only as a "storm patrol," but also as a potentially revolutionary new tool with global capabilities.

> Since the satellite will be the first vehicle contrived by man which will be entirely out of the influence of weather it may at first glance appear rather startling that this same vehicle will introduce a revolutionary chapter in meteorological science–not only by improving global weather observing and forecasting, but by providing a better understanding of the atmosphere and its ways. There are many things that meteorologists do not know about the atmosphere, but of this they can be sure: that the atmosphere is indivisible, and that meteorological events occurring far away will ultimately affect local weather. This global aspect of meteorology lends itself admirably to an observation platform of truly global capability – the Earth satellite.[1]
>
> *(Wexler, 1957)*

Taking up Rossby's call for more information about interactions between hemispheres, Wexler accepted the added challenge involved in serving as chief scientist for the United States expedition to the Antarctic for the International Geophysical Year (1957–58). By doing so he could add critical new information about both the South Pole and the Southern Hemisphere and integrate it into a global picture of circulation and dynamics of the *entire* atmosphere (Fleming, 2010b).

[1] Harry Wexler, "The Satellite and Meteorology," *Journal of Astronautics* 4 (Spring 1957): 1–6, on 1; Wexler had made considerable progress since 1954 in his thinking about the role of meteorological satellites. See "Meteorological Satellites," *Exploring The Unknown: Selected Documents in the History of the U.S. Civil Space Program*, vol. III: *Using Space*, ed. John M. Logsdon (Washington, DC: NASA History Office, 1998), p. 156 and passim, http://history.nasa.gov/SP-4407/vol3/cover.pdf.

4.6 Climate change and the general circulation

As documented above, Wexler was deeply interested in modeling the general circulation of the atmosphere and issues related to climate change. In 1952 Wexler attended a conference on climatic change at the American Academy of Arts and Sciences, convened by the distinguished astronomer Harlow Shapley. Wexler noted in his travel report, this is

> "the first meeting I have attended where this important subject has received the concentrated effort of scientists from many different fields."

Wexler's paper, presented early in the program, was entitled "The radiation balance of the Earth as a factor in climatic change." Here, according to his abstract, he considered the incoming (solar) and outgoing (terrestrial and atmospheric) radiation budgets of the Earth "in the light of their possible influence on climatic change."

> Effects on this budget caused by changes in amount and composition of solar radiation, in the amount of absorbing gases in the atmosphere, and in the Earth's albedo (reflecting power) are considered. Loss of radiation by volcanic dust in the atmosphere is given especial attention.
>
> *(Wexler, 1952a)*

Wexler's article resulting from this conference, a clear and comprehensive explanation of what was known at the time, follows his abstract closely. In a section on changes in the composition of the Earth's atmosphere, he examined the roles of trace absorbing gases, citing the work of G. S. Callendar on rising levels of carbon dioxide and the possibility of anthropogenic warming (Fleming, 2007b). He later discussed human-caused changes in atmospheric turbidity. Although purposeful climate engineering was not part of this article, in his discussion of changes in albedo Wexler used language that did not rule out purposeful manipulation.

> "Apparently, changes in the Earth's albedo, *no matter how brought about,* can be an important factor in changing the mean temperature of the earth and its atmosphere, and consequently world climate".
>
> *(Wexler, 1953, emphasis added)*

Wexler returned to this idea in his 1962 lectures on climate control, where he examined theoretical questions concerning natural and anthropogenic forcings, both inadvertent and purposeful, using the latest results from climate modeling and satellite measurements as applied to studies of the Earth's heat budget.

4.7 Gilbert Plass and CO_2 modeling

About this time an associate professor of physics at the Johns Hopkins University, Gilbert N. Plass (1920–2004) was drawing connections between the physics of infrared absorption, the geochemistry of the carbon cycle, and computer modeling. In 1954–55, with funds from the Office of Naval Research and using recent measurements of the infrared flux in the region of the 15-micron absorption band of CO_2, he constructed a one-dimensional model of radiative transfer. Plass calculated a 3.6 °C surface temperature increase for doubling of atmospheric carbon dioxide and a 3.8 °C decrease if the concentration were halved. He used these results to argue for the applicability of the carbon dioxide theory of climate change for geological epochs and in recent decades (Fleming, 2010c).

Stressing the intrinsic role carbon dioxide plays in our atmosphere, Plass discussed the danger of anthropogenic carbon release. He estimated that fossil fuel burning was adding six billion tons of carbon per year to the atmosphere, with more being added by deforestation and other activities. He estimated that this amount was sufficient to increase the CO_2 content of the atmosphere by thirty percent by the year 2000, causing noticeable changes in the Earth's radiation balance and thus the climate. His estimate was a 1.1 °C climate warming per century due to human influence. According to Plass, the oceans would be able to sequester only a small amount of the anthropogenic carbon, leaving the majority in the atmosphere. He noted that if humanity consumes the Earth's fossil fuel resources over the course of the next millennium, the CO_2 content of the atmosphere would quadruple from its present value and the planet would warm by at least 7 °C (Plass, 1956a, b).

Pre-dating the more widely cited warning of Roger Revelle and Hans Suess (1957), Plass pointed out in 1956 that humanity was conducting a large-scale experiment on the atmosphere, the results of which would not be available for several generations:

> "If at the end of this century, measurements show that the carbon dioxide content of the atmosphere has risen appreciably and at the same time the temperature has continued to rise throughout the world, it will be firmly established that carbon dioxide is an important factor in causing climatic change."

Plass also noted, prophetically, that the accumulation of CO_2 in the atmosphere from human activity will become increasingly important in the near future and will remain a problem through the centuries. Building on the work of G. S. Callendar and presaging the work of C. D. Keeling which began two years later, Plass called for new accurate measurements of the increasing CO_2 concentration in the atmosphere, which he rightly estimated should be on the order of 0.3 percent per year (Plass, 1956c). Wexler incorporated this work into the weather bureau's GCM and climate modeling efforts and established radiation, ozone, and notably, CO_2 measurements at the Mauna Loa Observatory starting just before the International Geophysical Year of 1957–58 (Fleming, 1998, 2008).

4.8 Harry Wexler's article and final lectures on climate control

In 1958 Wexler published a paper in *Science* that examined some of the consequences of tinkering with the Earth's heat budget. After a brief examination of albedo changes that would be caused by blacking the deserts and polar icecaps, Wexler turned to the notion, probably originating with Edward Teller, that detonating ten really "clean" hydrogen bombs in the Arctic Ocean would produce a dense ice cloud at high latitudes and would likely result in the removal of the sea ice much more quickly than the Soviet proposal of P. M. Borisov to dam the Bering Strait, divert Atlantic waters into the Pacific, and melt the Arctic sea ice. The balance of Wexler's paper is an examination of the radiative, thermal, and meteorological consequences of such outrageous acts, not only for warming the Polar regions, but also for the equatorial belt and middle latitudes. Noting that

> "the disappearance of the Arctic ice pack would not necessarily be a blessing to mankind,"

and implying that a nation like the Soviet Union already had the wherewithal to try such an experiment, Wexler concluded with a paragraph whose relevance has not been diminished by time:

> When serious proposals for large-scale weather modification are advanced, as they inevitably will be, the full resources of general-circulation knowledge and computational meteorology must be brought to bear in predicting the results so as to avoid the unhappy situation of the cure being worse than the ailment.
>
> *(Wexler, 1958)*

In 1962, armed with new information from computer models and satellite heat budget measurements, Wexler lectured to technical audiences "On the Possibilities of Climate Control," at the Boston Chapter of the American Meteorological Society, the Traveler's Research Corporation, and the UCLA Department of Meteorology. In his lectures he addressed inadvertent and purposeful damage to the ozone layer involving catalytic reactions of chlorine and bromine; he also examined climate engineering through purposeful manipulation of the Earth's heat budget. Wexler discussed increasing pollution from industry and recent developments in science, including computing and satellites, that led him to believe that manipulating and controlling large-scale phenomena in the atmosphere were distinct possibilities. He cited rising carbon dioxide emissions as an example of indirect control and explained how the general circulation research of the weather bureau was in part informed by this:

> We have for decades been releasing huge quantities of carbon dioxide and other gases and particles to the lower atmosphere. It is recognized that this atmospheric pollution may have serious effect not only on health but on global radiation or heat balance which is the cause of our present pattern of climate and weather".
>
> *(Wexler, 1962b)*

Wexler told his audiences that he was concerned not with the long and checkered history of rain making, but with planetary-scale manipulation of the environment that would result in

"rather large-scale effects on general circulation patterns in short or longer periods, even approaching that of climatic change."

He assured them he did not intend to cover all possibilities

"but just a few . . . *limited primarily to interferences with the Earth's radiative balance on a rather large scale* [original emphasis].
I shall discuss in a purely hypothetical framework those atmospheric influences that man might attempt deliberately to exert and also those which he may now be performing or will soon be performing, perhaps in ignorance of its consequences. We are in weather control *now* whether we know it or not".

(Wexler, 1962a)

Wexler was aware that any intervention in the Earth's heat budget would change the atmospheric circulation patterns, the storm tracks, and the weather itself, so, as he pointed out, weather and climate control are not two different things. He was interested in both inadvertent climatic effects, such as might be created by rocket exhaust gases or space experiments gone awry, and purposeful effects, whether peaceful or done with hostile intent. After presenting some twenty technical slides on the atmosphere's radiative heat budget and discussing means of manipulating it, Wexler concluded with a grand summary of various techniques to:

(a) *increase* global temperature by 1.7 °C by injecting a cloud of ice crystals into the polar atmosphere by detonating 10 H-bombs in the Arctic Ocean (a reprise of his 1958 *Science* article);
(b) *lower* global temperature by 1.2 °C by launching a ring of dust particles into equatorial orbit, a modification of an earlier Soviet proposal to warm the Arctic;
(c) *warm* the lower atmosphere and *cool* the stratosphere by artificial injections of water vapor or other substances; and
(c) *destroy all stratospheric ozone,* raise the tropopause, and cool the stratosphere by up to 80 °C by an injection of a catalytic agent such as chlorine or bromine. He estimated that a mere 0.1 MT of bromine would destroy all ozone in Polar regions and 0.4 MT would be needed near the equator (Wexler, 1962a).

Wexler was concerned that inadvertent damage to the ozone layer might occur if increased rocket exhaust polluted the stratosphere. He noted a 1961 study to this effect by the Geophysics Corporation of America (1961) on modification of the Earth's upper atmosphere by rocket exhaust. He was also concerned that future near-space experiments could go awry, citing Operation Argus (nuclear blasts in near space, 1958), Project West Ford (a ring of small copper dipole antennas in orbit,

1961), and Project Highwater (ice crystals injected into the ionosphere, 1962) as recent significant interventions with unknown risks. Purposeful damage was also not out of the question. In 1934 the noted British geophysicist Sydney Chapman had proposed making a temporary "hole in the ozone layer" using a yet-to-be-identified catalytic "deozonizer" (Chapman, 1934). According to Chapman, a small hole cut at a remote location might enable astronomers to make observations at ultra-violet wavelengths where radiation was otherwise absorbed by ozone. Much more sinister and relevant to the Cold War was possible military interest in waging geophysical warfare by attacking the ozone layer over a rival nation.

Seeking advice on how to cut a "hole" in the ozone layer, Wexler turned to chemist Oliver Wulf at Caltech who suggested that chlorine or bromine atoms might act in a catalytic cycle with atomic oxygen to destroy thousands of ozone molecules. In a handwritten note composed in January 1962 Wexler scrawled the following:

"UV decomposes $O_3 \rightarrow O$ in presence of halogen like Br. $O \rightarrow O_2$ recombines and so prevents more O_3 from forming".

On another slip of paper:

"$Br_2 \rightarrow 2$ Br in sunlight destroys $O_3 \rightarrow O_2 + BrO$."

These are essentially the basis of the modern ozone depleting chemical reactions. Using the radiation model of S. Manabe and F. Möller, Wexler was able to calculate an 80 °C stratospheric cooling with no ozone layer (Manabe and Möller, 1961).

In the summer of 1962 Wexler accepted an invitation from the University of Maryland Space Research and Technology Institute to lecture on "The Climate of Earth and Its Modifications," and might, under normal circumstances, have prepared his ideas for publication. However, Wexler was cut down in his prime by a sudden heart attack on 11 August 1962 during a working vacation at Woods Hole, Massachusetts. The documents relating to his career – from his early work at MIT, his work as liaison to the IAS meteorology project, his research into all sorts of new technologies, to his final speeches on ozone depletion and climate control – headed into the archives, probably not to be seen and certainly not to be fully re-evaluated until today.

The well-known and well-documented Supersonic Transport (SST) and ozone depletion issues developed about a decade later. The idea that bromine and other halogens could destroy stratospheric ozone was published in 1974, while CFC production expanded rapidly and dramatically between 1962 and its peak in 1974 (Crutzen, 1971; Molina and Rowland, 1974). Had Wexler lived to publish his ideas, they would certainly have been noticed and could have led to a different outcome and perhaps an earlier coordinated response to the issue of stratospheric ozone depletion. The received history of stratospheric ozone depletion dates only to the 1970s and certainly does not include Wexler's role. Recently, I have been in correspondence with three notable ozone scientists about Wexler's early work: Nobel Laureates Sherwood Rowland, Paul Crutzen, and current National Academy of Sciences President Ralph Cicerone. They are uniformly interested and quite amazed by this story.

Remarkable too, is the fact that with all his sophistication and the leading roles he played in the development of computer modeling, satellite monitoring, and many, many other technical fields, Wexler still opened his 1962 lectures by quoting extensively from Vladimir Zworykin's 1945 "Outline of Weather Proposal" and von Neumann's response to it. Remember it was Wexler, about 50 years ago, who first claimed climate control was now "respectable."

4.9 Conclusion

In a 1971 interview Joseph Smagorinsky identified meteorology as the field having

"probably the longest continuous history of the use of computers for scientific problems"

and indicated that, at least in the early decades, most computer applications were defense oriented:

"Up until about [1962] we were about the only ones using a top-of-the-line computer that was unclassified."

He also alluded to the program of his mentor, Harry Wexler, who played a central role in the development of computer models and used them to examine the possibilities of climate control:

> There have been proposals not only on questions regarding our inadvertent tampering with the atmosphere, but also on what people could agree they would like the climate to really be like. Is there something we could do to change it?... I doubt whether everyone would agree as to what they really would like. But one could test proposals, and these proposals would have scientific value too, because it tells you something about the stability of climate.
>
> *(Smagorinsky, 1971)*

The possibility of manipulating global climate through planetary-scale engineering is currently being actively debated, although its feasibility and desirability are highly questionable if not contentious. Most of the debate centers on back-of-the-envelope calculations (which are not good enough) or basic climate models (which also are not good enough). Still, the checkered history of the subject is only now receiving scholarly attention (Fleming, 2010a). On the other hand, accounts of the early history of computer use in meteorology follow a well-rehearsed script, identifying Vilhelm Bjerknes and Lewis Fry Richardson as early pioneers and emphasizing progress after 1946 through the work of a now-familiar cast of characters and technical breakthroughs. Through the career of Harry Wexler we can now see that the two histories, the familiar and the (until now) unwritten, are closely interrelated, and that climate control is not so much a newcomer in the era of global warming, but something that has been in the air for quite a long time.

References

Aydelotte, F. (1946). Proposal [from the Institute of Advanced Study] to the U.S. Navy's Office of Research and Inventions, 8 May 1946, copy in Wexler Papers, Box 2.

Bulletin of the American Meteorological Society (BAMS), 1947, vol. 28, p. 51.

Byers, H. (1974). History of Weather Modification. In *Weather and Climate Modification*, Hess, W. N. ed., New York, John Wiley.

Chapman, S. (1934). Presidential Address. *Quarterly Journal of the Royal Meteorological Society*, 60, 133–135.

Charney, J. G. (1951). Dynamic forecasting by numerical process. In *Compendium of Meteorology*. Boston, Mass.: American Meteorological Society, pp. 470–482.

Chicago Conference (1946). Notes on Problems of Meteorological Research, 9–13 Dec., Wexler Papers, Box 3.

Crutzen, P. J. (1971). *Quarterly Journal of the Royal Meteorological Society*, 96, 320–325.

Fleming, J. R. (1998). *Historical Perspectives on Climate Change*. New York, Oxford University Press.

Fleming, J. R. (2007a). A 1954 color painting of weather systems as viewed from a future satellite. *Bulletin of the American Meteorological Society* 88, 1525–1527.

Fleming, J. R. (2007b). *The Callendar Effect: The Life and Work of Guy Stewart Callendar (1898–1964)*. Boston, American Meteorological Society.

Fleming, J. R. (2008). *Climate Change and Anthropogenic Greenhouse Warming: A Selection of Key Articles, 1824–1995, with Interpretive Essays*. Classic Articles in Context. National Science Foundation, National Science Digital Library, http://wiki.nsdl.org/index.php/PALE:ClassicArticles/GlobalWarming

Fleming, J. R. (2010a). *Fixing the Sky: The Checkered History of Weather and Climate Control*. New York, Columbia University Press.

Fleming, J. R. (2010b). Polar and Global Meteorology in the Career of Harry Wexler, 1933–1962. In *Globalizing Polar Science: Reconsidering the Social and Intellectual Implications of the International Polar and Geophysical Years*, ed Launius, R. D. DeVorkin, D. K. and Fleming, J. R. New York, Palgrave Studies in the History of Science and Technology.

Fleming, J. R. (2010c). Gilbert N. Plass: Climate Science in Perspective. *American Scientist*, 98, 60–61.

General Electric (1946). Press Release, 13 Nov., GE News Bureau Binders, Schenectady Museum.

General Electric (1947). Press Release, 30 Jan., GE News Bureau Binders, Schenectady Museum.

Geophysics Corporation of America (1961). Modification of the Earth's upper atmosphere by missiles, Unpublished report, 84 p. + appendices, copy in Wexler Papers, Box 18.

Harper, K. C. (2008). *Weather by the Numbers: The Genesis of Modern Meteorology*. Cambridge, MA, MIT Press.

Jaw, J.-J. (1946). The formation of the semipermanent centers of action in relation to the horizontal solenoidal field. *Journal of Meteorology* 3, 103–114.

Kaempffert, W. (1945). Julian Huxley Pictures the More Spectacular Possibilities that Lie in Atomic Power, *New York Times* (9 Dec.), 77.

Kennedy, J. F. (1961). Address in New York City Before the General Assembly of the United Nations, 25 Sept. John F. Kennedy Presidential Library and Museum, http://www.jfklibrary.org/

Manabe, S. and Möller, F. (1961). On the radiative equilibrium and balance of the atmosphere. *Monthly Weather Review* 89, 503–532.

Mauchly, J. W. (1945). "Note on Possible Meteorological Use of High Speed Sorting and Computing Devices," dated April 14, 1945, copy marked "Confidential: copied from copy loaned by Prof. Mauchly 24 Jan. 1946," Wexler Papers, Box 2.

Molina, M. J. and Rowland, F. S. (1974). *Nature* 249, 890.

Nebeker, F. (1995). *Calculating the Weather: Meteorology in the 20th Century.* San Diego: Academic Press.

New Yorker (1946). "Talk of the Town" (12 Oct.), 23.

New York Times (1946a). "Blasting Polar Ice" (2 Feb.), 11.

New York Times (1946b). "Sarnoff Predicts Weather Control and Delivery of the Mail by Radio" (1 Oct.), 1.

New York Times (1946c). "Scientist Creates Real Snowflakes" (14 Nov.), 33.

New York Times (1947a). "Storm Prevention Seen by Scientist" (31 Jan.), 16; and "Weather to Order" (1 Feb.), 14.

New York Times (1947b). "$28,000,000 Urged to Support M.I.T." (15 June), 46.

Phillips, N. A. (1956). The general circulation of the atmosphere: a numerical experiment. *Quarterly Journal of the Royal Meteorological Society*, 82, 123–164.

Plass, G. N. (1956a). The carbon dioxide theory of climatic change. *Tellus* 8, 140–154.

Plass, G. N. (1956b). The influence of the 15μ band on the atmospheric infra-red cooling rate. *Quarterly Journal of the Royal Meteorological Society*, 82, 310–329.

Plass, G. N. (1956c). Effect of carbon dioxide variations on climate. *American Journal of Physics* 24, 376–387.

Reichelderfer, F. W. (1945). Letter of 29 Dec. to V. K. Zworykin, copy in Wexler Papers, Box 2.

Reichelderfer, F. W. (1946a). Letter of 11 Jan. to V. K. Zworykin and reply, 14 Jan., copies in Wexler Papers, Box 2.

Reichelderfer, F. W. (1946b). Memorandum for the Secretary, 18 Feb, copy in Wexler Papers, Box 2.

Revelle, R. and Suess H. E. (1957). Carbon dioxide exchange between atmosphere and ocean and the question of an increase in atmospheric CO2 during the past decades. *Tellus*, 9, 18–27.

Richardson, L. F. (1922). *Weather Prediction by Numerical Process*. Cambridge, Cambridge University Press.

Rigby, M. and Keehn, P. A. (1963). Bibliography of the publications of Harry Wexler. *Monthly Weather Review*, 91, 477–481.

Rossby, C. G. (1946a). Letter of 16 April to F. W. Reichelderfer, Wexler Papers, Box 2.

Rossby, C. G. (1946b). Letter of 23 April to John von Neumann, Wexler Papers, Box 2.

Schaefer, V. (1946). The production of ice crystals in a cloud of supercooled water droplets. *Science*, 104, 459.

Shalett, S. (1946). Electronics to Aid Weather Figuring. *New York Times* (11 Jan.), 12.

Smagorinsky, J. (1971). Interview with Richard R. Mertz, 19 May, Computer Oral History Collection, Archives Center, National Museum of American History, Smithsonian Institution.

Smagorinsky, J. (1983). The beginnings of numerical weather prediction and general circulation modeling: Early recollections. *Advances in Geophysics*, 25, 3–37.

Sverdrup, H. U. (1946). Letter of 2 June to F. W. Reichelderfer, Wexler Papers, Box 2.
Thompson, P. D. (1983). A history of numerical weather prediction in the United States. *Bulletin of the American Meteorological Society*, 64, 755–769.
United Nations General Assembly (1961). Resolution on the Peaceful Uses of Outer Space, 20 Dec., http://www.unoosa.org/
U.S. Joint Numerical Weather Prediction Unit (USJNWPU) (1955). Facts sheet, Rare book, NOAA Central Library.
Von Neumann, J. (1946). Letter of 28 June to Harry Wexler, Wexler Papers, Box 2.
Washington, W. M. (2006). Computer modeling the twentieth- and twenty-first-century climate. *Proceedings of the American Philosophical Society*, 150, 414–427.
Wexler, H. (1946–1962). Papers. Manuscript Division, Library of Congress (abbreviated throughout as Wexler Papers).
Wexler, H. (1946a). "Weather Proposal by V. K. Zworykin," Air Coordinating Committee, Subcommittee on Aviation Meteorology, ACC/SAM 37, 22 Jan., copy in Wexler Papers, Box 2.
Wexler, H. (1946b). Notes from telephone conversation with C. G. Rossby, 12 April, Wexler Papers, Box 2.
Wexler, H. (1946c). Trip Report – Chicago, Illinois and Iowa City, Iowa – 23–25 April, Wexler Papers, Box 2.
Wexler, H. (1946d). Trip Report – Princeton, New Jersey, 7 May, Wexler Papers, Box 2.
Wexler, H. (1946e). Trip Report – Princeton, N.J. and New York, N.Y., 23–25 May, Wexler Papers, Box 2.
Wexler, H. (1946f). Notes from telephone conversation with J. von Neumann, 24 June, Wexler Papers, Box 2.
Wexler, H. (1946g). Letter of 6 June to John von Neumann, Wexler Papers, Box 2.
Wexler, H. (1946h). Letter of 19 July to "Meteorology Computing Project," Wexler Papers, Box 2.
Wexler, H. (1946i). Draft agenda, agenda, hand-written meeting notes, typed meeting notes, and trip report, 29–31 Aug., dated 12 Sept., Wexler Papers, Box 2.
Wexler, H. (1946j). Trip Report – Princeton, N.J., 14–15 Oct., Wexler Papers, Box 2.
Wexler, H. (1946k). Trip Report – Princeton, N.J., 30 Oct., Wexler Papers, Box 2.
Wexler, H. (1946l). Trip Report – Princeton, N.J., 13 Nov., Wexler Papers, Box 2.
Wexler, H. (1946m). Trip Report – Princeton, N.J., 22 Nov., Wexler Papers, Box 2.
Wexler, H. (1946n). Trip Report – Princeton, N.J., 7 Dec., Wexler Papers, Box 2.
Wexler, H. (1947a). Trip Report – Princeton, N.J., and New York, N.Y. – Camp Upton, L.I., 28–30 Jan., Wexler Papers, Box 3.
Wexler, H. (1947b). Trip Report – Princeton, N.J., 7 Feb., Wexler Papers, Box 3.
Wexler, H. (1947c). Trip Report – Princeton, N.J., 17 Mar., Wexler Papers, Box 3.
Wexler, H. (1947d). Report of Travel to Princeton, N.J., 19 Nov., Wexler Papers, Box 3.
Wexler, H. (1947e). Structure of hurricanes as determined by radar. *Annals of the New York Academy of Sciences*, 48, 821–844.
Wexler, H. (1950). The great smoke pall-September 24–30, 1950, *Weatherwise* 3, Dec., 129–134, 142.
Wexler, H. (1951). Meteorology and air pollution. *American Journal of Public Health*, Part II, *Yearbook* 41.
Wexler, H. (1952a). Travel accomplished under T.O. No 1050, 12 May, Wexler Papers, Box 5.
Wexler, H. (1952b). "Reinstatement of Joseph Smagorinsky in the Weather Bureau," 3 Sept., Wexler Papers, Box 5.
Wexler, H. (1953). Radiation balance of the Earth as a factor in climatic change. In *Climatic Change*, Harlow Shapley, ed. Cambridge: Harvard University Press, pp. 73–105.

Wexler, H. (1956a). Meteorological aspects of atomic radiation. *Science*, 124, 105–112.
Wexler, H. (1956b). Variations in insolation, general circulation, and climate. *Tellus*, 8, 480–494.
Wexler, H. (1957). The satellite and meteorology. *Journal of Astronautics*, 4, Spring, 1–6.
Wexler, H. (1958). Modifying weather on a large scale. *Science*, n.s. 128, 1059–1063.
Wexler, H. (1962a). On the Possibilities of Climate Control. Unpublished manuscript and notes, Wexler Papers, Box 18.
Wexler, H. (1962b). Further justification for the General Circulation Research Request for FY 63, Draft, 9 Feb. Wexler Papers, Box 18.
Wexler, H. and Namias, J. (1946). Trip report to RCA Laboratories, 8 Feb. 1946, copy in Wexler Papers, Box 2.
Wexler, H., Schrenk, H. H., Heimann, H., Clayton, G. D., and Gafafer W. M. (1949). Air Pollution in Donora, Pa.: Epidemiology of the Unusual Smog Episode of October 1948, *Public Health Bulletin* No. 306 (Washington, D.C.: U.S. Public Health Service, 173 p.)
Yalda, S. (2007). Harry Wexler. *Complete Dictionary of Scientific Biography*. Vol. 25: 273–276. Detroit: Scribner's/Thomson Gale.
Zworykin, V. K. (1945). Outline of Weather Proposal, October 1945. Princeton, NJ: RCA Laboratories, 1945, 12 p. + appendices, copy in Wexler Papers, Box 18. Reproduced in full in *History of Meteorology* 4 (2008): 57–78.

5

Synergies between numerical weather prediction and general circulation climate models

CATHERINE A. SENIOR, ALBERTO ARRIBAS, ANDREW R. BROWN, MICHAEL J. P. CULLEN, TIMOTHY C. JOHNS, GILLIAN M. MARTIN, SEAN F. MILTON, STUART WEBSTER AND KEITH D. WILLIAMS

In this chapter we will explore the value of taking a seamless approach to model development and prediction across a range of time and space scales. General circulation models were originally distinct from numerical weather prediction models due to the need for more advanced atmospheric physics and extended length runs and hence the requirement for lower resolution. However with the availability of more detailed observations and improved NWP capability, it is clear that model representation of "fast" physical processes might best be advanced from testing and evaluating against real-time observations in a strongly dynamically constrained environment provided by the first few days of a short-range numerical weather prediction. In parallel, it has become obvious that detailed representation of physical processes, previously considered of less importance on short time-scales, can provide considerable additional skill for short-range forecasts.

We will explore the motivation for the development of a unified modeling approach at some centers and initialization techniques at others aimed at exploiting the synergies between numerical weather prediction and climate modeling. We will consider the benefits of a unified approach including: studying systematic errors in climate models, using initial tendencies from short-range forecasts, and comparison with real-time observations; diagnostic capabilities applicable across time-scales; cross-fertilization of physics development; initialization and ensemble techniques, among others. We will illustrate the benefits by discussing the move towards seamless predictions and traceability of mechanisms across a range of space and time-scales. Finally we will explore how to further exploit synergies in the future.

5.1 Background and motivation

This introduction surveys the historical background which has led to the modern realization of the synergy between weather forecast and climate models. It is inevitably a brief survey of modeling developments world-wide. The history of numerical weather prediction (NWP) is documented more comprehensively in Spekat (2000) and good reviews of early climate models are contained in Arakawa (2000) and Edwards (2000). Many of the numerical aspects are surveyed in the papers presented at the André Robert Memorial Symposium (Lin *et al.*, 1997).

Many national weather services and research institutes have been running state-of-the-art primitive equation models for both weather forecasting and climate applications since the mid 1960s. In the early stages, weather forecasting was limited to short-range prediction over small areas, and was a clearly distinct activity from climate prediction. Thus in most cases weather forecasting and climate research were not the responsibility of the same institute. Exceptions to this included the U.K. Met Office and Meteo France. Climate models have inevitably been global general circulation models (GCMs) from the outset. One of the earliest successful climate models was the Geophysical Fluid Dynamics Laboratory (GFDL) model of Smagorinsky, Manabe, and collaborators (Smagorinsky *et al.*, 1967). After considerable experimentation with alternatives, finite difference models were written on latitude–longitude grids, and filtering of small scales near the poles was used to maintain a reasonable timestep. Conservation forms of the equations were used to aid the construction of numerically stable approximations. Weather forecast models were at this stage invariably limited area models. Examples were the limited area fine-mesh model used in the USA (Howcroft, 1966), and the rectangle model based on the work of Sawyer and Bushby used in the UK (Bushby and Timpson, 1967). This was natural, because the idea of a global forecast model was not seen as practical. The scope of national weather services did not extend world-wide, and the scarcity of observations would have made the forecasts of little practical use in the tropics. The advantage of this approach was that the equations could be written on a map projection, avoiding the difficulties of designing a satisfactory grid on the sphere. The numerical methods were chosen for efficiency, rather than conservation properties. Many of the other differences between the formulations of weather forecast and climate models were the natural result of different teams of scientists, usually in different institutes, working with a different focus.

Developments in the 1970s brought the requirements and capabilities closer together. The Global Atmospheric Research Program's (GARP) Atlantic Tropical Experiment (GATE) provided data which was used to show that tropical, and hence global, data assimilation was feasible.[1] This experience led to the 1979 First GARP Global Experiment (FGGE) year when observations were collected and global analyses were produced on a routine basis. The realization that extended-range

[1] Data assimilation techniques allow analyses to be prepared from limited observational data by using a forecast model to carry information forward in time.

forecasting required both accurate initial data and short-range evolution, together with realistic longer-term climatology, led to the thinking that the necessary computing resources could not be afforded on a national basis. This was the main argument leading to the establishment of the European Centre for Medium-range Weather Forecasts (ECMWF) (Woods, 2006). ECMWF's mission was global forecasting from the outset, and several existing climate models were imported to guide the design of the initial ECMWF model. There were also major technical developments, which meant that high-resolution modeling became much more affordable. In particular, the spectral method (Bourke, 1974),[2] was demonstrated as a very efficient approach to global modeling, avoiding the difficulties with the poles, and use of semi-implicit time integration (Robert *et al.*, 1972), allowed the use of much larger timesteps. By the end of the 1970s nearly all global weather forecasting and climate models used the spectral semi-implicit technique. This was exploited in designing very efficient time integration schemes, which allowed high-resolution forecasts to be produced in a timely manner. Spectral methods also naturally conserve quadratic quantities, and are very stable. In the meantime, climate modeling efforts became more focused on modeling of surface and atmospheric physical processes; and the treatment of these effects had become much more sophisticated than that of typical forecasting models.

In the 1980s development of global weather forecast models was more incremental in nature, but it was recognized that the more sophisticated representation of physical processes used in climate models was necessary for improved forecasting in at least the medium range (6–10 days). Separately, it was becoming possible to run limited area short-range forecast models at much higher resolution, allowing realistic simulation of precipitation patterns and type. Developments in numerical methods, in particular semi-Lagrangian advection (Staniforth and Côté, 1991),[3] made a significant contribution to the affordability of high resolution, though the loss of conservation meant that these schemes were not adopted in most climate models. Exploiting the high resolution required the same sophistication in the representation of physical processes as already recognized as necessary for climate models. Forecasting convective precipitation, for instance, is very dependent on forecasting of surface conditions, and thus surface schemes developed for climate models were introduced into weather forecasting models. A further development in this direction was the first demonstrations of local forecasting using very high-resolution local non-hydrostatic models. These models had been developed for

[2] In the spectral method, the data are represented by coefficients of spherical harmonics rather than by values at grid points.

[3] In semi-Lagrangian methods, the Lagrangian formulation rather than the Eulerian formulation of the equations is discretized, so that advection is represented by transferring the value of a variable from a departure point to the arrival point.

idealized research in the 1970s, (Tapp and White, 1976). It became clear very quickly that the usefulness of such models would lie in the forecasting of parameters such as low cloud, visibility, and precipitation, rather than in forecasting pressure patterns. Again the physics and prediction of land surface properties became a strong focus.

Meanwhile, the focus in climate modeling was moving towards coupled atmosphere/ocean/land surface/sea-ice modeling, which was clearly essential and required a major commitment of effort from climate scientists. In particular, it required investment in software structure and management as well as in scientific formulation. This had a strong synergy with attempts at extended-range forecasting, typically on a monthly time-scale, by dynamical models (Miyakoda *et al.*, 1983). Computer power was now sufficient for credible simulations to be run. This initially exposed model systematic errors as a major issue, and resulted in many improvements to the models. It also exposed the need for coupled modeling (specifically atmosphere to land surface) in weather forecasting applications as well as climate simulations.

In the mid 1980s, it was becoming clear that the expectation that a new computer code would be written from scratch for each new generation of supercomputers was becoming untenable because of the increasing complexity of the models. The convergence of model formulations in the 1980s made it sensible to combine the models and create unified systems. It was also realized that the overall capabilities of the software would have to be much greater than just a unified atmospheric model. The need for coupling to the ocean and land surface meant that a considerable investment was required in software design and data handling. Three such large projects were the introduction of a unified forecast/climate model by the U.K. Met Office (Cullen, 1993), the creation of the Integrated Forecast System (IFS) by ECMWF and Meteo France (Woods, 2006), and the Flexible Modeling System at GFDL (http://www.gfdl.noaa.gov/fms). The Met Office system was designed as a modeling system, including data assimilation as a component. The ECMWF system was seen as an integration of observation processing, data assimilation, and forecasting. Another example is the Community Climate System Model (Boville and Gent, 1998), which is hosted by the U.S. National Center for Atmospheric Research (NCAR) and provides a coupled ocean–atmosphere–land-ice model that can be used for both research and educational purposes.

The 1990s saw this evolution coming to fruition with large, general-purpose modeling systems coming into production use. Major IT developments in the 1990s greatly enhanced the ability to manage, document, and disseminate such complex models. A large amount of work was therefore devoted to software design and management to exploit the new tools available. Scientifically, the benefits of integrated systems soon became obvious with new capabilities, such as regional climate modeling, being exploited. Dynamical seasonal forecasting became a firmly established routine activity, for instance at the U.S. National Center for Environmental Prediction (NCEP; Saha *et al.*, 2006), and ECMWF, (Woods, 2006). The ability to run much higher-resolution models for routine weather forecasting meant that useful forecasts of many more quantities, such as cloud amount and distribution of pollutants, became possible. The strong synergy between issues relevant to climate change and opportunities for obtaining additional information from weather forecasting models continued. Scientifically, demonstrations of what could be achieved with cloud-resolving models led to the realization that similar very high resolution could greatly enhance the fidelity of climate models, particularly in the tropics.

In the 2000s, the benefits of unified systems were apparent, and so the development of numerical methods which were both efficient and conservative became a priority. Much of the issue could be addressed by moving away from the latitude–longitude grid, and considerable work was done on resolving the numerical accuracy issues associated with such grids. If the grid resolution is approximately the same at all points on the sphere, the use of semi-Lagrangian advection is less compelling for efficiency, and conservative finite volume schemes on such grids are gaining in popularity. The GFDL climate model (Putman and Lin, 2007), now uses such a scheme. At the same time, the U.K. Met Office implemented a mass-conserving but otherwise semi-Lagrangian climate model (Davies *et al.*, 2005). Though the semi-Lagrangian treatment does not give formal conservation of tracers, it was found to give advantages in the representation of the transport of chemical species. At this time, the idea of seamless model prediction or the use of a single modeling framework across all relevant spatial and temporal scales was gaining considerable momentum in the international community (Shapiro *et al.*, 2007; Hurrell *et al.*, 2009). This recognized both the multi-scale nature of many phenomena and issues of predictability but also identified a route to greater progress on improving models. Work on understanding and evaluating the ability of climate models to simulate key processes accurately was often hampered by a lack of relevant observations, sampling limitations, compensating errors, and the combination of local and remotely forced errors at equilibrium. Research at centers running a single model across a range of time-scales (Palmer, 1999; Martin *et al.*, 2010) suggested that many of the systematic errors in these key processes were common to integrations with the same physical models run from well-balanced initial states after only a few days. These centers were able to use these synergies to analyze sources of longer time-scale systematic errors. Climate modeling centers without the availability of data assimilation techniques in their models also recognized the potential benefits and have looked to develop initialization methods (e.g., Phillips *et al.*, 2004).

At the present time, it is clear that these synergies will develop further, and the use of integrated modeling systems to exploit these opportunities will be essential. Though the cost and complexity of such systems is large, it appears to be the only way of meeting the increasing range and sophistication of user requirements. Concerns about environmental quality are leading to requirements for forecasting air quality and atmospheric composition on short forecast time-scales as well as on climate time-scales. Flexible software implementations are essential so that systems can be widely used. An example is the use of the Met Office climate model in the climateprediction.net project (Stainforth *et al.*, 2005), where it was set up to be used by the general public. It is increasingly difficult to create such a system from scratch, so a policy of continuous development is required to ensure that the systems remain of the highest standard.

In the rest of this paper we discuss in more detail the scientific and technical benefits of exploiting the synergies between NWP and climate models. The term "seamless prediction" has been widely used, but often with different meanings. There is an acknowledgment that traditional boundaries between weather and climate prediction are artificial and that the accurate representation of the multi-scale processes within a single modeling framework is necessary. This encompasses both space- and time-scale interactions and questions of initialization and predictability. In this paper, we restrict our discussion to the exploitation of common aspects of weather and climate modeling to accelerate research, reduce model systematic errors, and improve capability in both disciplines. We have tried to include a broad range of examples from activities world-wide, but inevitably many come from the authors' home institution (the U.K. Met Office). The motivation here is simply to illustrate techniques.

Section 5.2 focuses on unified prediction across space scales and discusses the benefits of high resolution. Section 5.3 focuses on unified prediction across time-scales. We look at the value of investigating errors in processes at different time ranges, from data assimiliation increments to longer-term evolution of errors and through comparison against high-quality satellite or *in situ* data. Section 5.4 focuses on the other shared benefits of development across time-scales such as development of existing or new capability in models, diagnostics, and metrics and software engineering infrastructure.

5.2 Seamless prediction across a range of space scales

The governing equations of dynamics and thermodynamics cannot be solved exactly on today's computers. The equations have to be approximated in space and time, and the effect of unresolved scales modeled. In practice, the sub-grid modeling is mostly done implicitly within the numerical algorithms, while

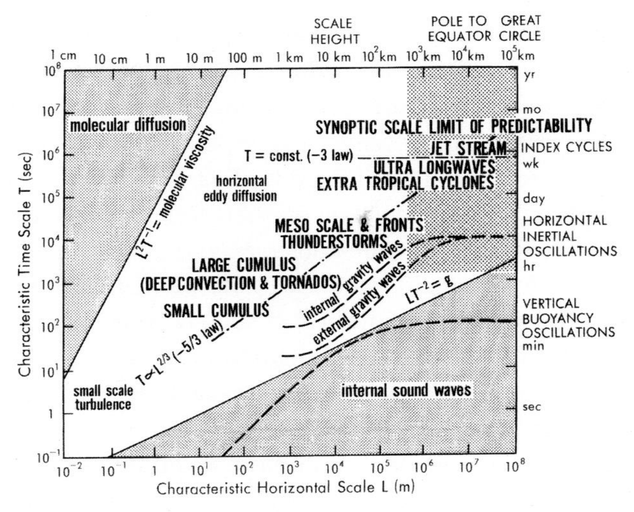

Figure 5.1 Typical space and time-scales of atmospheric phenomena following Smagorinsky (1974).

sophisticated sub-grid models are used in areas of organized unresolved activity such as the atmospheric boundary layer and regions of cumulus convection. Very different model resolutions (and effective averaging scales for the governing equations of motion) are used in different applications. These are usually chosen to be as high as is affordable given the length of the simulations required and the time available to complete them. On computers presently available, typical global climate models are using resolutions of order 100 km or more, global NWP models are using a few tens of km, and local weather forecasting models are moving towards resolutions of around 1 km. Different configurations can also be used at different forecast lead times reflecting the loss of deterministic predictability at smaller scales and the need to use coupled ocean/atmosphere models for longer lead times. Thus in 2008, ECMWF introduced a variable-resolution ensemble prediction system with T399 resolution (50 km grid spacing) from 0 to 10 days with persisted sea surface temperature (SST) anomalies, and T255 (80 km grid spacing) coupled to an ocean model for days 10–30. Variable resolution can be used where particular regions require higher resolution. For example, the ocean component of the Met Office HadGEM1 model and the Modular Ocean Model (MOM) of GFDL (Johns *et al.*, 1997; Delworth *et al.*, 2006) use higher resolution in the tropics.

Clearly these various configurations can explicitly represent different features, and different processes are important across scales (e.g. Figure 5.1). This can be accommodated into a seamless modeling system by modifying or removing certain parts of the sub-grid model, as necessary, as the resolution is increased.

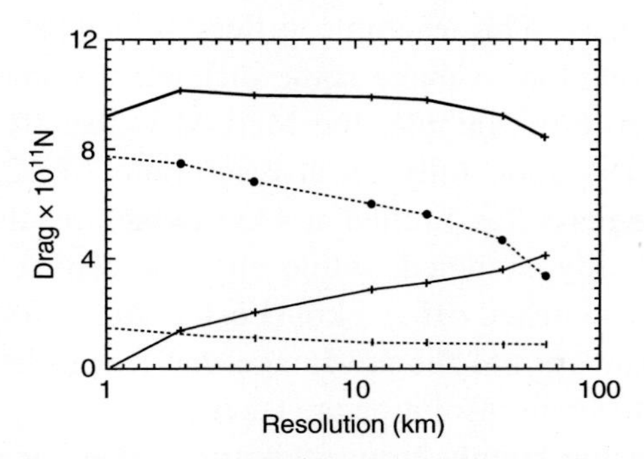

Figure 5.2 Drag on the Alps as a function of model resolution from case study of 8 November 1999. Uppermost solid: total; heavy dotted: resolved orographic; Lower solid: parameterized orographic; light dotted: boundary layer.

5.2.1 Horizontal resolution

As an example of seamless prediction across a range of space scales we show the way in which models represent the effect of orography on the large-scale flow. In reality, the tendency to produce high pressure on the upwind side of mountains and low pressure on the downwind side results in a net force on the mountains. There is a corresponding decelerating drag force on the atmospheric flow which is a significant term in the atmospheric angular momentum budget. Its accurate representation is important in obtaining realistic forecasts of the evolution of the flow and weather systems. However, while the drag due to mountains which can be explicitly modeled at the chosen resolution may be predicted directly, that due to mountains which are too small to be resolved has to be parameterized.

Figure 5.2 shows the resolved and parameterized drag on a region encompassing the Alps from a series of Met Office Unified Model (MetUM) forecasts for 8 November 1999 as a function of model resolution. Between 60 and ~1 km resolution the resolved orographic drag (that resulting from a correlation between pressure perturbation and orographic slope) in the model increases from around 3.3 x 10^{11} N to 7.8 x 10^{11} N. At the same time the sub-grid drag provided by the Webster *et al.* (2003) scheme steadily reduces (primarily because the amount of unresolved orography decreases). Hence the parameterization scheme is automatically switching itself off as the model resolution improves, and there is a smooth handover from parameterized to resolved drag. Note also that the extent to which the total drag is independent of resolution gives a test of the performance of the parameterization which would not be available without the ability to run the same model across a range of resolutions.

Another example is in the modeling of cumulus convection – a process which is parameterized in large-scale models, but which can be explicitly modeled in kilometer-scale models. This example differs somewhat from the orographic one, in that it is common to make some different parameterization choices at different resolutions. For example, the MetUM is usually configured with the convection parameterization fully on at resolutions of 12 km and coarser, but with the convective mass flux limited at 4 km (which results in deep convection being largely explicitly modeled, while shallow convection continues to be parameterized), and switched off at 1 km (Lean *et al.*, 2008). Even for cases like this, a unified framework has the advantage that it ensures that decisions to do different things at different resolution are taken consciously and explicitly (rather than differences arising simply from arbitrary choices being taken in different models).

Figure 5.3, from Lean *et al.* (2008), shows an example case study over the southern UK run with the MetUM at 12, 4, and 1 km resolution. The verifying radar data shows an area of very heavy rain near point B. This was generated by a squall line associated with a cold pool at the surface, which itself was generated by previous convection. This chain of processes cannot be predicted without explicit modeling of at least the large-scale aspects of the convection. In the 12 km model, the area of rain near the squall line is no heavier than in the other areas of rain predicted by the model. Since the sub-grid model only uses the local stratification and moisture content, the interaction between convection and dynamics which occurred in reality cannot be represented. The 4 km model shows a cluster of heavy convective cells near the squall line and in the 1 km model, an area of organized heavy rain is predicted. This illustrates that the explicit modeling possible with a 1 km grid can give significant benefits. Note however that none of the simulations give a particularly convincing simulation of the showery activity behind (northwest of) the squall line (Lean *et al.*, 2008).

As well as high-resolution convection-resolving simulations having direct application in forecasting systems, their results are often used to develop understanding of physical processes involved, and to develop parameterizations (e.g. of cloud and convection) for use in coarser-resolution models (e.g. Randall *et al.*, 2003). While the high-resolution simulations are often done with separate models covering relatively small domains, this is not always the case. For example, Shutts and Palmer (2007) have performed and analyzed idealized simulations with 2 km by 10 km resolution with a model domain large enough to cover a significant fraction of the tropics. The ongoing Cascade project (a consortium of U.K. academic institutes and the Met Office studying scale interactions in the tropical atmosphere) also uses large domains, but is using the MetUM in cloud-resolving mode. Other related work uses the so-called super-parameterization or multi-scale modeling approach, in which cloud-resolving models are embedded within a coarser-scale model (e.g. Khairoutdinov and Randall, 2001). Global climate models at resolutions

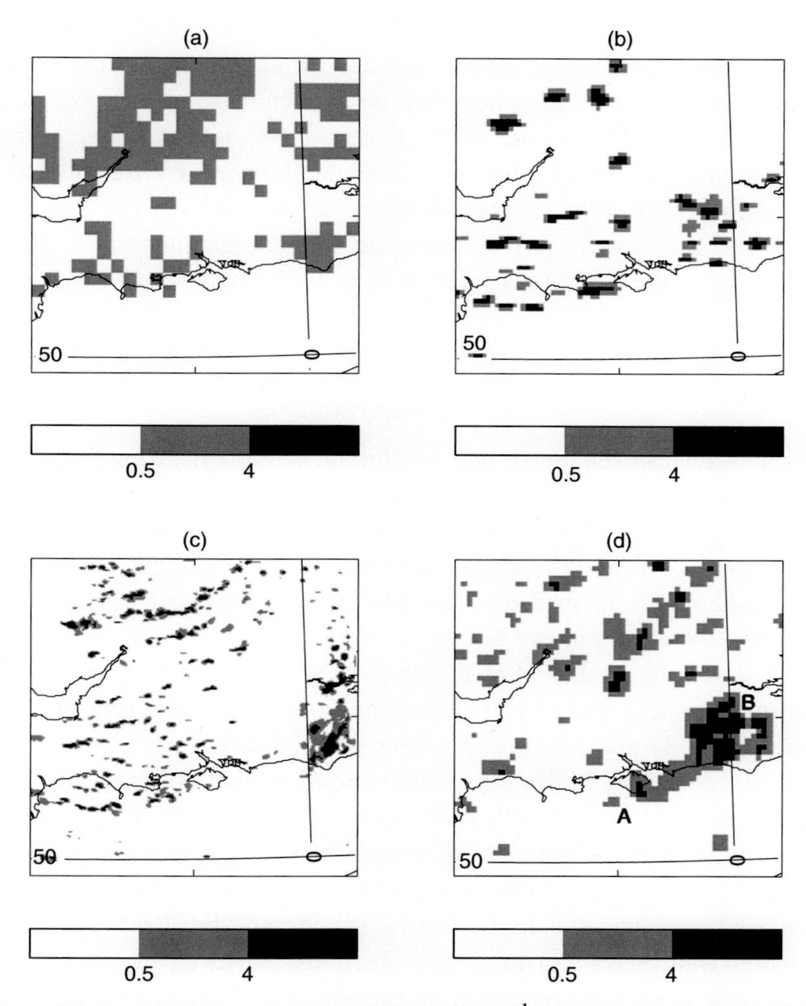

Figure 5.3 Instantaneous rainfall rates in mm hour^{-1} for 13 UTC 25 August 2005 with model runs started at 07UTC using 06UTC data. (a) 12 km model, (b) 4 km model, (c) 1 km model, and (d) 5 km radar data. The main squall line is the feature between the letters 'A' and 'B'. From Lean *et al.* (2008)

more typical of weather forecasting models are increasingly being developed, e.g. the high atmosphere and ocean resolution HiGEM model (Shaffrey *et al.*, 2009), the eddy-permitting ocean resolution model, HadCEM (Roberts *et al.*, 2004), and high-resolution ocean versions of the GFDL model (Lau and Ploshay, 2009)) and used for understanding the role of enhanced resolution on the simulation of physical processes. Perhaps the current limiting case of this high-resolution, global domain work are global simulations at 3.5 km resolution for months to seasons run with the Non-hydrostatic ICosahedral Atmospheric Model (NICAM) (e.g. Satoh *et al.*, 2008). At these resolutions, the NICAM model is able to partially resolve cumulus convection and can simulate a large-scale cloud envelope propagating slowly eastward that is characteristic of the Madden–Julian oscillation (MJO) (Miura *et al.*, 2007). Computer limitations currently mean that it is impractical to perform the long simulations necessary for climate change runs at such high resolutions, but these initial studies may point to the future direction and can already provide valuable insights into important climate processes, such as the MJO, that will inform parameterization development for lower-resolution models.

5.2.2 Vertical resolution

The trend for increased horizontal resolution in numerical models enabled by increased computing power has not always been matched by increases in vertical resolution in either NWP or climate configurations. This is in spite of evidence of the importance of vertical resolution and of consistent choices of horizontal and vertical resolution published over the last two decades (e.g. Lindzen and Fox-Rabinovitz, 1989; Persson and Warner, 1991).

Model sensitivity studies of changing vertical resolution have been carried out with climate models, global and mesoscale forecast models, and single column models (e.g. Colle and Mass, 2000; Lane *et al.*, 2000; Tompkins and Emanuel, 2000; Byrkjedal *et al.*, 2008). Most studies confirm that the benefits of increasing resolution are only seen if both horizontal and vertical resolution are increased consistently. Bushell and Martin (1999) applied a threefold enhancement of vertical resolution in the boundary layer of the MetUM climate model version HadAM3, and achieved improvements in the vertical structure of the cloud-topped boundary layer in both well-mixed and decoupled situations. However, closer examination of the model results revealed an underlying sensitivity to vertical resolution in model interactions between boundary layer and convection processes which appeared unrealistic. Inness *et al.* (2001) showed that the appearance of congestus clouds in a model can be improved by increasing the vertical resolution around the melting level. Similar improvement in the representation of cumulus congestus was also seen in the ECHAM5 model by Roeckner *et al.* (2006). Inness *et al.* (2001) found that representation of such clouds improved their simulation of the Madden–Julian oscillation. However, they also showed that sensitivities to vertical resolution in the convective parameterization resulted in noise in the atmospheric profiles which also altered the nature of the convection in the model. These examples also illustrate the difficulties which are often encountered when vertical resolution is increased: physical parameterizations often have such sensitivities built in and they can be difficult both to find and to remove.

The facility to develop and test parameterizations in a seamless modeling environment offers the potential to uncover and remove dependencies on vertical resolution during the development stage and thus allow both NWP and climate models to benefit from improved parameterizations at an earlier stage. An example, is the development of a new, non-local, boundary layer parameterization (Lock *et al.*, 2000; Martin *et al.*, 2000) in which the eddy diffusivities for unstable conditions are, at a given height within the boundary layer, determined not by any local properties of the mean profiles at that height, but solely by the magnitude of the turbulence forcing applied to the layer and the height within the layer. Tests were carried out in the single column version, 'mesoscale' limited area United Kingdom model and the climate version (HadAM3; Pope *et al.*, 2000) of the MetUM with two different vertical resolutions. In both configurations, the new boundary layer scheme was able to diagnose different boundary layer types that appeared to be consistent with the observed conditions, and the boundary layer structure was improved in comparison with observations. However, there were also differences at the different resolutions, and these were attributed to vertical resolution dependencies in the model's convection scheme. The MetUM subsequently adopted the higher vertical resolution in the lower troposphere when the new boundary layer scheme was implemented in both NWP and climate configurations, and parallel development work was carried out in both configurations to reduce level dependencies in order that the benefits of the improved parameterization could be realized.

An additional area of synergy is in the development of vertically extended models in order to represent stratospheric processes and variability and their impacts on the troposphere. For example, Mohankumar and Pillai (2008) showed that the quasi-biennial oscillation (QBO), a regular oscillation of the zonal wind in the equatorial stratosphere with a periodicity of about 26 months, has a significant association with the Asian summer monsoon on a biennial time-scale. Representing phenomena such as the QBO requires that models have an adequate vertical resolution to represent the shear zones and the vertical wavelengths as well as a generation mechanism for a tropical wave spectrum (e.g. Boville and Randel, 1992; Takahashi, 1999; Scaife *et al.*, 2000). The model vertical resolution is also important for modeling stratosphere–troposphere exchange on a range of time-scales (e.g. Pope *et al.*, 2001; Land *et al.*, 2002; Gray, 2003). The location of the upper model boundary can affect the representation of vertically propagating wave modes. Boville and Baumhefner (1990) suggested that there may be no reasonable location for the model upper boundary until all significant wave energy propagating upwards from the troposphere has been absorbed by damping processes. Using the NCAR Community Climate Model, Boville and Baumhefner (1990) showed that the location of the upper model boundary can affect the simulation of the lower stratosphere within the first few days of a forecast, while the impact on tropospheric systematic errors is manifest after about 20 days.

Many forecasting models have been extended into the mesosphere in order to benefit from assimilation of the full range of satellite data (e.g. Untch and Simmons, 1999; Polavarapu *et al.*, 2005; Hoppel *et al.*, 2008). This has been found to improve extended-range weather forecasts as well as improving understanding of the middle atmosphere. For example, Untch and Simmons (1999) showed that the ECMWF model gave better mean verification scores for 500 hPa height forecasts throughout the medium range following its vertical extension and increase in stratospheric resolution. Hoppel *et al.* (2008) found that medium-range stratospheric forecasts from the high-altitude (up to 100 km) extended version of the Navy's Operational Global Atmospheric Prediction System (NOGAPS) global forecast model showed reduced root-mean-square errors due to the assimilation of stratospheric temperature measurements from satellites. Furthermore, Jung and Leutbecher (2007) showed that the observed downward propagation of stratospheric polar vortex anomalies into the troposphere is well captured in seasonal integrations of the vertically extended ECMWF model. The development and evaluation of vertically extended climate models can also benefit from comparison with daily operational meteorological analyses obtained using a data assimilation system such as that described by Swinbank and O'Neill (1994), which utilizes the mix of observations available in the troposphere and stratosphere. For example, Amodei *et al.* (2001) compared the tropical oscillations and planetary-scale Kelvin waves[4] in four troposphere–stratosphere climate models with the assimilated dataset produced by the Met Office, and demonstrated more realistic wave activity in the models with better vertical resolution in middle and lower stratospheres.

Increasing attention is now also being directed towards improved modeling of the diurnal cycle, which is obviously a fundamentally important mode of variability in the atmosphere–land–ocean system on time-scales of interest for weather prediction and also an essential building block of intraseasonal to climate time-scale phenomena. There is some evidence (Bernie *et al.*, 2005) linking observations and idealized modeling that the presence of air–sea interaction on diurnal time-scales may be a significant factor in explaining the variance of SST on intraseasonal time-scales, for instance related to the MJO in the tropical warm pool region. The idea of simulating diurnal air–sea variability interactively in operational climate and NWP models has been explored to a limited extent so far in published work (e.g. McCulloch *et al.*, 2004; Zeng and Beljaars, 2005; Danabasoglu *et al.*, 2006). Modeling evidence suggests that to truly capture the diurnal response of SSTs certainly requires explicit two-way air–sea coupling involving either a mixed layer-type ocean model or a full

[4] Equatorially trapped eastward propagating internal gravity waves.

ocean GCM with a high near-surface ocean vertical resolution (1 m or less), and correspondingly high frequency air–sea coupling of 3 hourly or less together with the appropriate physics (Bernie *et al.*, 2007).

5.3 Seamless prediction across a range of time scales

Predicting across a range of time-scales offers substantial benefits for understanding the physics and dynamics of phenomena in the coupled atmosphere–ocean–land–cryosphere system and ultimately improving their prediction. At time-scales of hours to days, routine predictions from several operational centers provide high-quality analyses and forecasts of the weather on both global and regional scales. Using the NWP framework we can explore deficiencies in model parameterizations with fast time-scales (e.g. clouds and precipitation) by evaluating analyses and forecasts of individual weather elements against detailed *in situ* (surface, profiling, and aircraft data) and satellite observations. As we move into the medium range (6–15 days) and deterministic predictability declines, then the slower physical processes play a larger role (e.g. land surface, ocean, cryosphere, aerosols, and biogeochemical forcing), and ensemble prediction becomes vital for assessing uncertainty. For sub-seasonal to seasonal/decadal time-scales, the promise of improving regional climate prediction capability lies in the ability to capture low-frequency phenomena such as the MJO (Zhang, 2005), El-Niño Southern Oscillation (ENSO) (Wallace *et al.*, 1998; Trenberth *et al.*, 1998), and North Atlantic Oscillation (NAO) (Hurrell, 1995).[5] This includes capturing global teleconnections and associated high impact weather events such as drought, flood, blocking, and tropical/extratropical cyclones (Barsugli *et al.*, 1999). However, these phenomena are inherently multi-scale in nature. The MJO modulates the large-scale convective activity across the Indian Ocean and Pacific on 30–60 day time-scales, but is made up of phenomena on numerous space and time-scales from individual cloud clusters to planetary-scale circulation anomalies. In turn the MJO plays a role in modulating extratropical synoptic and low-frequency variability through excitation of Rossby wave trains (Ferranti *et al.*, 1990) and also influences Asian and Australian monsoons and ENSO. Investigating the ability of atmospheric models to predict these phenomena across several time (and space) scales will hopefully shed light on these multi-scale interactions and the role they play in the physics, dynamics, and predictability at weather and climate time-scales.

Evaluating the models against high-quality observations is a key component of a unified or seamless modeling strategy (Randall *et al.*, 2003). Observational datasets

[5] Fluctuations in the difference of atmospheric pressure at sea level between the Icelandic low and the Azores high.

are available for model evaluation across a range of time (and space) scales. Field campaigns provide a wealth of information but typically only last a period of weeks to months, whilst climate monitoring datasets like the global temperature record extend over a century. Intensive observation sites such as those run by the U.S. Department of Energy ARM (Atmospheric Radiation Measurement) program (Ackerman and Stokes, 2003) provide detailed information at a number of point locations whereas satellite observations often achieve near global coverage. A unified modeling framework can help exploit the different scales of these observations by comparing climate model errors identified by comparison against long-term datasets, with errors in shorter-range forecasts of individual events (such as case studies from a field campaign).

The advantages of using an NWP framework to assess physical parameterizations in climate models are that

(i) locally forced rather than remotely forced errors are emphasized,
(ii) errors in the large-scale flow are minimized in running from initial states generated by state-of-the-art data assimilation systems, and
(iii) as discussed above, detailed comparison with observational datasets focusing on key physical processes and individual weather systems is possible.

This strategy has been advocated by Phillips *et al.* (2004) and Rodwell and Palmer (2007), and recently formalized by the Working Group on Numerical Experimentation (WGNE) as an international model inter-comparison exercise (Transpose AMIP – Williamson *et al.*, 2008), with a preliminary study evaluating model parameterizations over the Southern Great Plains ARM site in spring and summer. Implicit in this strategy is that short-term NWP systematic errors and longer-term "climate drift" show similarities. Figure 5.4 is an example of systematic errors in zonally averaged temperature and zonal wind from predictions of the MetUM at a variety of time-scales and horizontal resolutions. The similarity between the climate and NWP systematic errors is striking, for example, the tropical mid-tropospheric warm bias and upper-tropospheric westerly zonal wind biases. This suggests that a proportion of the systematic errors appear on very short time-scales and are probably associated with deficiencies in the fast physics. A comparison of precipitation errors in the NCAR CAM and GFDL models shows a very similar agreement at 3 days and over a season (Moncrieff *et al.*, 2010).

However, model systematic errors do evolve in time as local and remotely forced errors interact and deficiencies in "slow physics" processes (e.g. land surface) come into play at longer ranges. The study of the evolution of errors from very short-range to monthly, seasonal, and decadal time-scales should inform these interacting error sources. The following sections provide some examples of these approaches across the various time-scales from data assimilation, initial tendencies, and short-range

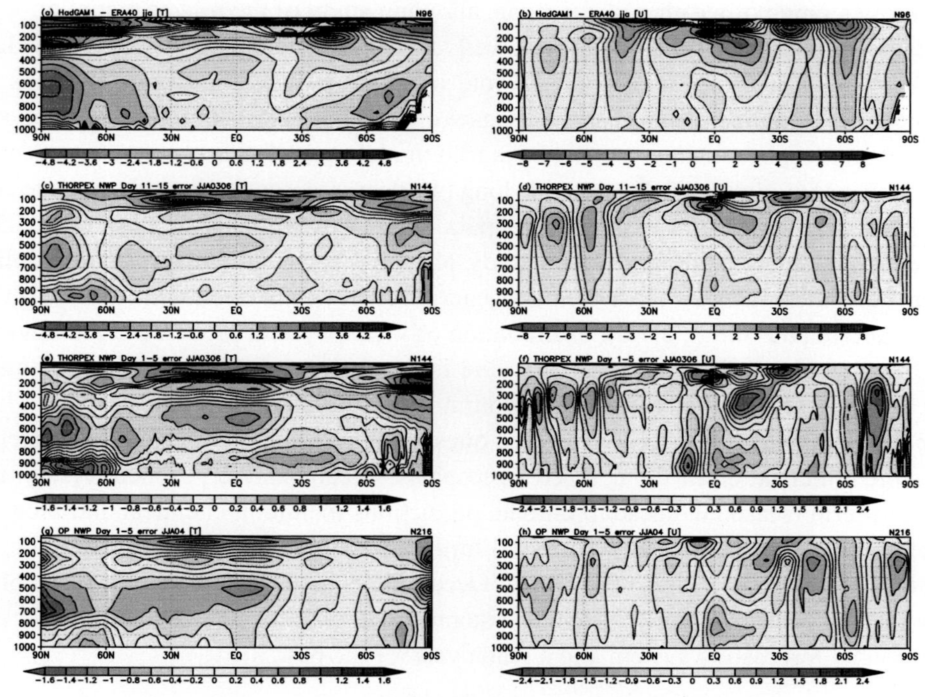

Figure 5.4 Zonally averaged cross-sections of temperature and zonal wind systematic errors (model–analyses) for model forecasts/simulations during June–August performed across a range of temporal scales and at differing horizontal resolutions. (a) & (b) HadGAM1 climate model 20 year AMIP run–ERA40 re-analysis, N96 (135 km) horizontal resolution (c) & (d) MOGREPS-15 (THORPEX) days 11–15 forecast – Met Office analyses for JJA 2003 and 2006, N144 (90 km) horizontal resolution (e) & (f) as (c) & (d) but for days 1–5, (g) & (h) days 1–5 from the operational deterministic global NWP model (circa 2004) – Met Office analyses for JJA 2004, N216 (60 km) resolution. See also color plate.

forecasts to the evolution and reduction of systematic errors in weather and climate predictions.

5.3.1 Data assimilation and reanalyses

In recent years the development of a new generation of variational assimilation systems (overview by Rabier, 2005) and ensemble Kalman filter techniques (Evensen, 1994), combined with a rapid increase in the number and quality of satellite observations, have provided a huge scientific opportunity for understanding atmospheric circulation/composition and evaluating atmospheric models. The daily observation–background (O–B) statistics provide continuous monitoring of the fit of 6-hour model atmospheric background state to quality-controlled observations across a range of weather phenomena, allowing study of the role of model error in data assimilation (Dee, 2005), as well as regime-dependent biases in models (ECMWF, 2007). A review of the role of NWP model-assimilation systems in model development is given in Hollingsworth (2000). Efforts to provide reanalyses of the atmosphere using the most advanced model-assimilation systems have seen a number of modeling centers provide long (40-year) records of atmospheric state and circulation (Kalnay *et al.*, 1996; Uppala *et al.*, 2005; Onogi *et al.*, 2007). The reanalysis datasets allow detailed study of global and regional climate variability in quantities such as the water and energy cycles (e.g. Trenberth *et al.*, 2007; Trenberth and Smith, 2008), examination of local physical processes in the model (e.g. Betts *et al.*, 2003; Betts, 2004), and full four-dimensional datasets to evaluate climate model predictions. Finally, within the seamless prediction paradigm it has been recognized that a major research question for future data assimilation systems is the initialization of the coupled atmosphere–ocean–land–cryosphere system for short-term (seasonal–decadal) climate predictions including chemical and biogeochemical processes. This is discussed further in Section 5.4.3 but, for example, at NCEP (Saha *et al.*, 2006) the Global Ocean Data Assimilation System is coupled with the Climate Forecast System to generate both real-time initial conditions for the seasonal forecasting system and a reanalysis for the past.

5.3.2 Initial tendency diagnosis of model errors

Klinker and Sardeshmukh (1992) attempted to take the NWP approach to error diagnosis to its ultimate limit by considering just the tendencies in the very first forecast timestep. The idea is that initial tendencies are generated from a state as close as possible to the true state and so errors should be directly associated with deficiencies in the model processes, and not with the application of erroneous atmospheric states to these processes. Errors have no time to interact or propagate and so it may be possible to identify which model process(es) lead to the erroneous total systematic tendency or budget residuals. Klinker and Sardeshmukh (1992) examined the budget residuals in the time–mean momentum budget and identified problems with the ECMWF gravity wave drag scheme. Other studies have also applied this technique to various problems. Milton and Wilson (1996) examined the time–mean initial momentum balance in the NWP version of the MetUM demonstrating benefits from revisions to gravity wave drag and inclusion of a new orographic roughness scheme representing a previously missing process. Evaluation of the gravity wave drag scheme was also carried out in climate simulations with the MetUM (Gregory *et al.*, 1998). Kanamitsu and Saha (1996) investigated systematic initial tendency errors in the NCEP global model and noted that ascribing the source of model error to a given parameterization from time-mean budgets still remained challenging due to interactions amongst dynamics and physics and potential errors in the data assimilation process. The initial tendency diagnosis can potentially be extended to take account of the co-variability of initial tendencies and errors in an automated search for model error sources (this was explored to some extent by Klinker and Sardeshmukh, 1992). Schubert and Chang (1996) and Krishnamurti *et al.* (2004) developed the tendency method further along these lines, applying statistical regression techniques to determine linear changes to the GCM forcing terms which produce the best model in a least squares sense (compared to observed tendencies). These studies provided quantitative estimates of both systematic and random error components partitioned amongst the various models' parameterized forcing terms (e.g. radiative heating, convection, etc.). Bergman and Sardeshmukh (2003) have used the single column model (SCM) framework to provide an economical strategy for diagnosing errors in the NCAR model parameterizations. They used six-hour forecasts and initial tendencies from SCM runs to explore both the linear error growth and the nonlinear interactions between parameterization schemes, highlighting the interaction of boundary layer vertical diffusion and convection as a potential source of tropical temperature biases. The use of SCM 6-hour runs overcomes the problem of unrealistic drift seen in long-term SCM integrations that traditionally hampers their use for error diagnosis (Randall *et al.*, 2003). Bergman and Sardeshmukh (2003) also proposed a technique to couple adiabatic and diabatic tendencies in the SCM which could partially overcome the model drift problem and make SCM studies of climate variability on longer time-scales more tractable.

Rodwell and Palmer (2007) have recently used the initial tendency approach in order to better quantify the uncertainty in climate change forecasts arising from model error. The central idea is that the initial tendency budget residual, which is a manifestation of systematic errors in the model's fast physics processes, can be used to produce probability weightings for each model that could be used in the construction of probability distribution functions of climate change. For example, if their initial tendency results are applicable to the base model used by Stainforth *et al.* (2005), then they suggest that the high climate sensitivity models in that study can be rejected or given a low probability weighting. One advantage of this approach is that it typically costs 5% of the cost of a 100-year coupled model simulation that might otherwise be used to assess the simulation of present-day climate. Rodwell and Palmer (2007) demonstrated that the initial tendencies combine in a near-linear fashion (owing to linearity and/or spatial orthogonality in the effects of different perturbations), showing that the fast physics of a model with any combination of parameter perturbations can be approximately assessed from the initial tendencies associated with each individual perturbation alone. This approximate linearity could further reduce the computational expense of assessing model fast physics by several orders of magnitude and thus help greatly in the quantification of climate change uncertainty.

5.3.3 Evaluating short-range forecasts against observations

Another advantage of the NWP framework is that short-range (e.g. 12–36-hour) model predictions of individual weather systems can be evaluated against detailed *in situ* and satellite observations. An example is the U.S. DOE sponsored link between climate modelers in the Climate Change Prediction Program (CCPP) and the parameterization specialist community in the ARM program. This so-called CAPT initiative (CCPP–ARM Parameterization Testbed) aims to support implementation of the NWP methodology in climate GCMs and to facilitate the necessary scientific collaborations (Phillips *et al.*, 2004; Boyle *et al.*, 2008; Xie *et al.*, 2008).

The Transpose AMIP project is a WGNE initiative related to CAPT which aims to compare several climate models (run on NWP time-scales) against detailed observational data such as that from the ARM sites. Figure 5.5 provides an example from the case study at the ARM Southern Great Plains (SGP) site during the June–July intensive observing period (IOP) in 1997. It shows error growth in both MetUM global NWP model (at 60 km resolution) and NCAR CAM2 climate model vertical temperature structures (Williamson *et al.*, 2005). MetUM shows a large-scale warming in the lower troposphere and stratosphere, and a cooling in the upper troposphere. The temperature tendencies from the individual physical parameterizations (K/day) in MetUM show a close correspondence between the budget residual (total tendency) and the overall error growth, although ascribing the contributions of the individual processes to this "residual" is still challenging. Looking in more detail at diurnal time-scales can give further clues to error sources. Superimposed on the large-scale temperature drift we see, (i) cooling in the lowest kilometer between sunrise and midday that is associated with too slow an evolution of the growing convective mixed layer, (ii) a strengthening warm bias between 1 km and 8 km that grows during daytime. Further diagnosis suggests that some of this excessive warming comes from shortwave (SW) heating from a too-absorbing aerosol climatology currently used in the lowest layers in the NWP version of the MetUM. Work is underway to improve the aerosol climatology and ultimately to replace the climatologies with fully prognostic aerosols as used in the climate version (see Section 5.4.1 for more discussion), (iii) a cold bias in the upper troposphere which is largest just before sunrise. The CAM2 also shows a warm bias but with quite different structure, having the largest warming in the upper troposphere.

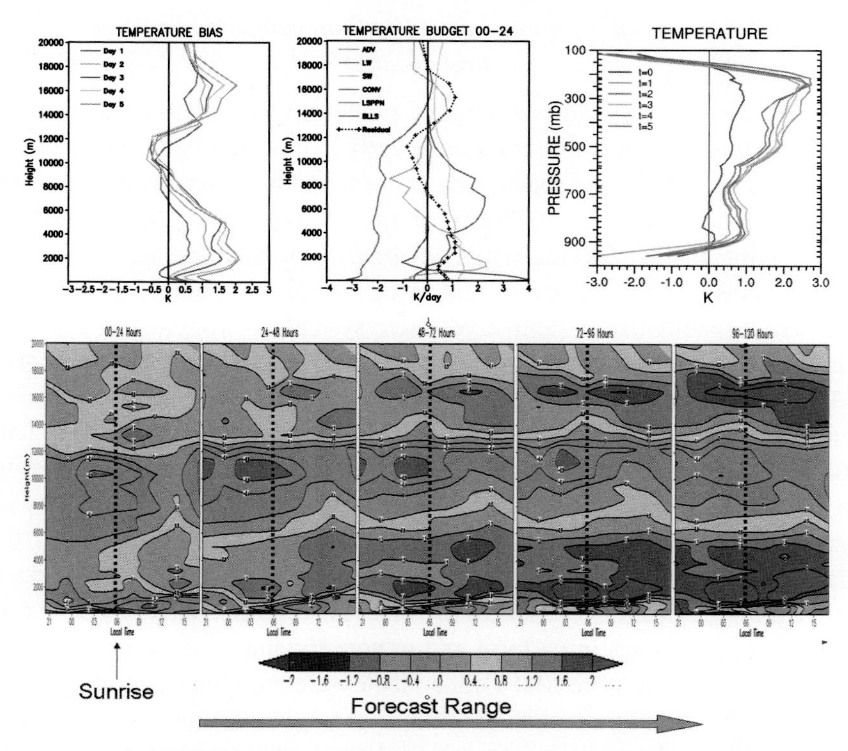

Figure 5.5 Diagnosing MetUM and NCAR-CAM2 systematic temperature biases at ARM-SGP during IOP June–July 1997 as part of Transpose-AMIP intercomparisons (see text). Results from the MetUM are shown in the bottom, upper left, and upper center panels. Results from NCAR-CAM2 are shown in the upper right panel. The bottom shows the evolution of MetUM vertical temperature biases (model-sonde) as a function of forecast range (0 to 5 days, showing 5 diurnal cycles). Results are averaged over all forecasts during the IOP. The upper left panel shows the evolution of MetUM temperature bias integrated over a diurnal cycle (and over all forecasts), and the middle panel shows MetUM mean 00–24 hour tendencies from the individual physics routines. Note the strong similarity between evolving temperature bias and the residual amongst the tendencies and also how the contributions to the temperature bias come from different parts of the diurnal cycle point to specific errors in physical processes (see text for more details). The upper right panel shows the NCAR-CAM2 (from Williamson *et al*, 2005) temperature biases as a function of forecast range which also shows an evolving warm bias but differs from MetUM in that larger errors are seen in the upper troposphere. See also color plate.

Klein *et al.* (2006) also studied this summertime warm and dry bias over the SGP in the GFDL AM2 model. They concluded that deficient precipitation and feedbacks between land-surface and atmosphere were responsible for the warm/dry bias. The lack of precipitation may have been associated with the failure of the coarse-resolution model to capture the observed rain-bearing nocturnal mesoscale convective systems (MCS) triggered over the Rockies.

5.3.4 Evolution of errors – medium range to seasonal and decadal

Finally, we can consider how systematic model errors develop into the medium range (6–10 days) and beyond, from those early in model integrations to the mature errors in the model climate mean state, in order to gain insight into the physical processes involved in the establishment of the long-term "climate drift." This evolution typically involves the interaction of errors in both fast and slow processes, local and remotely forced errors, and errors in the separate components of the coupled atmosphere–ocean–land–cryosphere system.

Jung (2005) examined the evolving systematic error growth in operational ECMWF model wintertime forecasts from 1–2 days out to seasonal time-scales and for the period 1983–2003. He demonstrated a number of interesting features. While some systematic errors had clearly been reduced in medium-range forecasts in recent model cycles, consideration of the seasonal time-scale showed very similar systematic errors. This suggested that the growth of error had slowed but the fundamental error structure remained. He also noted that the systematic errors underwent a significant evolution in the first few days but were more invariant at extended ranges, suggesting a local and remote forcing of the errors respectively. Another example, from the Met Office Hadley Centre climate model HadGEM1 (Johns *et al.*, 2006) are precipitation errors in the Asian summer monsoon, with too little rainfall over the Indian peninsula and too much over the equatorial Indian Ocean to the southwest of the peninsula (Martin *et al.*, 2006). Development of the monsoon precipitation errors in Met Office NWP forecasts to 15 days show that different parts of the error pattern evolve on different time-scales. Excessive rainfall over the equatorial Indian Ocean to the southwest of the Indian peninsula develops rapidly, over the first day or two of the forecast, while a dry bias over the Indian land area takes approximately 15 days to develop.

Accurate simulation of cloud is crucial for both climate and shorter-range forecasts. The radiative response from cloud accounts for much of the variation in climate sensitivity between GCMs, impacts via diabatic processes on the evolution of a forecast, and has a direct effect on key forecast products such as surface temperature, precipitation, and visibility. Following Jakob and Tselioudis (2003), Williams and Brooks (2008) applied a clustering technique on International Satellite Cloud Climatology Project (ISCCP) data and comparable diagnostics from MetUM to group together grid-boxes with similar cloud top pressures, cloud optical depths, and grid-box cloud covers in order to identify a small number of "cloud regimes." They found a number of errors in the climate model such as the shortwave cloud forcing of the stratocumulus regime being excessive (i.e. the cloud reflects too much sunlight), and the extratropical cirrus regime occurring too infrequently with an

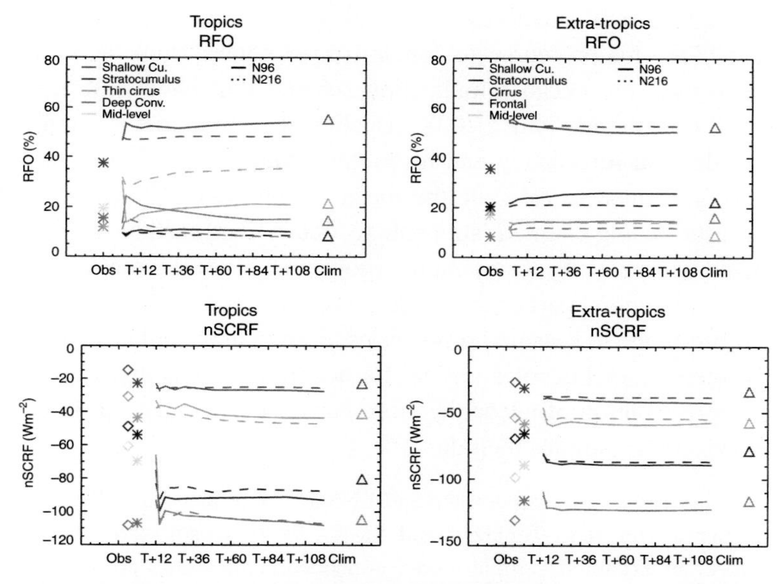

Figure 5.6 Evolution of cloud-regime relative frequency of occurrence (RFO) and normalized shortwave cloud radiative forcing (nSCRF) through the model forecast. N96 is shown solid; N216 is shown dashed. Observed climatology from ISCCP is shown with an asterisk; from Earth radiation budget experiment (ERBE) data is shown with a diamond; climate model climatology is shown with a triangle. Note that the ISCCP tropical stratocumulus and deep convective RFO are identical (from Fig. 3 of Williams and Brooks, 2008). See also color plate.

excessive longwave cloud forcing (i.e. the cloud is too high and/or too opaque in the longwave). Williams and Brooks (2008) investigated the time-scale over which these errors occurred by running a series of five-day forecasts, initialized with the four-dimensional variational data assimilation system used operationally with this model. The initialization had the effect of constraining the dynamics and, to a large extent, the thermodynamic structure of the atmosphere to be close to that observed, however cloud is not currently assimilated in this system. The authors showed that errors noted in the climate model climatology were also present in the NWP forecasts, usually from the first timestep (Figure 5.6). This indicates that the biases are local in nature, being present even in the well-constrained situation at the start of the forecast, rather than evolving as a result of impact of a remote error on the dynamics. The study also showed that increasing the resolution to the operational forecast resolution of the time had little effect on these errors.

For prediction on seasonal time-scales the advent of new observations and more sophisticated assimilation systems is likely to improve the performance of coupled models on short- to medium-range time-scales, but in many cases this may be overshadowed by the systematic errors in the models. This is the case in seasonal forecasting, now a routine activity in several operational centers, where it has been shown that if the major source of forecast error comes from the coupled model systematic biases, the ocean initialization scheme will have little impact on the forecast skill (Balmaseda *et al.*, 2008). Dealing with this drift is a major issue for seasonal-to-decadal forecasting and historical reforecasts (or hindcasts) have been traditionally generated to estimate the mean model biases. Potentially, the use of these hindcasts – using coupled (atmosphere–ocean–land surface–cryosphere) models initialized with realistic observations – goes well beyond forecast calibration and they are now an integral part of the model development strategies at the Met Office.

On medium-range to seasonal time-scales, correct representation of the physics–dynamics interactions becomes critical for accurate forecasting. For example, the improper representation of teleconnections can be due to missing processes in our modeling system. Examples include

(i) the link between European climate and ENSO, where there is evidence for an active stratospheric role (e.g. Toniazzo and Scaife, 2006; Ineson and Scaife, 2009) and, therefore, an extended vertical model is needed. This shows an interesting synergy with NWP, where an increased number of vertical levels has been sought in order to better assimilate satellite radiance information,
(ii) the intraseasonal interaction between the Madden–Julian oscillation and the North Atlantic oscillation (Cassou, 2008),
(iii) the positive impact of coupling wave models to atmospheric models (Janssen, 1992).

5.4 Benefits of shared development

As models become increasingly complex, running on a variety of platforms and with the need to encompass initialization and boundary forcing, maximizing the utility of the scientific and technical development becomes essential. In this section we discuss synergies across time and space scales in developing and evaluating existing and new parameterizations of physical and biogeochemical processes; in maximizing the benefits of new diagnostic techniques; in developing initialization techniques for slower-scale processes and the need for coupled initialization; the use of ensemble systems for understanding and representing uncertainty; and the benefits and costs of a shared software engineering infrastructure.

5.4.1 Complexity

The development and evaluation of parameterizations across a range of temporal and spatial scales is a key element in the NWP and climate synergy. The different diagnostic/evaluation approaches from each area help identify strengths and weaknesses of the parameterization schemes at the testing stage. Past examples of this approach in the MetUM have been in the development of the boundary layer scheme (Martin *et al.*, 2000) and in the orographic drag schemes (Milton and Wilson, 1996; Gregory *et al.*, 1998).

It is also true that the atmosphere–ocean–land–cryosphere physical system is generally modeled with more complexity in coupled climate models than in NWP models, in order to capture feedbacks important on long time-scales within the climate system. In recent years the Met Office and other centers have looked to exploit their unified modeling strategy to implement some of these physical processes within the global and regional NWP systems moving towards an environmental prediction capability. For example, there is ongoing work to improve the representation of aerosols in NWP systems. The ability to account for the transport and radiative properties of mineral dust in short-range to seasonal weather forecasts has been explored at various modeling centers (Greed *et al.*, 2008; Milton *et al.*, 2008; Rodwell and Jung, 2008), and comparisons of real-time predicted dust with satellite, ground, and aircraft observations (Haywood *et al.*, 2008) give the opportunity to improve the mineral dust parameterization for both weather and climate prediction. Another example of this cross fertilization is the UK Climate and Aerosols (UKCA) model (Morgenstern *et al.*, 2009) developed for climate prediction, which is being used as the basis to develop an air quality prediction system for the UK based on the MetUM. A strong synergy was also found between meeting the climate modeling requirement for more detailed description of surface processes and the mesoscale modeling requirement for predicting weather elements. In the UK, during anticyclonic conditions, the weather is almost entirely determined by boundary layer processes. Prediction of moisture availability at the surface is essential for forecasting of convective rainfall amounts. When the MetUM was introduced, an enormous step up was required in the complexity of the land-surface model, which was rewarded by a much greater ability to predict screen-level temperature, fog, etc. The use of the model for climate simulations ensures reasonable performance when there is a need to create a limited area forecast model at short notice over a new part of the world.

5.4.2 Sharing of diagnostic techniques across time-scales

Many traditional diagnostics are used by both the weather and climate communities for model evaluation (e.g. pressure at mean sea level, screen-level temperature, etc.). However, both communities are also evolving new innovative diagnostic techniques which can offer the potential to exploit more of the information within observations than would otherwise be possible. Within the climate community there has been an ongoing effort to develop satellite simulators (Klein and Jakob, 1999; Webb *et al.*, 2001; Chepfer *et al.*, 2008; Bodas-Salcedo *et al.*, 2008). These aim to provide model diagnostics which emulate the satellite product, taking account of cloud overlaps, using the satellite retrieval algorithm, etc. Such forward models are an essential component of NWP data assimilation systems, but there is also growing interest in the use of simulator diagnostics from the NWP community. For example Bodas-Salcedo *et al.* (2008) use the CloudSat simulator to evaluate a mid-latitude frontal case study and produce statistical summary information to show how the errors relate to general model biases (Figure 5.7). Such statistical

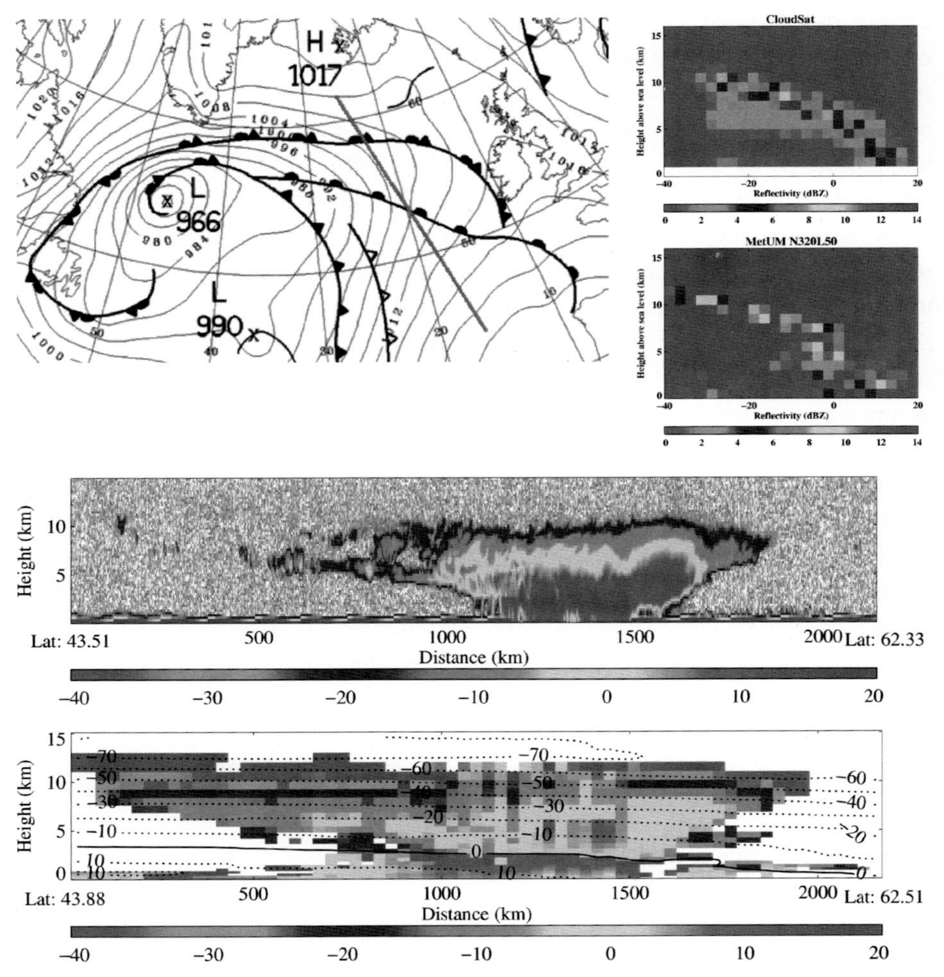

Figure 5.7 Top left: Met Office analysis chart for North Atlantic at 1200 UTC on 26 February 2007. The CloudSat track is shown in red. Bottom: radar reflectivity (dBZ) observed by CloudSat (upper) and simulated by the MetUM global forecast model (lower) along this track. Top right: joint height–reflectivity hydrometeor frequency of occurrence (%) observed by CloudSat (upper) and simulated by the MetUM global forecast model (lower) for this case study. Taken from Bodas-Salcedo *et al.* (2008). See also color plate.

summaries can then be used to compare biases at differing time-scales. In contrast, diagnostic methods developed in the weather forecasting community to objectively identify synoptic features (e.g. Hewson, 1998) are starting to be used for climate and climate change analysis. Inclusion of new observations in an NWP data assimilation system yields information on the characteristics of the observations which can also be valuable for climate monitoring purposes. An example is radio occultation data, which are now routinely assimilated for NWP (e.g. Healy and Thepaut, 2006). Based on the knowledge gained in that setting, radio occultation data has recently been proposed for climate change fingerprinting (Ringer and Healy, 2008).

Recently, there have been a number of studies which have aimed to isolate particular processes in order to provide a more process-based diagnostic. For example, compositing cloud variables like cloud amount or cloud radiative forcing by 500 hPa vertical velocity (e.g. Bony *et al.*, 2004), or surface pressure (e.g. Tselioudis *et al.*, 2000), or simply clustering together similar cloud types (e.g. Williams and Tselioudis, 2007) in both the model and observations allows the statistics of the cloud variables in different regimes to be evaluated. In a unified modeling framework, these techniques can be used to compare model errors between, say, a short-range forecast and a climate model climatology (Williams and Brooks, 2008), enabling some of the studies described in the previous section. Case studies of systematic errors evaluate an individual event over a limited region, often where there are a large number of *in situ* observations. Compositing techniques can be used to relate such case studies to the more general model errors occurring over large regions and over multiple forecasts or model climatology.

There is increasing interest amongst some groups to introduce more internationally agreed metrics for models in order to objectively demonstrate their usefulness and how they have improved over time. For NWP this builds on the long-standing internationally agreed metrics which are regularly exchanged, and for climate comprises a new set of measures (e.g. Gleckler *et al.*, 2008; Reichler and Kim, 2008). Metrics based on diagnostics which can be applied across time-scales, such as the compositing techniques discussed above, can be particularly valuable in providing a consistent set of metrics between the different time-scales.

5.4.3 Initialization

Model initialization is a shared problem at all time-scales. In the case of NWP, atmospheric initialization is critical to predict weather changes over the coming days. Much progress in both observational networks and data assimilation techniques has been made in recent years, substantially improving forecast skill.

For longer time-scales, from medium-range to seasonal and centennial, the initialization of other slower parts of the climate system (land surface, oceans, cryosphere) is fundamental given that consistently small influences can have a significant effect if applied for a long enough time.

As in the atmosphere, much progress has been made in the initialization of these slower systems. However, the lack of quality global observations remains a significant problem. For example, there are few global observations of soil moisture so initialization of the land surface variables has typically only been possible through indirect information about low-level temperatures and humidity. An assumption of linearity is made and the soil moisture is "nudged" towards the optimum value based on this indirect information (Douville *et al.*, 2000; Best and Maisey, 2002). As we move forward, it is likely that more progress will be made through exploitation of independent information on soil moisture, either directly from satellites or indirectly from, for example, stand-alone models of the land surface forced with observed precipitation and evaporative fluxes (e.g. the Global Soil Wetness Project – Dirmeyer *et al.*, 1999, the Land Information System – Kumar *et al.*, 2006).

A similar situation is found for the initialization of sea ice. Satellite data of sea-ice extent, concentrations, and motion are routinely assimilated using a simple nudging scheme in the Met Office Forecast Ocean Assimilation Model (Martin *et al.*, 2007). Methods of assimilating satellite and field observations of thickness change and ice-flow of ice sheets are now being explored (Arthern and Hindmarsh, 2003). The observation of the ocean's surface and subsurface has improved significantly from the late 90s with the deployment of the Argo floats and new and better satellite instruments.

The assimilation of these slower-scale processes has clearly benefitted from advances in atmospheric assimilation for NWP. Thus, several centers are now developing systems for ocean initialization (Pham *et al.*, 1998; Sugiura *et al.*, 2008; Fu *et al.*, 2009). In the same way, there is growing evidence that short- and medium-range predictions benefit from coupling of the atmosphere to the ocean, sea ice, and waves which improves the simulation of momentum transfer at the air–sea interface.

With the potential application of coupled atmosphere–ocean models across time-scales, a current limitation is that the individual model components are initialized independently, with the optimal data assimilation theory now becoming established for linearized models. One of the shortcomings of this is the introduction of initialization shocks given that the coupling of systems such as the atmosphere, ocean, and land-surface is largely nonlinear. The idea of truly coupled initialization systems has been put forward in recent years (Zhang *et al.*, 2007). However, such a system will need to address issues such as the spin-up of nonlinear processes, model errors and biases, and the inconsistent fluxes required to balance different model components of the coupled system. Although progress has been made (see for example the soil moisture nudging described in Best and Maisey (2002)) a usable theory for this is not yet available – and may not exist given the very different time-scales involved. It is necessary to combine these requirements with the optimal assimilation of observations in variational data assimilation systems such as 4D-Var. One way of dealing with this is the use of time-averaged values for the atmosphere as done by Sugiura *et al.* (2008) using a 4D-Var scheme or Zhang *et al.* (2005) using an ensemble square root filter approach. However, it may be more practical to seek optimal assimilation by each model component of its own observations, together with pragmatic bias correction, coupling, and initialization methods for the non-linear processes and fluxes between systems.

5.4.4 Ensemble generation

It is now fully recognised that it is essential to account for the uncertainties that limit predictability at all time-scales. In order to do so, ensemble prediction systems are run routinely in various centers for: short-range (including the use of regional models as pioneered at NCEP, Du and Tracton (2001); see also Bowler *et al.* (2008) and references therein); medium-range (Toth and Kalnay, 1993; Buizza and Palmer, 1995); monthly (Buizza *et al.*, 2006); seasonal (Graham *et al.*, 2005; Saha *et al.*, 2006); decadal (Smith *et al.*, 2007), and centennial (Murphy *et al.*, 2004).

The main sources of uncertainty can be classified within three categories depending on whether they affect the initial conditions, model, or boundary conditions. However, these uncertainties are not fully independent – for example, the initial conditions are contaminated by model error through the data assimilation systems – and their importance varies depending on the time-scale of interest.

Medium range is perhaps the most mature area of ensemble prediction with operational systems running routinely since 1992 in both the USA and Europe (Toth and Kalnay, 1993; Buizza and Palmer, 1995). In these systems, reflecting the fact that weather is a chaotic system, the representation of initial condition uncertainties was given priority, resulting in the development of singular and breeding vectors, both aiming to capture the fastest-growing perturbations in the atmosphere. However, it was soon recognized that, even for medium-range and shorter time-scales, model uncertainties had to be included. Not surprisingly, given that model errors develop continuously during the simulation, the role of model uncertainties is even bigger at longer time-scales. One technique for addressing model error is the use of stochastic physics schemes which are now being jointly developed between NWP and GCM groups. For example, ECMWF and the Met Office have both developed schemes based on Shutts (2005) aiming to backscatter energy excessively dissipated by advection and physical parameterizations. These schemes are being used in short-, medium-range and seasonal ensemble prediction systems.

The sampling of the uncertainties due to the subjective parameters in physical parameterizations has also been attempted in different ways. For example, for short- and medium-range prediction, Lin *et al.* (2000); Bright and Mullen (2002) and Bowler *et al.* (2008) treat certain parameters as stochastic variables which are allowed to change during the simulation following a first-order auto-regression function. For centennial prediction Murphy *et al.* (2004) created different model versions, each with a different set of parameter values kept constant during the simulation, that were run as an ensemble.

It is important to remark that the schemes to represent model uncertainties are not simply intended to generate spread, but also to reduce model errors. Berner *et al.* (2008) show a reduction in some tropical wind biases in the ECMWF model with the introduction of a backscatter scheme and the MetUM scheme generates a better representation of power spectra. In the case of Murphy *et al.* (2004), the climate quality of the different model versions constructed can be evaluated, offering a valuable insight into the physical processes behind the systematic model biases.

5.4.5 Software engineering infrastructure

In the case of the MetUM , originally developed in the early 1990s, a large part of the rationale for unification was the perceived "economy of scale" in designing and operating a common software infrastructure to support both types of application (Cullen, 1993), taking an end-to-end view (from job set-up through a user interface right through to output processing stages). It is extremely rare that such an opportunity arises fundamentally to examine common design requirements starting from a fairly blank sheet of paper in a large operational software system, as was the case here.

The subsequent longevity and scientific success of the MetUM owes much to the decisions made originally and as the system has evolved regarding the architecture and practical realization of the software infrastructure and management practices. This success rests on foundations such as good coding practice and code review, efficient source code management and release cycles, and methodological approaches to specific model development projects.

Effective communication at all levels is also a vital ingredient, and an important factor in the case of the MetUM has undoubtedly been the synergy within the same organization between well-established centers of scientific and technical expertise in both forecast and climate applications, allowing close technical as well as scientific coordination of the software development cycles amongst the full range of developers and users of the system, aided by the fact that the coding is largely done by scientific domain experts, who are also generally code users. This has helped to minimize any communication gaps allowing an "agile" and yet controlled approach to the overall software development even in such a large and complex code (Easterbrook and Johns, 2009). These organizational synergies have also mitigated against the temptation to "fork" from a single unified system back to separate (possibly multiple) systems tailored to individual forecast and climate applications (there is an interesting analogy here in that in open source projects forking/deunifying, while always a possibility, is acknowledged to be a retrograde step for the overall user community (Fogel, 2005).

A similar software infrastructure design philosophy can be seen, for example, in the Flexible Modeling System (FMS) of GFDL in Princeton (http://www.gfdl.noaa.gov/fms) which has been in use for the past decade. In this case, the driving goal is to provide flexibility in the coupling of sub-component models rather than a seamless unification between NWP and GCM applications. The implementation, tools, user interfaces, and working practices associated with the MetUM and FMS systems naturally differ as a result of differences in design goals and requirements. Individuals involved in the design and development process, and their organizational cultures, also influence the detail of the software systems (Easterbrook and Johns, 2009), but common "good practice" features can nonetheless be identified.

Several other formal and informal tools have been introduced to aid the communication and change cycles in the MetUM. These include the use of newsgroup channels for disseminating information, and the use of wiki-based pages for recording changes as model configurations evolve – divorced from any individual's ownership.

The flexible/unified software engineering picture painted above is not a complete utopia of course. To meet short- and medium-term performance improvement goals, one model development path frequently needs to run ahead of the other, or schemes in different applications diverge to the extent that the actual model code in them is not identical. It is crucial for the long-term health of the unified system as a whole that developments feeding into the forward trunk of the common model development path can be easily tested and verified in any model applications to which they apply or could impact at the next software release cycle.

Integration testing and code review mechanisms are therefore crucial in providing the appropriate checks against undesirable changes appearing in the code that have a beneficial impact in one application but a deleterious effect in another, for instance. A centrally coordinated process is in place to manage software release cycles of the MetUM, with both pre-and post-release testing of changes across a range of model configurations an important part of ensuring that changes are fit for overall purpose.

In summary, striking the right balance between rapid code development, adequate testing, and long-term integrity of all the model applications in a unified system is difficult and requires considerable resources. Given those resources, the potential synergistic benefits – including a solid framework for building a seamless prediction system – arguably outweigh the costs.

5.5 Conclusion and future exploitation

In this chapter we have given a brief history of the evolution of numerical modeling across a range of time-scales. The challenges for both weather and climate models have evolved to global scales. However, there has also been a requirement for more detailed regional predictions, extremes, and predictions of environmental factors. Such demands have meant that models across scales have moved towards more common infrastructures, dynamics, and physics. A number of centers have developed or adopted a unified modeling strategy in which the same model is applied to both weather forecasting and seasonal, decadal, and centennial climate prediction. We have shown that there are considerable benefits to such an approach, although there can also be costs. Notably these are associated with the potential need for compromises to aspects of performance or cost on one time-scale in favor of maintaining a common system and the technical difficulties of maintaining a highly complex system ranging from data assimilation to Earth system components. In practice, with the MetUM we have encountered relatively few occasions where compromise was required, and more typical is delayed implementation of a change because of lack of performance at a particular time-scale. However, this provides a rigorous process for testing changes and can ensure proper evaluation. Some climate modeling centers without the availability of data assimilation techniques have also recognized the potential benefits to model improvement of running at shorter time-scales and have looked to develop methods to enable initialization for climate models to run from reanalyses. In both cases, a clear benefit is the potential to use a much broader range of observational data (e.g. the newest satellite missions, aircraft, or surface measurements) for evaluation, development of metrics of performance, and long-term improvement of models. Application of sophisticated diagnostics (such as the model–to–satellite approaches) across time and space scales will further benefit this effort. A unified modeling structure encompasses transitions between the traditional, if artificial, boundaries of weather and climate. The different requirements and importance of processes at different scales (from climate models with resolutions of order 100 km, global NWP models with resolutions of a few tens of km, to regional models with resolutions of around 1 km) is accounted for within the single modeling structure and as the models develop, parameterized representation of sub-grid processes (e.g. orography or convection) is removed as the processes become resolved. Such benefits will become more apparent as the need for accurate climate prediction at km scales develops. We have also highlighted the "cross fertilization" benefits of a unified modeling structure. Here we mean use and testing of new physics, chemistry, or biology in applications across time and space. In many cases this will allow earlier adoption of new functionality than would otherwise be the case. A good example is development of air quality forecasting systems from chemistry modeling developed for climate applications.

As we move to higher resolution for both weather and climate modeling the synergies to be gained from a seamless transition between these modeling capabilities will increase. Climate models will benefit from accurate mesoscale modeling of detailed regional processes at extremely high resolution (e.g. convection, clouds, vegetation, land-use change, river flows, hydrology, etc.) and increasingly sophisticated methods of initialization of these processes. Weather modeling will be able to use existing coupled modeling techniques to simulate the impact of tides, waves, and mixed-layer ocean processes, and modes of variability on short- to medium-range forecasts. We will be able to easily develop forecasting and assimilation capability for air quality, dust prediction, and other interactions with atmospheric chemistry from the increasingly large community of coupled climate–Earth-system models. Arising from this we would expect the ability to assimilate conventional observations to improve as the physical capability of models increases, e.g. through representation of the direct and indirect effects of aerosols. Challenges will include the need to improve the basis for parameterization based on such high-resolution modeling and comparison to observations; to develop fully coupled assimilation schemes and to enable the accurate representation of the multi-scale nature of many phenomena.

The ultimate manifestation of a seamless system is a single unified model running across all time-scales. The MetUM system developed at the U.K. Met Office has a growing number of adopters around the world (Australia, India, Korea, South Africa, Norway), most of whom are using the system for forecasting from NWP to climate. ECEarth is a collaboration of ten European countries developing an Earth-system version of the ECMWF weather forecasting model whose goal is to provide a fully coupled system for use across seasonal-decadal to centennial time-scales. The Weather Research and Forecasting (WRF) model developed at NCAR has a growing community of research and operational users including NCEP and the U.S. Air Force Weather Agency. A challenge will be to retain a fully unified software system while opening up the science code development to these wider collaborations.

Acknowledgments

Catherine Senior, Alberto Arribas, Timothy Johns, Gillian Martin, and Keith Williams were supported by the Joint DECC and Defra Integrated Climate Programme, DECC/Defra (GA01101).

References

Ackerman, T. P. and Stokes, G. M. (2003). The Atmospheric Radiation Measurement program. *Physics Today*, 56, 38–44.

Amodei, M., Pawson, S., Scaife, A. A., Langematz, U., Lahoz, W., Li, D. M., and Simon, P. (2001). The SAO and Kelvin waves in the EuroGRIPS GCMS and the UK Met. Office analyses. *Annales Geophysicae*, 19, 99–114.

Arakawa, A. (2000). Chapter 1: A Personal Perspective on the Early Years of General Circulation Modeling at UCLA. In *General Circulation Model Development – Past, Present, and Future*, ed. Randall, D. A., International Geophysics Series; Vol. 70, San Diego, CA, Academic Press, pp. 1–65.

Arthern, R. J. and Hindmarsh, R. C. A. (2003). Optimal estimation of changes in the mass of ice sheets. *Journal of Geophysical Research-Earth Surface*, 108.

Balmaseda, M. A., Vidard, A., and Anderson, D. L. T. (2008). The ECMWF ocean analysis system: ORA-S3. *Monthly Weather Review*, 136, 3018–3034.

Barsugli, J. J., Whitaker, J. S., Loughe, A. F., Sardeshmukh, P. D., and Toth, Z. (1999). The effect of the 1997/98 El Niño on individual large-scale weather events. *Bulletin of the American Meteorological Society*, 80, 1399–1411.

Bergman, J. W. and Sardeshmukh, P. D. (2003). Usefulness of single column model diagnosis through short-term predictions. *Journal of Climate*, 16, 3803–3819.

Berner, J., Doblas-Reyes, F. J., Palmer, T. N., Shutts, G., and Weisheimer, A. (2008). Impact of a quasi-stochastic cellular automaton backscatter scheme on the systematic error and seasonal prediction skill of a global climate model. *Philosophical Transactions of the Royal Society A–Mathematical Physical and Engineering Sciences*, 366(1875), 2561–2579.

Bernie, D. J., Woolnough, S. J., Slingo, J. M., and Guilyardi, E. (2005). Modeling diurnal and intraseasonal variability of the ocean mixed layer. *Journal of Climate*, 18, 1190–1202.

Bernie, D. J., Guilyardi, E., Madec, G., Slingo, J. M., and Woolnough, S. J. (2007). Impact of resolving the diurnal cycle in an ocean-atmosphere GCM. Part 1: a diurnally forced OGCM. *Climate Dynamics*, 29, 575–590.

Best, M. J. and Maisey, P. E. (2002). *A physically based soil moisture nudging scheme,* Hadley Centre technical note 35, Met Office, Bracknell, UK.

Betts, A. K. (2004). Understanding hydrometeorology using global models. *Bulletin of the American Meteorological Society*, 85, 1673–1687.

Betts, A. K., Ball, J. H., and Viterbo, P. (2003). Evaluation of the ERA-40 surface water budget and surface temperature for the Mackenzie River basin. *Journal of Hydrometeorology*, 4, 1194–1211.

Bodas-Salcedo, A., Webb, M. J., Brooks, M. E. *et al.* (2008). Evaluating cloud systems in the Met Office global forecast model using simulated CloudSat radar reflectivities. *Journal of Geophysical Research-Atmospheres*, 113, D00A13.

Bony, S., Dufresne, J. L., Le Treut, H., Morcrette, J. J., and Senior, C. (2004). On dynamic and thermodynamic components of cloud changes. *Climate Dynamics*, 22, 71–86.

Bourke, W. (1974). Multilevel spectral model. 1. Formulation and hemispheric integrations. *Monthly Weather Review*, 102, 687–701.

Boville, B. A. and Baumhefner, D. P. (1990). Simulated forecast error and climate drift resulting from the omission of the upper-stratosphere in numerical-models. *Monthly Weather Review*, 118, 1517–1530.

Boville, B. A. and Gent, P. R. (1998). The NCAR Climate System Model, Version One. *Journal of Climate*, 11, 1115–1130.
Boville, B. A. and Randel, W. J. (1992). Equatorial waves in a stratospheric GCM – effects of vertical resolution. *Journal of the Atmospheric Sciences*, 49, 785–801.
Bowler, N. E., Arribas, A., Mylne, K. R., Robertson, K. B., and Beare, S. E. (2008). The MOGREPS short-range ensemble prediction system. *Quarterly Journal of the Royal Meteorological Society*, 134, 703–722.
Boyle, J., Klein, S., Zhang, G., Xie, S., and Wei, X. (2008). Climate model forecast experiments for TOGA COARE. *Monthly Weather Review*, 136, 808–832.
Bright, D. R. and Mullen, S. L. (2002). Short-range ensemble forecasts of precipitation during the southwest monsoon. *Weather and Forecasting*, 17, 1080–1100.
Buizza, R. and Palmer, T. N. (1995). The singular-vector structure of the atmospheric global circulation. *Journal of the Atmospheric Sciences*, 52, 1434–1456.
Buizza, R., Bidlot, J.-R., Wedi, N., Fuentes, M., Hamrud, M., Holt, G., and Vitart, F. (2006). The new ECMWF VarEPS (Variable Resolution Ensemble Prediction System). *Quarterly Journal of the Royal Meteorological Society*, 133, 681–695.
Bushby, F. H., and Timpson, M. S. (1967). A 10-level atmospheric model and frontal rain. *Quarterly Journal of the Royal Meteorological Society*, 93, 1–17.
Bushell, A. C. and Martin, G. M. (1999). The impact of vertical resolution upon GCM simulations of marine stratocumulus. *Climate Dynamics*, 15, 293–318.
Byrkjedal, O., Esau, I., and Kvamsto, N. G. (2008). Sensitivity of simulated wintertime Arctic atmosphere to vertical resolution in the ARPEGE/IFS model. *Climate Dynamics*, 30, 687–701.
Cassou, C. (2008). Intraseasonal interaction between the Madden-Julian Oscillation and the North Atlantic Oscillation. *Nature*, 455, 523–527.
Chepfer, H., Bony, S., Winker, D., Chiriaco, M., Dufresne, J. L., and Seze, G. (2008). Use of CALIPSO lidar observations to evaluate the cloudiness simulated by a climate model. *Geophysical Research Letters*, 35, L157O4.
Colle, B. A. and Mass, C. F. (2000). The 5–9 February 1996 flooding event over the Pacific Northwest: sensitivity studies and evaluation of the MM5 precipitation forecasts. *Monthly Weather Review*, 128, 593–617.
Cullen, M. J. P. (1993). The unified forecast climate model. *Meteorological Magazine*, 122, 81–94.
Danabasoglu, G., Large, W. G., Tribbia, J. J., Gent, P. R., Briegleb, B. P., and McWilliams, J. C. (2006). Diurnal coupling in the tropical oceans of CCSM3. *Journal of Climate*, 19, 2347–2365.
Davies, T., Cullen, M. J. P., Malcolm, A. J. *et al.* (2005). A new dynamical core for the Met Office's global and regional modelling of the atmosphere. *Quarterly Journal of the Royal Meteorological Society*, 131, 1759–1782.
Dee, D. P. (2005). Bias and data assimilation. *Quarterly Journal of the Royal Meteorological Society*, 131, 3323–3343.
Delworth, T. L., Broccoli, A. J., Rosati, A. *et al.* (2006). GFDL's CM2 global coupled climate models. Part I: Formulation and simulation characteristics. *Journal of Climate*, 19, 643–674.
Dirmeyer, P. A., Dolman, A. J., and Sato, N. (1999). The pilot phase of the Global Soil Wetness Project. *Bulletin of the American Meteorological Society*, 80, 851–878.
Douville, H., Viterbo, P., Mahfouf, J. F., and Beljaars, A. C. M. (2000). Evaluation of the optimum interpolation and nudging techniques for soil moisture analysis using FIFE data. *Monthly Weather Review*, 128, 1733–1756.

Du, J. and Tracton, M. (2001). Implementation of a real-time short-range ensemble forecasting system at NCEP: an update, Preprints, 9th conference on mesoscale processes, Ft. Lauderdale, Florida, Amer. Meteor. Soc., 355–356.

Easterbrook, S. M. and Johns, T. C. (2009). Engineering the Software for Understanding Climate Change. *Computing in Science & Engineering*, 11, 64–74.

ECMWF (2007). ECMWF Workshop on flow-dependent aspects of data assimilation, ECMWF workshop proceedings – available from ECMWF.

Edwards, P. (2000). Chapter 2: A Brief History of Atmospheric General Circulation Modelling. In *General Circulation Model Development – Past, Present, and Future*, ed. Randall, D. A., International Geophysics Series; Vol. 70, San Diego, CA, Academic Press, pp. 67–90.

Evensen, G. (1994). Sequential data assimilation with a nonlinear quasi-geostrophic model using monte-carlo methods to forecast error statistics. *Journal of Geophysical Research-Oceans*, 99, 10 143–10 162.

Ferranti, L., Palmer, T. N., Molteni, F., and Klinker, E. (1990). Tropical extratropical interaction associated with the 30–60 day oscillation and its impact on medium and extended range prediction. *Journal of the Atmospheric Sciences*, 47, 2177–2199.

Fogel, K. (2005). *Producing Open Source Software: How to Run a Successful Free Software Project*. Cambridge, MA, O'Reilly Media Inc., 304 pp.

Fu, W. W., Zhu, J., and Yan, C. X. (2009). A comparison between 3DVAR and EnOI techniques for satellite altimetry data assimilation. *Ocean Modelling*, 26, 206–216.

Gleckler, P. J., Taylor, K. E., and Doutriaux, C. (2008). Performance metrics for climate models. *Journal of Geophysical Research-Atmospheres*, 113.

Graham, R. J., Gordon, M., McLean, P. J. *et al.* (2005). A performance comparison of coupled and uncoupled versions of the Met Office seasonal prediction general circulation model. *Tellus Series A-Dynamic Meteorology and Oceanography*, 57, 320–339.

Gray, S. L. (2003). A case study of stratosphere to troposphere transport: The role of convective transport and the sensitivity to model resolution. *Journal of Geophysical Research–Atmospheres*, 108(D18).

Greed, G., Haywood, J. M., Milton, S. *et al.* (2008). Aerosol optical depths over North Africa: 2. Modeling and model validation. *Journal of Geophysical Research–Atmospheres*, 113, D00C05.

Gregory, D., Shutts, G. J., and Mitchell, J. R. (1998). A new gravity-wave-drag scheme incorporating anisotropic orography and low-level wave breaking: Impact upon the climate of the UK Meteorological Office Unified Model. *Quarterly Journal of the Royal Meteorological Society*, 124, 463–493.

Haywood, J. M., Pelon, J., Formenti, P. *et al.* (2008). Overview of the Dust and Biomass-burning Experiment and African Monsoon Multidisciplinary Analysis Special Observing Period-0. *Journal of Geophysical Research–Atmospheres*, 113, D00C17.

Healy, S. B. and Thepaut, J. N. (2006). Assimilation experiments with CHAMP GPS radio occultation measurements. *Quarterly Journal of the Royal Meteorological Society*, 132, 605–623.

Hewson, T. D. (1998). Objective fronts. *Meteorological Applications*, 5, 37–65.

Hollingsworth, A. (2000). Chapter 11: Prospects for Development of Medium-Range and Extended-Range Forecasts. In *General Circulation Model Development – Past, Present, and Future*, ed. Randall, D. A., International Geophysics Series; Vol. 70, San Diego, CA, Academic Press, pp. 327–354.

Hoppel, K. W., Baker, N. L., Coy, L. *et al.* (2008). Assimilation of stratospheric and mesospheric temperatures from MLS and SABER into a global NWP model. *Atmospheric Chemistry and Physics*, 8, 6103–6116.

Howcroft, J. (1966). Fine-mesh limited area forecasting model, Tech. Report 188, U.S. Air Service.
Hurrell, J., Meehl, G. A., Bader, D., Delworth, T. L., Kirtman, B., and Wielicki, B. (2009). A unified modeling approach to climate system prediction. *Bulletin of the American Meteorological Society*, 90, 1819–1832.
Hurrell, J. W. (1995). Decadal trends in the North Atlantic Oscillation: regional temperatures and precipitation. *Science*, 269, 676–679.
Ineson, S. and Scaife, A. A. (2009). The role of the stratosphere in the European climate response to El Niño. *Nature Geoscience*, 2, 32–36.
Inness, P. M., Slingo, J. M., Woolnough, S. J., Neale, R. B., and Pope, V. D. (2001). Organization of tropical convection in a GCM with varying vertical resolution; implications of the Madden-Julian Oscillation. *Climate Dynamics*, 17, 777–793.
Jakob, C. and Tselioudis, G. (2003). Objective identification of cloud regimes in the Tropical Western Pacific. *Geophysical Research Letters*, 30.
Janssen, P. (1992). Experimental-evidence of the effect of surface-waves on the air-flow. *Journal of Physical Oceanography*, 22, 1600–1604.
Johns, T. C., Carnell, R. E., Crossley, J. F. *et al.* (1997). The second Hadley Centre coupled ocean-atmosphere GCM: Model description, spinup and validation. *Climate Dynamics*, 13, 103–134.
Johns, T. C., Durman, C. F., Banks, H. T. *et al.* (2006). The new Hadley Centre Climate Model (HadGEM1): Evaluation of coupled simulations. *Journal of Climate*, 19, 1327–1353.
Jung, T. (2005). Systematic errors of the atmospheric circulation in the ECMWF forecasting system. *Quarterly Journal of the Royal Meteorological Society*, 131, 1045–1073.
Jung, T. and Leutbecher, M. (2007). Performance of the ECMWF forecasting system in the Arctic during winter. *Quarterly Journal of the Royal Meteorological Society*, 133, 1327–1340.
Kalnay, E., Kanamitsu, M., Kistler, R. *et al.* (1996). The NCEP/NCAR 40-year reanalysis project. *Bulletin of the American Meteorological Society*, 77, 437–471.
Kanamitsu, M. and Saha, S. (1996). Systematic tendency error in budget calculations. *Monthly Weather Review*, 124, 1145–1160.
Khairoutdinov M. F. and Randall, D. A. (2001). A cloud resolving model as a cloud parameterization in the NCAR Community Climate System Model: Preliminary Results. *Geophys. Res. Lett.*, 28, 3617–3620.
Klein, S. A. and Jakob, C. (1999). Validation and sensitivities of frontal clouds simulated by the ECMWF model. *Monthly Weather Review*, 127, 2514–2531.
Klein, S. A., Jiang, X. N., Boyle, J., Malyshev, S., and Xie, S. C. (2006). Diagnosis of the summertime warm and dry bias over the U.S. Southern Great Plains in the GFDL climate model using a weather forecasting approach. *Geophys. Res. Lett.*, 33, L18805.
Klinker, E. and Sardeshmukh, P. D. (1992). The diagnosis of mechanical dissipation in the atmosphere from large-scale balance requirements. *Journal of the Atmospheric Sciences*, 49, 608–627.
Krishnamurti, T. N., Sanjay, J., Mitra, A. K., and Kumar, T. (2004). Determination of forecast errors arising from different components of model physics and dynamics. *Monthly Weather Review*, 132, 2570–2594.
Kumar, S. V., Peters-Lidard, C. D., Tian, Y. *et al.* (2006). Land information system: An interoperable framework for high resolution land surface modeling. *Environmental Modelling & Software*, 21, 1402–1415.
Land, C., Feichter, J., and Sausen, R. (2002). Impact of vertical resolution on the transport of passive tracers in the ECHAM4 model. *Tellus Series B-Chemical and Physical Meteorology*, 54, 344–360.

Lane, D. E., Somerville, R. C. J., and Iacobellis, S. F. (2000). Sensitivity of cloud and radiation parameterizations to changes in vertical resolution. *Journal of Climate*, 13, 915–922.

Lau, N. C. and Ploshay, J. J. (2009). Simulation of synoptic- and subsynoptic-scale phenomena associated with the East Asian summer monsoon using a high-resolution GCM. *Monthly Weather Review*, 137, 137–160.

Lean, H. W., Clark, P. A., Dixon, M. *et al.* (2008). Characteristics of high-resolution versions of the Met Office Unified Model for forecasting convection over the United Kingdom. *Monthly Weather Review*, 136, 3408–3424.

Lin, C., Laprise, R., and Ritchie, H. (1997). *Numerical Methods in Atmospheric and Oceanic Modelling*. Ottawa, Canada, Canadian Meteorological and Ocean Society, 581pp.

Lin, J. W. B., Neelin, J. D., and Zeng, N. (2000). Maintenance of tropical intraseasonal variability: Impact of evaporation-wind feedback and midlatitude storms. *Journal of the Atmospheric Sciences*, 57, 2793–2823.

Lindzen, R. S. and Fox-Rabinovitz, M. (1989). Consistent vertical and horizontal resolution. *Monthly Weather Review*, 117, 2575–2583.

Lock, A. P., Brown, A. R., Bush, M. R., Martin, G. M., and Smith, R. N. B. (2000). A new boundary layer mixing scheme. Part I: Scheme description and single-column model tests. *Monthly Weather Review*, 128, 3187–3199.

Martin, A. J., Hines, A., and Bell, M. J. (2007). Data assimilation in the FOAM operational short-range ocean forecasting system: A description of the scheme and its impact. *Quarterly Journal of the Royal Meteorological Society*, 133, 981–995.

Martin, G. M., Bush, M. R., Brown, A. R., Lock, A. P., and Smith, R. N. B. (2000). A new boundary layer mixing scheme. Part II: tests in climate and mesoscale models. *Monthly Weather Review*, 128, 3200–3217.

Martin, G. M., Ringer, M. A., Pope, V. D., Jones, A., Dearden, C., and Hinton, T. J. (2006). The physical properties of the atmosphere in the new Hadley Centre Global Environmental Model (HadGEM1). Part I: Model description and global climatology. *Journal of Climate*, 19, 1274–1301.

Martin, G., Milton, S., Senior, C. *et al.* (2010). Analysis and reduction of systematic errors through a seamless approach to modelling weather and climate. *Journal of Climate*, **23**, doi: 10.1175/2010JCLI3541.1.

McCulloch, M. E., Alves, J. O. S., and Bell, M. J. (2004). Modelling shallow mixed layers in the northeast Atlantic. *Journal of Marine Systems*, 52, 107–119.

Milton, S. F. and Wilson, C. A. (1996). The impact of parameterized subgrid-scale orographic forcing on systematic errors in a global NWP model. *Monthly Weather Review*, 124, 2023–2045.

Milton, S. F., Greed, G., Brooks, M. E. *et al.* (2008). Modeled and observed atmospheric radiation balance during the West African dry season: Role of mineral dust, biomass burning aerosol, and surface albedo. *Journal of Geophysical Research-Atmospheres*, 113, D00C02.

Miura, H., Satoh, M., Nasuno, T., Noda, A. T., and Oouchi, K. (2007). A Madden-Julian Oscillation event realistically simulated by a global cloud-resolving model. *Science*, 318, 1763–1765.

Miyakoda, K., Gordon, T., Caverly, R., Stern, W., Sirutis, J., and Bourke, W. (1983). Simulation of a blocking event in January 1977. *Monthly Weather Review*, 111, 846–869.

Mohankumar, K. and Pillai, P. A. (2008). Stratosphere-troposphere interaction associated with biennial oscillation of Indian summer monsoon. *Journal of Atmospheric and Solar-Terrestrial Physics*, 70, 764–773.

Moncrieff, M., Waliser, D. E., and Shapiro, M. (2010): The multiscale organization of tropical convection and its interaction with the global circulation: Year of Tropical Convection (YOTC). *Bulletin of the American Meteorological Society*. Submitted.

Morgenstern, O., Braesicke, P., O'Connor, F. M. *et al.* (2009). Evaluation of the new UKCA climate-composition model – Part 1: The stratosphere. *Geoscientific Model Development*, 2, 43–57.

Murphy, J. M., Sexton, D. M. H., Barnett, D. N., Jones, G. S., Webb, M. J., and Collins, M. (2004). Quantification of modelling uncertainties in a large ensemble of climate change simulations. *Nature*, 430, 768–772.

Onogi, K., Tslttsui, J., Koide, H. *et al.* (2007). The JRA-25 reanalysis. *Journal of the Meteorological Society of Japan*, 85, 369–432.

Palmer, T. N. (1999). A nonlinear dynamical perspective on climate prediction. *Journal of Climate*, 12, 575–591.

Persson, P. O. G. and Warner, T. T. (1991). Model generation of spurious gravity-waves due to inconsistency of the vertical and horizontal resolution. *Monthly Weather Review*, 119, 917–935.

Pham, D. T., Verron, J., and Roubaud, M. C. (1998). A singular evolutive extended Kalman filter for data assimilation in oceanography. *Journal of Marine Systems*, 16, 323–340.

Phillips, T. J., Potter, G. L., Williamson, D. L. *et al.* (2004). Evaluating parameterizations in general circulation models – Climate simulation meets weather prediction. *Bulletin of the American Meteorological Society*, 85, 1903–1915.

Polavarapu, S., Ren, S. Z., Rochon, Y. *et al.* (2005). Data assimilation with the Canadian middle atmosphere model. *Atmosphere-Ocean*, 43, 77–100.

Pope, V. D., Gallani, M. L., Rowntree, P. R., and Stratton, R. A. (2000). The impact of new physical parametrizations in the Hadley Centre climate model: HadAM3. *Climate Dynamics*, 16, 123–146.

Pope, V. D., Pamment, J. A., Jackson, D. R., and Slingo, A. (2001). The representation of water vapor and its dependence on vertical resolution in the Hadley Centre Climate Model. *Journal of Climate*, 14, 3065–3085.

Putman, W. M. and Lin, S. H. (2007). Finite-volume transport on various cubed-sphere grids. *Journal of Computational Physics*, 227, 55–78.

Rabier, F. (2005). Overview of global data assimilation developments in numerical weather-prediction centres. *Quarterly Journal of the Royal Meteorological Society*, 131, 3215–3233.

Randall, D., Krueger, S., Bretherton, C. *et al.* (2003). Confronting models with data – The GEWEX cloud systems study. *Bulletin of the American Meteorological Society*, 84, 455–469.

Reichler, T. and Kim, J. (2008). How well do coupled models simulate today's climate? *Bulletin of the American Meteorological Society*, 89, 303–311.

Ringer, M. A. and Healy, S. B. (2008). Monitoring twenty-first century climate using GPS radio occultation bending angles. *Geophysical Research Letters*, 35, L05708.

Robert, A., Henderson. J., and Turnbull, C. (1972). An implicit time integration scheme for baroclinic models of atmosphere. Monthly Weather Review, 100, 329–335.

Roberts, M. J., Banks, H., Gedney, N. *et al.* (2004). Impact of an eddy-permitting ocean resolution on control and climate change simulations with a global coupled GCM. *Journal of Climate*, 17, 3–20.

Rodwell, M. J. and Jung, T. (2008). Understanding the local and global impacts of model physics changes: An aerosol example. *Quarterly Journal of the Royal Meteorological Society*, 134, 1479–1497.

Rodwell, M. J. and Palmer, T. N. (2007). Using numerical weather prediction to assess climate models. *Quarterly Journal of the Royal Meteorological Society*, 133, 129–146.

Roeckner, E., Brokopf, R., Esch, M. *et al.* (2006). Sensitivity of simulated climate to horizontal and vertical resolution in the ECHAM5 atmosphere model. *Journal of Climate*, 19, 3771–3791.

Saha, S., Nadiga, S., Thiaw, C. *et al.* (2006). The NCEP Climate Forecast System. *Journal of Climate*, 19, 3483–3517.

Satoh, M., Matsuno, T., Tomita, H., Miura, H., Nasuno, T., and Iga, S. (2008). Nonhydrostatic icosahedral atmospheric model (NICAM) for global cloud resolving simulations. *Journal of Computational Physics*, 227, 3486–3514.

Scaife, A. A., Butchart, N., Warner, C. D., Stainforth, D., Norton, W., and Austin, J. (2000). Realistic quasi-biennial oscillations in a simulation of the global climate. *Geophysical Research Letters*, 27, 3481–3484.

Schubert, S. and Chang, Y. H. (1996). An objective method for inferring sources of model error. Monthly Weather Review, 124, 325–340.

Shaffrey, L. C., Stevens, I., Norton, W. A. *et al.* (2009). UK HiGEM: The new UK high-resolution global environment model – Model description and basic evaluation. *Journal of Climate*, 22, 1861–1896.

Shapiro, M., Hoskins, B. J., Shukla, J. *et al.* (2007). *The socio-economic and environmental benefits of a revolution in weather, climate and Earth-system analysis and prediction*, Tudor Rose, pp. 137–139.

Shutts, G. J. (2005). A kinetic energy backscatter algorithm for use in ensemble prediction systems. *Quarterly Journal of the Royal Meteorological Society*, 131, 3079–3102.

Shutts, G. J. and Palmer, T. N. (2007). Convective forcing fluctuations in a cloud-resolving model: Relevance to the stochastic parameterization problem. *Journal of Climate*, 20, 187–202.

Smagorinsky, J. (1974). *Weather and Climate Modification*, chapter: Global atmospheric modelling and the numerical simulation of climate, 633–686. John Wiley and Sons, New York.

Smagorinsky, J., Strickler, R., Sangster, W., Manabe, S., Holloway, J., and Membree, G. (1967). Prediction experiments with a general circulation model. *International Symposium – Dynamics of Large-Scale Atmospheric Processes (Moscow 1965)*, Izdatel'stvo Nauka, Moscow, 70–34.

Smith, D. M., Cusack, S., Colman, A. W., Folland, C. K., Harris, G. R., and Murphy, J. M. (2007). Improved surface temperature prediction for the coming decade from a global climate model. *Science*, 317, 796–799.

Spekat, A. (2000). *50th Anniversary of Numerical Weather Prediction. Book of Lectures from Commemorative Series.* Potsdam, Deutsche Meteorologische Gesellschaft.

Stainforth, D. A., Aina, T., Christensen, C. *et al.* (2005). Uncertainty in predictions of the climate response to rising levels of greenhouse gases. *Nature*, 433, 403–406.

Staniforth, A. and Côté, J. (1991). Semi-lagrangian integration schemes for atmospheric models – A review. *Monthly Weather Review*, 119, 2206–2223.

Sugiura, N., Awaji, T., Masuda, S. *et al.* (2008). Development of a four-dimensional variational coupled data assimilation system for enhanced analysis and prediction of seasonal to interannual climate variations. *Journal of Geophysical Research–Oceans*, 113(C10).

Swinbank, R. and O'Neill, A. (1994). A stratosphere troposphere data assimilation system. *Monthly Weather Review*, 122, 686–702.

Takahashi, M. (1999). Simulation of the quasi-biennial oscillation in a general circulation model. *Geophysical Research Letters*, 26, 1307–1310.

Tapp, M. C. and White, P. W. (1976). A non-hydrostatic mesoscale model. *Quarterly Journal of the Royal Meteorological Society*, 102, 277–296.

Tompkins, A. M. and Emanuel, K. A. (2000). The vertical resolution sensitivity of simulated equilibrium temperature and water-vapour profiles. *Quarterly Journal of the Royal Meteorological Society*, 126, 1219–1238.

Toniazzo, T. and Scaife, A. A. (2006). The influence of ENSO on winter North Atlantic climate. *Geophysical Research Letters*, 33, L24704.

Toth, Z. and Kalnay, E. (1993). Ensemble forecasting at NCM – The generation of perturbations. *Bulletin of the American Meteorological Society*, 74, 2317–2330.

Trenberth, K. E. and Smith, L. (2008). Atmospheric energy budgets in the Japanese reanalysis: Evaluation and variability. *Journal of the Meteorological Society of Japan*, 86, 579–592.

Trenberth, K. E., Branstator, G. W., Karoly, D., Kumar, A., Lau, N. C., and Ropelewski, C. (1998). Progress during TOGA in understanding and modeling global teleconnections associated with tropical sea surface temperatures. *Journal of Geophysical Research-Oceans*, 103, 14 291–14 324.

Trenberth, K. E., Smith, L., Qian, T. T., Dai, A., and Fasullo, J. (2007). Estimates of the global water budget and its annual cycle using observational and model data. *Journal of Hydrometeorology*, 8, 758–769.

Tselioudis, G., Zhang, Y. C., and Rossow, W. B. (2000). Cloud and radiation variations associated with northern midlatitude low and high sea level pressure regimes. *Journal of Climate*, 13, 312–327.

Untch, A. and Simmons, A. J. (1999). Increased stratospheric resolution. *ECMWF Newsletter*, No. 82, ECMWF, Reading, United Kingdom, 3–8.

Uppala, S. M., Kallberg, P. W., Simmons, A. J. *et al.* (2005). The ERA-40 re-analysis. *Quarterly Journal of the Royal Meteorological Society*, 131, 2961–3012.

Wallace, J. M., Rasmusson, E. M., Mitchell, T. P., Kousky, V. E., Sarachik, E. S., and von Storch, H. (1998). The structure and evolution of ENSO–related climate variability in the tropical Pacific: Lessons from TOGA. *Journal of Geophysical Research-Oceans*, 103, 14 241–14 259.

Webb, M., Senior, C., Bony, S., and Morcrette, J. J. (2001). Combining ERBE and ISCCP data to assess clouds in the Hadley Centre, ECMWF and LMD atmospheric climate models. *Climate Dynamics*, 17, 905–922.

Webster, S., Brown, A. R., Cameron, D. R., and Jones, C. P. (2003). Improvements to the representation of orography in the Met Office Unified Model. *Quarterly Journal of the Royal Meteorological Society*, 129, 1989–2010.

Williams, K. D. and Brooks, M. E. (2008). Initial tendencies of cloud regimes in the Met Office unified model. *Journal of Climate*, 21, 833–840.

Williams, K. D. and Tselioudis, G. (2007). GCM intercomparison of global cloud regimes: present-day evaluation and climate change response. *Climate Dynamics*, 29, 231–250.

Williamson, D. L., Boyle, J., Cedarwall, R. *et al.* (2005). Moisture and temperature balances at the Atmospheric Radiation Measurement Southern Great Plains Site in Forecasts with the Community Atmosphere Model (CAM2). *J. Geophys. Res.*, 110, D15516, doi:10.1029/2004JD005109.

Williamson, D., Nakagawa, M., Klein, S., Earnshaw, P., Nunes, A., and Roads, J. (2008). Transpose AMIP: a process oriented climate model intercomparison using model forecasts and field campaign observations, *Geophysical Research Abstracts*, 10, EGU2008–A–02919.

Woods, A. (2006). *Medium-Range Weather Prediction, the European Approach*, Springer, 270pp.

Xie, S. C., Boyle, J., Klein, S. A., Liu, X. H., and Ghan, S. (2008). Simulations of Arctic mixed-phase clouds in forecasts with CAM3 and AM2 for M-PACE. *Journal of Geophysical Research–Atmospheres*, 113, D04211.

Zeng, X. B. and Beljaars, A. (2005). A prognostic scheme of sea surface skin temperature for modeling and data assimilation. *Geophysical Research Letters*, 32, L14605.

Zhang, C. D. (2005). Madden-Julian oscillation. *Reviews of Geophysics*, 43, 1–36.

Zhang, S., Harrison, M. J., Wittenberg, A. T., Rosati, A., Anderson, J. L., and Balaji, V. (2005). Initialization of an ENSO forecast system using a parallelized ensemble filter. *Monthly Weather Review*, 133, 3176–3201.

Zhang, S., Harrison, M. J., Rosati, A., and Wittenberg, A. (2007). System design and evaluation of coupled ensemble data assimilation for global oceanic climate studies. *Monthly Weather Review*, 135, 3541–3564.

6

Contributions of observational studies to the evaluation and diagnosis of atmospheric GCM simulations

NGAR-CHEUNG LAU

6.1 Introduction

Throughout the past half-century, observational studies of the atmospheric circulation have played a pivotal role in the validation of output from general circulation model (GCM) simulations. The insights gained from examination of various observational datasets have also provided beneficial guidance to the design and diagnosis of model runs. Conversely, inferences from GCM experiments have opened new vistas for more incisive analyses of the available observational records. In this chapter, the historical evolution of this synergy between model and observational diagnostics is traced by describing selected studies based on data for both observed and GCM atmospheres. In view of the multitude of pertinent works on this theme, an all-inclusive survey is not attempted here. Attention is instead focused on a limited set of examples which serve to illustrate the close relationships between model- and observation-based studies in various stages of GCM development. Some of these choices are predicated upon the author's direct or indirect participation in the studies cited herein.

The following discussion begins with the early emphasis on model-observation comparisons of the nature of the atmospheric energy cycle, the atmospheric circulation averaged over longitude and time, and regional climatology (Section 6.2). The shift of interest towards the three-dimensional spatial characteristics and frequency dependence of the transient component of the circulation, and their dynamical interactions with the local, zonally varying quasi-stationary flow, is then documented (Sections 6.3 and 6.4). Applications of GCM experiments for understanding and attribution of observed atmospheric responses to anomalous forcings at the lower boundary, such as those associated with sea surface temperature (SST) changes, are highlighted in Section 6.5. The remarks in Section 6.6 are mainly concerned with the future prospects of joint diagnoses of observational and model data.

6.2 The early years – energy cycle, zonal-mean circulation, and regional climatology

Construction of the first GCMs in the 1950s and 1960s coincided with the advent of many observational analyses of the role of the general circulation in the energy cycle and in zonally averaged momentum and heat budgets of the atmosphere. The theoretical basis for these empirical studies was laid in the pioneering works of Lorenz (1955, 1967). The energy cycle offers a compact description of various forms of energy in the atmospheric system, and the role of the circulation in converting one form of energy to another. In the original framework, the energy reservoirs and energy transformation rates were deduced from spatial integrals of the terms in the energy budget over a large atmospheric domain.

The early formulation of the observed energy cycle also entailed the partition of atmospheric fields into a longitudinally averaged component ("zonal mean"), and the departure from this zonal-mean state ("eddy"). This strategy was partially motivated by the decomposition of the atmospheric flow into a "basic state" and a "perturbation" component in many theoretical studies. Partitioning of atmospheric structures into the zonal mean and eddy states exerted a strong influence on general circulation research in the middle of the twentieth century. Much attention was devoted to the pattern of the zonally averaged circulation, temperature, and water vapor fields in the meridional (latitude–height) plane. The role of various mechanisms, such as transports by the eddies, friction and heating, in maintaining the zonal-mean characteristics was also thoroughly investigated.

Considerable efforts were made to establish extensive observational data bases, so as to estimate individual terms in the energy cycle and in the zonal-mean budgets of various quantities for the real atmosphere (e.g. Starr *et al.*, 1970). The data source for such studies consists mainly of upper-air observations taken at a global network of rawinsonde stations. These data-gathering projects yielded voluminous compilations of atmospheric circulation statistics, such as those published by Oort and Rasmusson (1971) and Newell *et al.* (1972, 1974). Many of these data documentations bear the imprint of the domain or zonal averaged perspectives of the general circulation that were prevalent in that era. Due to the limited temporal coverage of the available observations (typically a few years), the analyses were mostly performed using time averages over the entire data record, with less attention being paid to variability of these quantities within subperiods of the record.

At the completion of the first GCM experiments, a natural question to ask was whether these models were capable of replicating the observed energy cycle. Such model validations were facilitated by the concurrent research activity aimed at quantifying different aspects of atmospheric energetics using observational data, as mentioned in the preceding paragraphs. The documentations of various early model runs (e.g. Smagorinsky, 1963; Smagorinsky, *et al.*, 1965; Manabe and Smagorinsky, 1967; Manabe *et al.*, 1970; Kasahara and Washington, 1971) contain many results on the observed and simulated energy cycles. A typical example of such comparisons is exhibited in Figure 6.1, which was presented by Smagorinsky *et al.* (1965). It shows the hemispherically averaged energy cycles under annual mean conditions for the observed (Figure 6.1a) and simulated (Figure 6.1b) atmospheres. The reservoirs for available potential energy (P) and kinetic energy (K) are partitioned into their respective zonal mean (subscript Z) and eddy (subscript E) components. Values in the black boxes represent conversion rates between various reservoirs, as well as generation or destruction rates of energy in individual reservoirs due to processes such as heating (Q) and subgrid-scale mixing (H, F).

The estimates displayed in Figure 6.1 and in similar diagrams for other models demonstrate that the first generation of GCMs reproduced the essential qualitative characteristics of the observed energy cycle. In this particular example, the simulated energy reservoirs P_Z and K_Z are noticeably larger than their observed counterparts; whereas the eddy component in the model atmosphere is considerably weaker than the corresponding observed values. Discrepancies between the model and observed estimates could be attributed to simplifications in model treatments of hydrological and oceanic transport processes, as well as orography and continent–ocean contrasts at the lower boundary. It is noteworthy that only some of the terms in the energy cycle could be estimated directly using observational data; whereas the remaining terms (e.g. $Q+H$, F) were deduced as "residues" by invoking budget constraints for individual reservoirs. On the other hand, GCM outputs enabled direct estimation of all terms in the energy cycle, and further yielded information on the separate contributions of Q and H to their sum, and of horizontal ($_HF$) and vertical ($_VF$) subgrid-scale mixing to F. Such model–observation comparisons served to validate the GCMs. Conversely, the model estimates could be utilized as independent assessments of the accuracy and physical consistency of those aspects of the energy cycle for which no direct observational information was readily available. These feedbacks between model and observational diagnostics have proven to be mutually beneficial not only for budget calculations, but also for many other general circulation studies.

The prominence of the zonal-mean perspective of the atmospheric circulation is also evident in the early GCM studies. The zonally averaged structure of the motion and thermodynamic fields in model atmospheres received much attention. These model results were checked in detail against the corresponding observations. A demonstration of this analysis approach is given in Figure 6.2, which is based on the work of Kasahara and Washington (1971). The diagram shows the latitude–height distributions of longitudinally averaged zonal [u] and meridional [v] wind components, as obtained from observations (left panels) and from GCM integrations

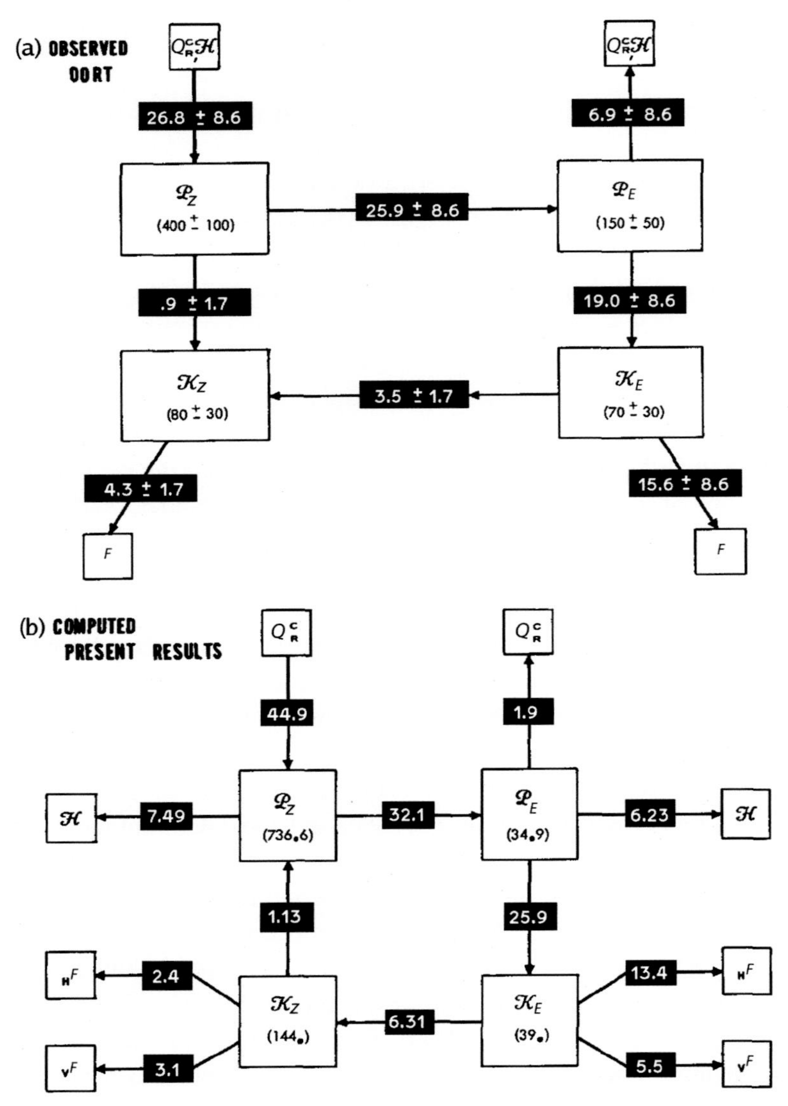

Figure 6.1 Estimates of the contents of various energy reservoirs of the atmosphere (large white boxes; units: J cm^{-2}) and the output/input of energy from/to the individual reservoirs due to conversions between reservoirs and other processes (small black boxes; units: 10^{-3} J cm^{-2} mb^{-1} d^{-1}), as computed using (a) observational data by Oort (1964) and (b) model output. P_Z and P_E denote the zonal and eddy available potential energy, respectively; K_Z and K_E the zonal and eddy kinetic energy, respectively; $Q_R{}^C$ and H the change in P_Z, P_E due to radiation/convection and horizontal mixing, respectively; $_HF$ and $_VF$ the change in K_Z, K_E due to horizontal and vertical mixing, respectively; $F = {}_HF + {}_VF$. The model data were extracted from a 70-day segment of an integration under annual mean conditions with a 9-level hemispheric GCM developed at the Geophysical Fluid Dynamics Laboratory (GFDL), with a horizontal resolution ranging from 320 km at the Equator to 640 km near the pole. From Smagorinsky *et al.* (1965).

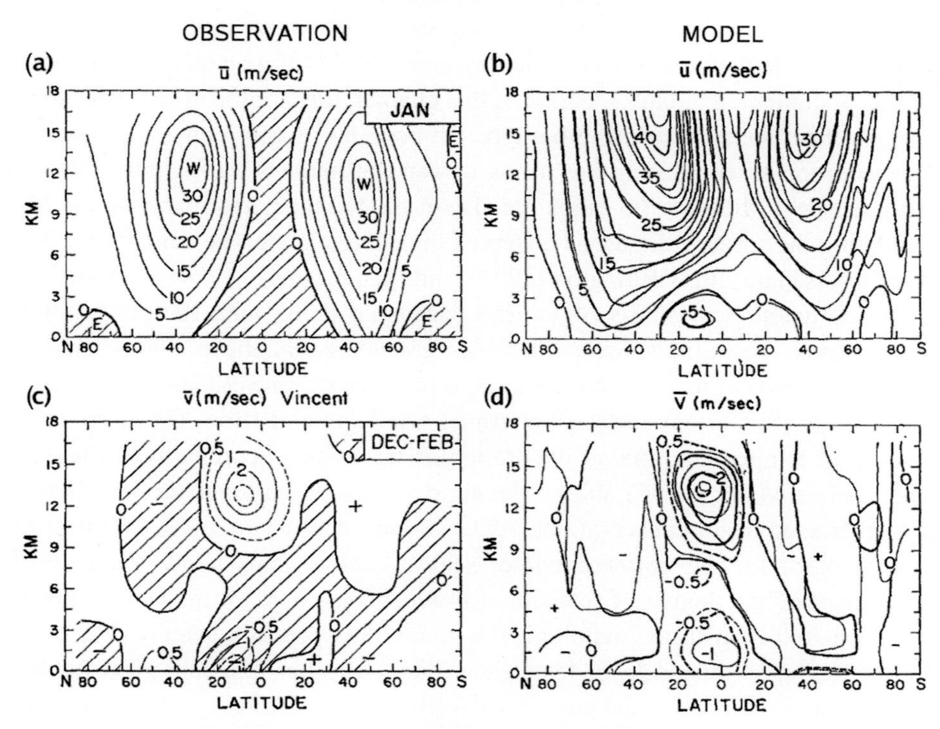

Figure 6.2 Latitude–height distributions during northern winter of zonally averaged (a) observed zonal wind based on data compilations of Crutcher (1961), Palmén (1964), and van Loon *et al.* (1971), (b) simulated zonal wind, (c) observed meridional wind based on estimates by Vincent (1969), and (d) simulated meridional wind. Units: m s^{-1}. The model data were obtained from 30-day segments of GCM integrations with a 6-level global GCM developed at the National Center for Atmospheric Research, with a horizontal resolution of 5° in latitude and longitude. Results based on model runs with and without orographic forcing are displayed in (b, d) using heavy and thin contours, respectively. From Kasahara and Washington (1971).

with and without orographic forcing (heavy and thin contours in right panels, respectively). Here the square parentheses denote zonal averages. The data presented in this example illustrate that the GCM reproduced the upper-level maxima in [u] at 30°N and 40°S (see upper panels). However, the placement of the observed jet cores at an altitude of 12 km was not properly simulated by the GCM. The lower panels indicate the presence in the model atmosphere of the tropical Hadley cell (with [v]>0 in the upper branch; [v]<0 in the lower branch), and the extratropical Ferrel cell ([v]<0 in the upper branch; [v]>0 in the lower branch), both in the Northern Hemisphere.

It is noteworthy that the model patterns in Figure 6.2 are based on two experiments, one with and the other without orographic forcing (see heavy and thin contours, respectively). These results illustrate an early attempt to delineate the role of mountains in the general circulation by using GCMs as an experimental tool. As will be documented in later sections of this chapter, this approach has been applied to study the effects of various other types of boundary forcing on the atmosphere.

In addition to computing the terms of various budgets in spatially integrated forms, some of the early investigators also evaluated the performance of GCMs in simulating the surface climate as observed in different geographical regions. Mintz (1965), Kasahara and Washington (1971), and Manabe *et al.* (1974) compared near-global climatological charts of various simulated fields with the corresponding observations. The atmospheric quantities considered in these evaluations were chosen by virtue of their strong linkages to regional surface climates, and included sea-level pressure, near-surface horizontal wind, precipitation, cloud cover, and surface air temperature. As an illustration of this type of comparison, Figure 6.3 (taken from Mintz, 1965) shows the time-averaged sea-level pressure patterns as constructed using observational data (Figure 6.3a) and a GCM integration (Figure 6.3b) for the northern winter season. These maps allow for a detailed assessment of the degree of realism with which the GCM simulated the position and intensity of various climatological high and low pressure centers. The evaluation of model-simulated local features in this and other climatological fields for different seasons is a crucial step in establishing the credibility of the GCMs for studying regional climates and for making climate projections for the future.

6.3 Shift of attention to longitudinal and frequency dependence of atmospheric variability

Commencing in the 1960s, various operational meteorological centers have routinely produced gridded analyses of the three-dimensional distributions of different atmospheric fields at sub-daily intervals. These products incorporate observations from multiple sources, including rawinsonde stations, ships of opportunity, aircraft, and satellites. In data sparse regions, predictions by numerical models are used to provide the "first-guess" fields in the analysis procedure. The new data resource offers an alternative to the observations taken at the global network of rawinsonde stations, which served as the backbone for the early general circulation studies described in the previous section. Circulation statistics as computed using the operational analysis products were compared in detail against the corresponding quantities based on the rawinsonde network by Lau and Oort (1981, 1982).

The gridded analysis products were utilized by Blackmon (1976) and Blackmon *et al.* (1977) to examine the geographical patterns of time-mean fields, as well as deviations from these time averages (hereafter referred as "transients"). Extensive use was made of temporal variance and covariance statistics, which serve as

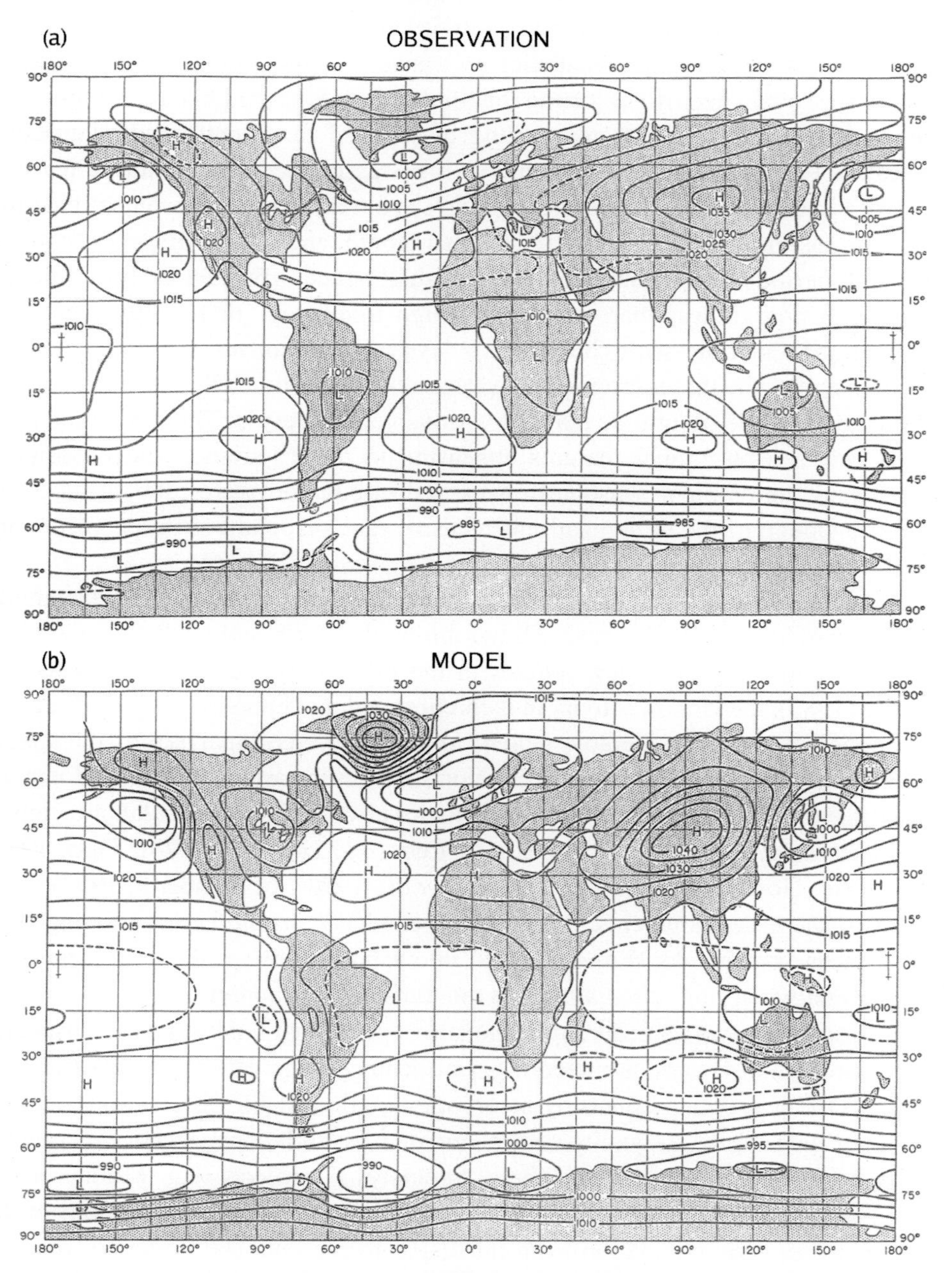

Figure 6.3 Horizontal distributions during northern winter of sea-level pressure, based on (a) observational data compiled by O'Conner (1961), Riehl (1954), and van Loon (1961), and (b) 30-day integration with a 2-level global GCM developed at University of California, Los Angeles, with a horizontal resolution of 7° in latitude and 9° in longitude. Units: mb. From Mintz (1965).

indicators of the local transient eddy activity and transport properties. Contributions to the second-moment statistics by transient fluctuations of different time-scales were evaluated by subjecting the time series of the gridded data to temporal filters with response functions peaking in selected frequency bands. Particular attention was devoted to atmospheric changes associated with synoptic weather with life-cycles ranging from about two days to a week, more persistent circulation episodes lasting between about 10 days to a month, and ultra-long period anomalies with periods longer than a month. We shall henceforth refer to those fluctuations with periods less than a week as "high-frequency" or HF transients, and fluctuations with time-scales between 10 days and a season as "low-frequency" or LF transients. The three-dimensional structure of these time-filtered statistics was plotted on latitude–longitude maps at various vertical levels. The spatial relationships between prominent features appearing in these variance and covariance charts and the ambient seasonally averaged circulation were interpreted using our knowledge of the dynamical mechanisms that shape atmospheric variability on different time-scales.

Considerable progress was made in our theoretical understanding of the temporal evolution of longitudinally varying flows during the 1970s and 1980s. As is evident from the collection of review articles in the monograph edited by Hoskins and Pearce (1983), the vibrant activity in theoretical and modeling investigations during this period was of considerable benefit to the concurrent analyses of the spatial behavior of the observed general circulation statistics in various frequency domains. The three-dimensional nature of these statistics also received particular attention in the observational data compilations in that era (e.g. Lau *et al.*, 1981; Oort, 1983; Lau, 1984; Hoskins *et al.*, 1989; Schubert *et al.*, 1990a, b; Trenberth, 1992). As a result of these developments, the previous emphasis on the domain-averaged or zonal-mean frameworks shifted to local interactions between the time-mean (but longitudinally varying) circulation and the transient fluctuations.

In view of the widespread applications of operational analyses in atmospheric circulation studies, various meteorological centers made efforts to generate new versions of these data products by incorporating all available observations, and by applying state-of-the-art quality control and data assimilation systems consistently throughout the analysis period. These "re-analysis" projects (e.g. Kalnay *et al.*, 1996; Uppala *et al.*, 2005) provide detailed and comprehensive descriptions of the global atmospheric circulation from the 1950s to present. The important roles of reanalysis in developing climate models and numerical weather prediction have been discussed in Chapters 3 and 5. Circulation statistics based on these reanalysis products can now be easily accessed through websites (e.g. see data atlases at http://www.cdc.noaa.gov/data/ncep_reanalysis/ and http://www.ecmwf.int/research/era/ERA-40_Atlas/docs/index.html).

In conjunction with the change in focus of the observational studies of the atmospheric general circulation, many GCM simulations were also diagnosed by partitioning the three-dimensional flow field into a time-mean component and transient fluctuations on various time-scales. An illustration of the model diagnosis based on this paradigm, as performed by Blackmon and Lau (1980), is given in Figure 6.4. This diagram shows the horizontal distributions of time-averaged geopotential height in the upper troposphere (top row), and root-mean squares (rms) of 500 mb height data subjected to timefilters retaining fluctuations between 2.5–6 days (middle row) and between 10 days and a season (bottom). Results based on observational and GCM data are presented in the left and right columns, respectively. In the following discussion, we shall first describe the salient features in the observed atmosphere (i.e. Figures 6.4a, 6.4c and 6.4e), and then compare these results with the model output.

The observed seasonally averaged upper-air circulation in the northern extratropics during winter is characterized by a wavy pattern (see solid contours in Figure 6.4a). The axes of the principal troughs in this stationary wave, as indicated by dotted lines in these panels, are situated over eastern Asia and North America. The climatological jetstreams, which correspond to maxima in the zonal wind speed (see dashed contours), are located in the vicinity of the stationary troughs.

Enhanced HF transient activity, as depicted by large rms values in the middle panels, occurs along zonally elongated belts over the North Pacific and North Atlantic at 40°–55°N in the observed atmosphere. These regions coincide with the preferred trajectories of extratropical weather disturbances. Comparison between the upper and middle panels in Figure 6.4 reveals that the "storm tracks" are displaced downstream (i.e. to the east) and poleward of the time-averaged jet cores. Other covariance statistics based on filtered data within the 2.5–6 day range indicate that the HF transients along the preferred storm tracks are active agents for poleward and upward transports of sensible heat and water vapor, and for convergence of horizontal momentum transports. These eddy signatures are consistent with the processes associated with the evolution of eastward-migrating baroclinic waves through their life-cycles (e.g. Simmons and Hoskins, 1978). The spatial relationship between the storm tracks and jetstreams indicates that the quasi-stationary circulation could exert a strong influence on the site of HF transient activity through modulation of the vertical and horizontal shear environment in the vicinity of the jetstreams.

The observed pattern for rms of 500 mb height fluctuations with periods between 10 days and a season exhibits maxima over the central North Pacific and North Atlantic at 50°–60°N, and northern Eurasia near 65°N 80°E. These centers are known to be preferred sites of prolonged circulation anomalies such as blocking highs and lows, and "centers of action" of recurrent modes of atmospheric

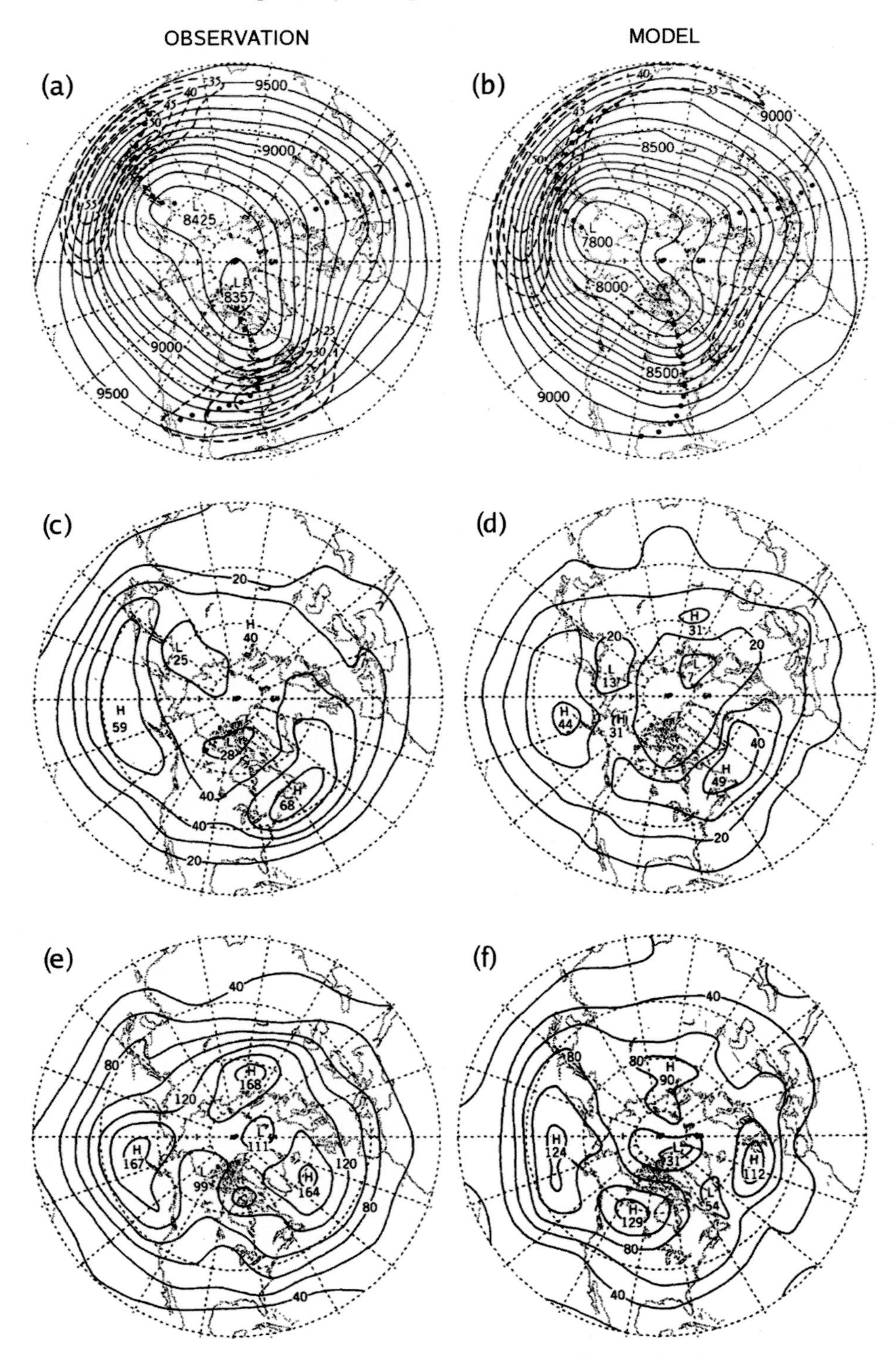

Figure 6.4 Horizontal distributions during northern winter of (a, b) time-averaged geopotential height (solid contours; interval: 100 m) and zonal wind (dashed

variability on time-scales longer than a month. The myriad contributors to the enhanced LF variance at these locations include long-period unstable modes of the longitudinally varying basic state, forcing due to anomalous conditions at the lower boundary, atmospheric energy dispersion, and dynamical interactions between multi-scale atmospheric perturbations, among others.

The geographical dependence of the time-mean and time-varying states of the observed atmosphere, as briefly summarized in the preceding paragraphs, offers a useful test bed for diagnosing and evaluating GCM output. Comparison between the temporal mean and variance maps based on model-generated data with their observed counterparts would reveal the strengths and deficiencies of a particular GCM in reproducing the essential regional characteristics of the atmospheric circulation. In the example presented in Figure 6.4, the specific GCM being examined is seen to reproduce some of these characteristics, such as the stationary trough and jetstream over East Asia, as well as the locations of the storm tracks and centers of enhanced LF variability over two ocean basins. However, it is also apparent that the simulated North American jetstream is too weak and displaced east of its observed position, and the simulated transient eddy amplitudes in both the HF and LF period ranges are lower than the observations. Additionally, there are notable discrepancies with regard to the positions of the extrema of LF variance.

LF atmospheric variability, such as that depicted in Figures 6.4e–f, is of considerable interest to both the academic and extended-range forecasting communities. Our empirical and theoretical knowledge of slowly varying phenomena has expanded in recent decades (e.g. see the review articles in the monograph edited by Cattle, 1987). These advances may partially be attributable to enhanced research activity on the atmospheric response to anomalous boundary forcing (such as that associated with El Niño–Southern Oscillation (ENSO) episodes), and on various internal atmospheric processes operating on long time-scales. This development is further aided by the deployment of innovative analysis techniques for delineating the space-time characteristics of different types of atmospheric fluctuations

---

Caption for Figure 6.4 (cont.)
contours; interval: 5 m s^{-1}) at 300 and 315 mb based on observational and model data, respectively, (c, d) root-mean squares of filtered 500 mb height retaining periods between 2.5 and 6 days (contour interval: 10 m), and (e, f) root-mean squares of filtered 500 mb height retaining periods between 10 days and a season. The dotted lines in (a, b) indicate the positions of the axes of the principal stationary troughs. The observational charts (left panels) were based on operational analyses produced by the National Meteorological Center (NMC) for a 9-year period. The model data (right panels) were obtained from a 120-day integration with a global GCM at GFDL with 11 vertical layers and a horizontal resolution of ~265 km (e.g. see Manabe *et al.* 1974). From Blackmon and Lau (1980).

(e.g. Blackmon *et al.*, 1984a, b; Wallace *et al.*, 1988). During the early years of GCM development, when only short-term integrations were conducted, investigations on LF variability in model atmospheres were impeded by inadequate sampling of the phenomena of interest. As simulations lasting for more than a decade became available (e.g. Manabe and Hahn, 1981), various modern statistical tools were applied to the model output, so as to evaluate the capability of GCMs in mimicking different aspects of the observed LF variability, and to assess the relative roles of various mechanisms in generating the long-period perturbations in the model.

An example of the comparative studies between observed and model-simulated variability on monthly time-scales is that undertaken by Lau (1981), who examined the output from a 15-year GCM integration subjected to climatological SST forcing at the lower boundary. The spatial distribution of the leading mode of low-frequency variability in that model, as identified by applying an eigenvector or empirical orthogonal function analysis (e.g. see Kutzbach, 1967) to monthly averaged 500 mb height for the northern winter, is shown in Figure 6.5b. This pattern may be compared with the corresponding observational result in Figure 6.5a. It is seen that the observed and model patterns exhibit considerable similarity in the North Pacific–North American sector, with the polarity of monthly mean height fluctuations over the northwestern portion of North America and eastern Siberia being opposite to those over the central North Pacific and southeastern United States. The presence of this spatial pattern in the observed atmosphere is well known (e.g. see Wallace and Gutzler, 1981). Its appearance in a model run with no interannual variability of SST forcing being prescribed suggests that it may be generated by internal atmospheric dynamics alone. In addition to the characteristic mode of variability shown in Figure 6.5, much attention has also been devoted to the observed and simulated behavior of other recurrent LF patterns, such as those associated with the North Atlantic Oscillation and the north–south seesaws in geopotential height in the western Pacific and western Atlantic sectors.

Another illustration of the comparison of time-space evolution of observed and simulated wintertime atmospheric fluctuations on different time-scales is shown in Figure 6.6. The results in this figure are taken from the study by Lau and Nath (1987), who applied a cross-spectral analysis to compute the phase differences and coherences between the time series of 500 or 515 mb height at a selected reference point (50°N, 170°W for observations; 47°N, 157°W for model data; hereafter referred to as RP), and the corresponding time series at all other gridpoints within the North Pacific–North American domain. The typical phase difference between a given location and RP is indicated by the orientation of the arrow plotted at that location. A poleward-pointing arrow denotes zero phase difference. The arrow rotates clockwise (counterclockwise) by one degree for each degree of phase lag (lead) of the local fluctuations relative to those at RP. Stippling indicates coherence

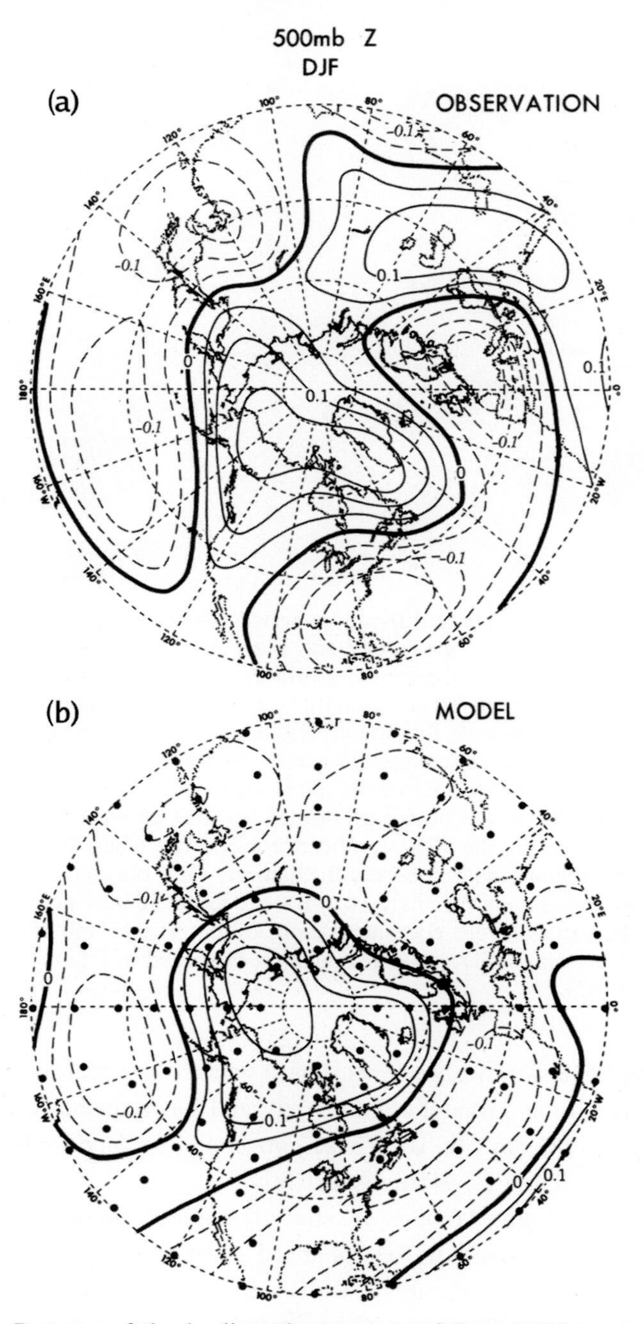

Figure 6.5 Patterns of the leading eigenvectors of the monthly mean 500 mb height field during northern winter, as computed using (a) observational NMC analyses for a 15-year period, and (b) output from a 15-year integration using a spectral global GCM at GFDL, with 9 vertical layers and rhomboidal truncation at 15 wavenumbers, which corresponds to a spatial resolution of 7.5° in longitude and 4.5° in latitude (see also Manabe and Hahn 1981). The dots in (b) represent the network of gridpoints used in the eigenvector analysis. From Lau (1981).

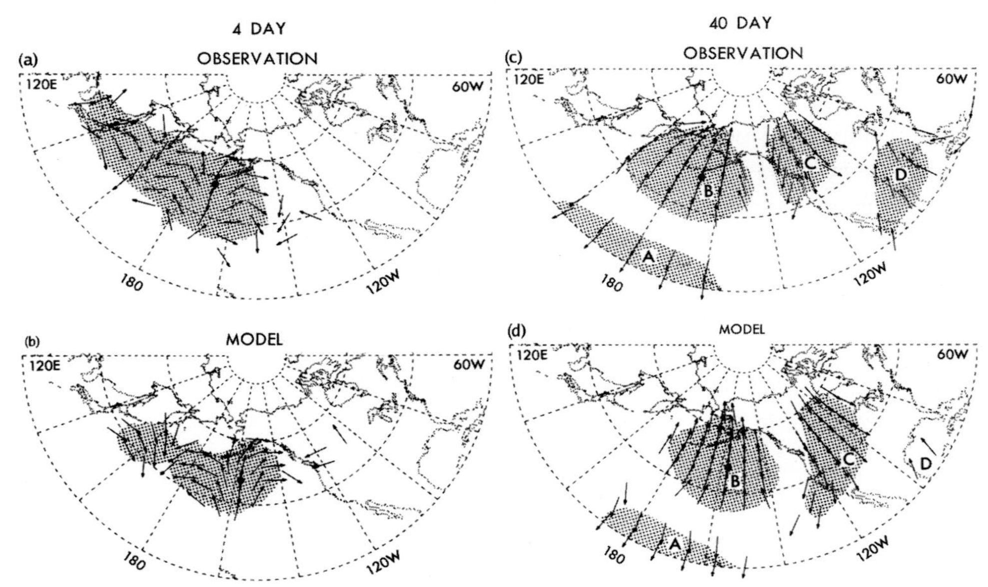

Figure 6.6 Patterns of phase differences between wintertime geopotential height fluctuations at individual gridpoints and the corresponding fluctuations at the reference point (RP) over the central North Pacific (see locations of solid dots), for frequency bands centered at the (a, b) 4-day and (c, d) 40-day periods. Arrows pointing due north indicate that the local fluctuations are in phase with those at RP; and rotate clockwise (counterclockwise) by one degree for each degree of phase lag (lead) relative to RP. Arrows are plotted only at gridpoints which exhibit coherence with RP at the 80% significance level or above. Stippling indicates coherences surpassing the 95% significance level. The labels A, B, C, and D in (c, d) denote regions with coherent fluctuations. The observational charts (left panels) were based on NMC analyses at 500 mb for an 18-year period. The model charts (right panels) were computed using output at 515 mb from a 12-year segment of the same GFDL model simulation examined in Figure 6.5. From Lau and Nath (1987).

surpassing the 95% significance level. Phase patterns are presented for the observations (panels a, c) and model output (b, d), and for frequency bands centered at 4 days (a, b) and 40 days (c, d). These two bands encompass fluctuations with periods of 3.8–4.2 d and 27–80 d, respectively.

HF disturbances (Figures 6.6a,b) are strongly coherent along a zonally elongated site that corresponds closely to the enhanced storm track activity over the North Pacific (see also maxima in Figures 6.4c, d), which is in turn located downstream of the time-mean jet over East Asia (Figures 6.4a,b). When the phase arrows are scanned from west to east, they are seen to rotate uniformly in a clockwise fashion, thus indicating continuous eastward phase propagation of the troughs and ridges along the storm track.

The phase structure for variations with 40-day periods (Figures 6.6c, d) is notably different from that for the HF transients. The LF pattern consists of three to four discrete coherent regions (labeled as A, B, C, and D in Figures 6.6c, d). The phase arrows within a given region have almost the same orientation, thus indicating in-phase variations at gridpoints in that region. On the other hand, out-of-phase relationships prevail between neighboring regions (e.g. note the poleward and equatorward arrows in B and A, respectively). Hence these long-period changes may be viewed as a combination of nodes and antinodes spread over a broad area. The dipole-like configuration of regions A and B over the North Pacific is particularly noteworthy. This pattern straddles the exit region of the East Asian jetstream (Figures 6.4a–b). The seesaw between the geopotential height at A and B is therefore accompanied by elongation or contraction of this jetstream over the North Pacific sector.

The observed phase patterns (Figures 6.6a, c) may be compared with the corresponding model results (Figures 6.6b, d). It is seen that the GCM is capable of reproducing the essential phase characteristics in both HF and LF bands.

6.4 Diagnoses of local interactions between transient eddies and the time-mean circulation

The geographical relationships between transient fluctuations and the stationary flow pattern, as delineated in the previous section, suggest that fresh insights on various types of local interactions between the time-averaged and time-varying components of the observed atmospheric circulation could be gained by examining the spatial distributions of the relevant statistics in some detail. The promising results from these empirical studies have motivated us to perform parallel diagnoses of GCM output, so as to determine whether these interactions are properly incorporated into the model atmosphere.

The baroclinic aspect of eddy-mean flow interaction is first considered by examining the relationships between heat fluxes by transient disturbances and the ambient mean temperature. These fields have been computed by Blackmon and Lau (1980) using unfiltered wintertime observational and GCM data at 850 and 835 mb, respectively, and are displayed in Figure 6.7. The heat transport by the transient fluctuations in the zonal and meridional directions $\boldsymbol{F}(T) = \overline{u'T'}\boldsymbol{i} + \overline{v'T'}\boldsymbol{j}$ is portrayed in this diagram using a vectorial format; whereas the contours represent the time-averaged temperature ($\overline{T}$). Here the overbar denotes time averaging, the prime indicates deviations from the time average, T is temperature, and $\boldsymbol{i}, \boldsymbol{j}$ represent unit vectors directed eastward and northward, respectively. In both the observed and model atmospheres, the magnitude of $\boldsymbol{F}(T)$ (as can be inferred from the length of the arrows in Figure 6.7, see scale at middle right) is largest over the western and central

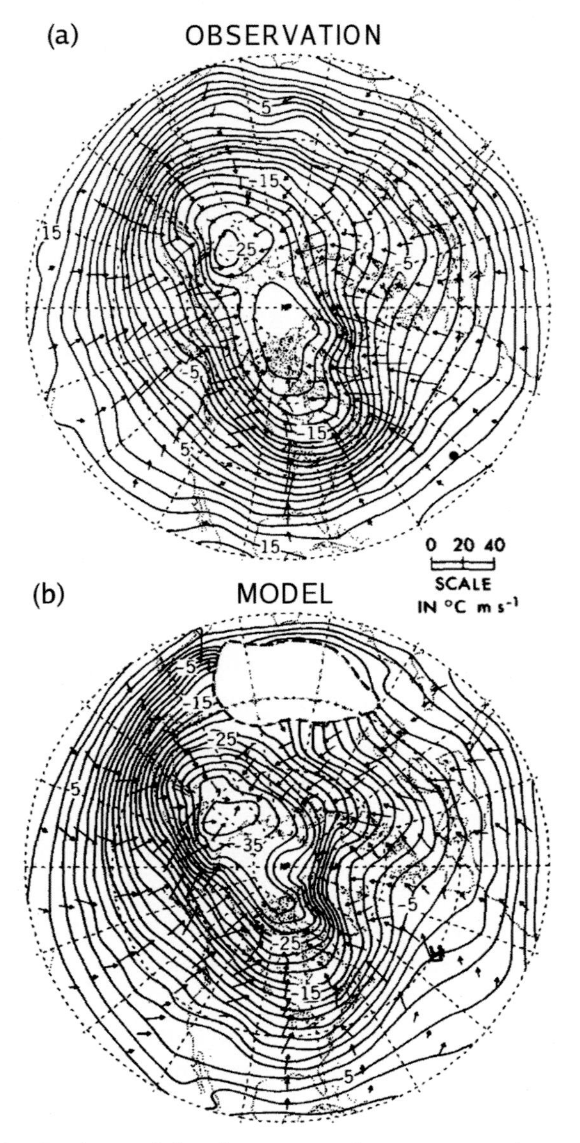

Figure 6.7 Distributions of the time-averaged temperature (contours; interval: 2 °C) and vectorial horizontal heat transport by transients ***F(T)*** (arrows; see scale at middle right) during the winter season, based on (a) NMC analyses at 850 mb for a 9-year period, and (b) output at 835 mb from the GFDL simulation examined in Figure 6.4. Data are not shown over the Tibetan Plateau in panel (b). From Blackmon and Lau (1980).

portions of the principal storm tracks (Figures 6.4c, d, 6.6a,b), where baroclinic growth of the transient disturbances preferentially occurs. Throughout the hemispheric domain, the transient fluctuations transport heat down the local, time-averaged temperature gradient, with the heat flux vectors being directed from high to low values of $\overline{T}$. These down-gradient transports lead to heat flux convergences (i.e. warming tendencies) in the vicinity of Hudson Bay–western Greenland, eastern Siberia and eastern Europe, where the lowest time-mean temperatures (relative to other locations along the same latitude circle) occur. Conversely, heat flux divergences (which result in cooling tendencies) prevail near Norwegian Sea–Greenland Sea and Gulf of Alaska, which correspond to the warmest sites within the zonal belt between 50°–70°N. By transporting heat away from the climatologically warm regions and towards the cold regions, the transient disturbances act to destroy the zonal asymmetries of the $\overline{T}$ pattern. From the viewpoint of atmospheric energetics, the down-gradient nature of $\boldsymbol{F(T)}$ facilitates the conversion of available potential energy of the stationary waves (which is related to the square of the deviations of $\overline{T}$ from the zonal mean) to that of the transient fluctuations (which is related to $\overline{T'^2}$). The agreement of the relationships between $\boldsymbol{F(T)}$ and $\overline{T}$ in the simulation (Figure 6.7b) with those in the observed atmosphere (Figure 6.7a) signifies that the GCM reproduces this important aspect of eddy-mean flow interaction.

Following the formulation of Hoskins *et al.* (1983), the local barotropic interactions between the transients and the time-mean circulation may be discerned using the horizontal components of the extended Eliassen–Palm flux vector $\boldsymbol{E} = (\overline{v'^2} - \overline{u'^2})i - \overline{v'u'}j$. The conversion of kinetic energy from the transients to the time-averaged flow may be approximated as $-\boldsymbol{E} \cdot \nabla\overline{u}$. It can also be shown that the pattern of the $\boldsymbol{E}$ field provides information on the characteristic horizontal shape of the transients as well as direction of their group velocity relative to the mean flow. These useful properties of the $\boldsymbol{E}$ vector have found many applications in diagnostic studies of various observed and simulated atmospheric circulation systems. For instance, the nature of barotropic energy conversion between the longitudinally varying stationary flow and transients of different time-scales was examined by Wallace and Lau (1985). Some sample results from that investigation are presented in Figure 6.8, which shows the distribution of $\boldsymbol{E}$ (arrows), as computed using filtered wintertime data retaining fluctuations between 2.5 and 6 days (panels a, b) and longer than 10 days (c, d). Superposed on these vector patterns are the distributions of the time-averaged zonal wind speed ($\overline{u}$, contours).

The typical shape of individual synoptic-scale ridges and troughs along the oceanic storm tracks has been described by Hoskins *et al.* (1983) and Blackmon *et al.* (1984a). These disturbances are elongated in the meridional direction, so that $\overline{v'^2} > \overline{u'^2}$ within the storm track regions. The ridges and troughs also exhibit a

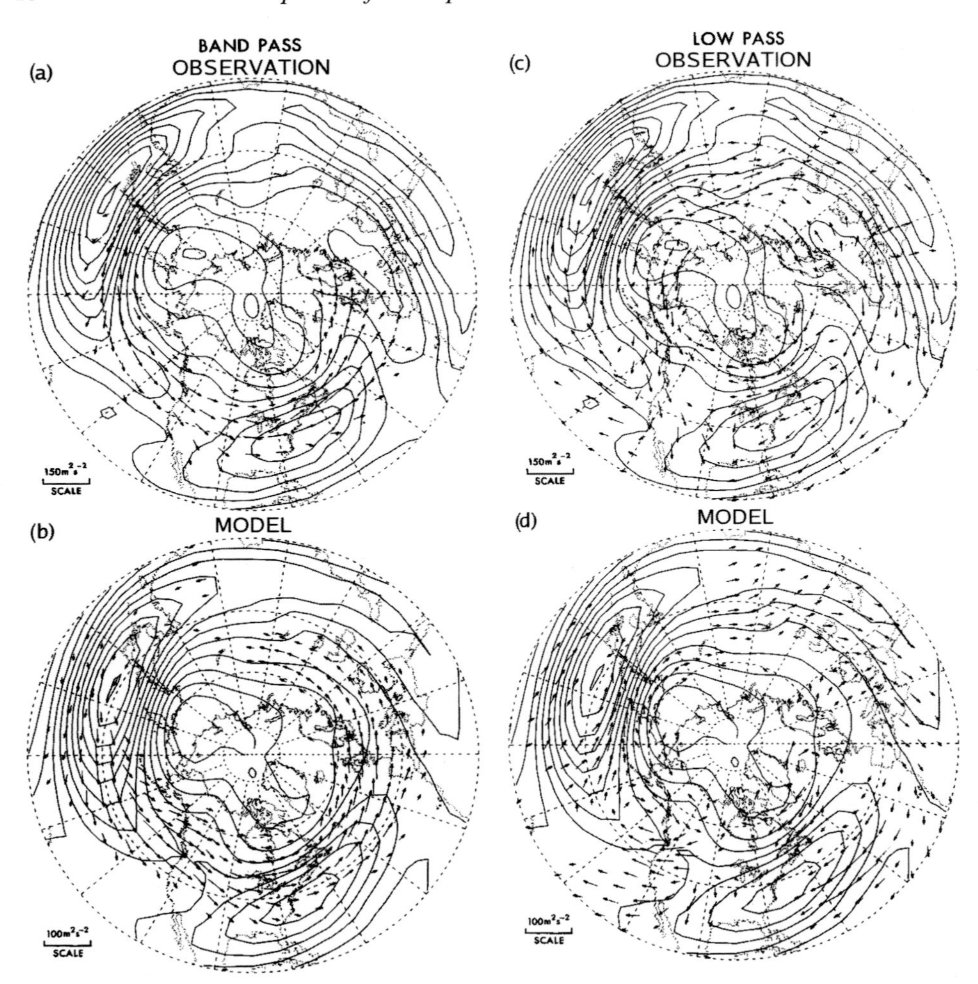

Figure 6.8 Distributions of the time-averaged zonal wind (contours; interval: 5 m s^{-1}) and $\boldsymbol{E}$ vector (arrows; see scale at lower left corner of each panel) at 300 mb, as computed using wintertime variance and covariance statistics for filtered wind data retaining periods of (a, b) 2.5–6 days and (c, d) between 10 days and a season. The observational charts (a, c) were based on NMC analyses for an 8-year period. The model charts (b, d) were computed using output from a 10-year segment of the same GFDL model simulation examined in Figure 6.5. From Wallace, J. M. and Lau, N.-C. (1985). Reprinted with permission from Elsevier.

southwest-to-northeast tilt equatorward of the storm track axis, thus leading to $\overline{v'u'} > 0$ at those sites. Conversely, $\overline{v'u'} < 0$ poleward of the storm track axis, due to the local prevalence of southeast-to-northwest tilt. By taking these characteristics of the HF transients into consideration, and recalling the definition of $\boldsymbol{E}$, it is readily seen that the $\boldsymbol{E}$ vectors for disturbances with synoptic time-scales would be oriented eastward near the storm track axis, and southeastward (northeastward) in regions lying equatorward (poleward) of the storm track axis. These inferences are supported by the observed and simulated patterns of $\boldsymbol{E}$ in Figures 6.8a,b in relation to the storm tracks (see maxima of HF variability in Figures 6.4c, d). Examination of the $\boldsymbol{E}$ field in relation to the local contours of $\overline{u}$ reveals that, throughout the storm tracks over the North Pacific and North Atlantic, the $\boldsymbol{E}$ vectors are directed down the local gradient of the $\overline{u}$ field in the jet exit regions, i.e. from high to low values of $\overline{u}$, so that the energy conversion term $-\boldsymbol{E} \cdot \nabla\overline{u} > 0$ (see Figures 4b and 10b of Wallace and Lau, 1985). The above chain of arguments illustrates that conversion of kinetic energy from the transients to the time-mean circulation near the storm tracks is closely linked to the typical horizontal structure of the HF disturbances in those locations.

In contrast to the HF transients, the LF fluctuations tend to be elongated in the zonal direction, so that $\overline{u'^2} > \overline{v'^2}$. This behavior is particularly evident for anomaly centers situated near the exits of the East Asian and North America jetstreams (e.g. see the shape of the extrema in the EOF patterns of Figure 6.5, and that of the coherent regions A and B in Figures 6.6c, d). Accordingly, the $\boldsymbol{E}$ vectors for low-frequency transients are mostly oriented westward (Figures 6.8c, d), from low to high values of the $\overline{u}$ field over the oceans. Thus $-\boldsymbol{E} \cdot \nabla\overline{u} < 0$ (see Figures 4c and 10c of Wallace and Lau, 1985), indicating that the momentum transports by the LF transients serve to extract kinetic energy from the time-averaged circulation.

The results in Figure 6.8 demonstrate that analysis tools based on the $\boldsymbol{E}$ vector, which were originally developed for analyzing observed atmospheric phenomena, also found fruitful applications in delineating the characteristic shape of GCM-simulated transients in various frequency bands. Moreover, these model diagnostics highlight the role of transients in the barotropic conversion of energy of the time-mean flow.

6.5 Experimentations on interactions with SST conditions at various sites

Considerable observational information has been collected on the covariability of the atmospheric circulation and different conditions at the lower boundary. The impacts of SST changes associated with ENSO episodes on the atmospheric flow pattern are of particular interest. These research activities have motivated numerous GCM experiments aimed at simulating and understanding the atmospheric responses to ENSO. The design of these experiments typically entails the prescription of idealized or observed SST anomalies related to ENSO at selected maritime sites. The outputs from these model runs were compared extensively with ENSO signals detected in the observed atmosphere. Various proposed mechanisms linking the SST forcing and the atmospheric responses were also critically examined on the basis of these GCM experiments.

Observational analyses of the global patterns of SST variability indicate that ENSO events are associated not only with oceanic anomalies in the tropical Pacific, but also with notable changes in other parts of the World Oceans, including the extratropical North and South Pacific, as well as the Atlantic and Indian Oceans (e.g. see discussion of Figure 6.10a in the latter part of this section). For a more complete understanding of the atmospheric responses to ENSO, it is necessary to evaluate the relative importance of SST anomalies at different sites in forcing the atmospheric circulation. This issue was addressed by Lau and Nath (1994) using a suite of three GCM experiments, with SST anomalies being prescribed in the following domains: the global ocean, the tropical Pacific only, and the mid-latitude North Pacific only. These three experiments were referred to as the GOGA, TOGA, and MOGA runs, respectively. In the course of each of these experiments, observed month-to-month SST variations during the 1946–1988 period were inserted into the corresponding forcing domain; whereas climatological SST conditions were prescribed at all maritime gridpoints lying outside of that forcing domain. Typical mid-tropospheric height responses in these experiments and in the observed atmosphere during ENSO episodes are summarized in Figure 6.9. Composites over El Niño (La Niña) events are presented in the left (right) columns. Results based on observations and the GOGA, TOGA, and MOGA experiments are shown in that order from the top to the bottom row.

The GOGA experiment generated geopotential height patterns (Figures 6.9c, d) that bear a qualitative resemblance to the observed response (Figures 6.9a, b), and also reproduced the reversal in polarity of the anomaly centers from warm to cold ENSO events. The amplitude of the simulated signals is notably lower than that of the observed responses, partially due to ensemble averaging over four independent samples of the GCM runs in this study, and to the relatively low resolution used in these model experiments. Investigations based on higher resolution yielded much stronger responses (Lau and Nath, 2001). There is a close correspondence between the responses in the GOGA and TOGA experiments (Figures 6.9c–f), thus indicating that the anomalous height pattern in these charts can mostly be attributed to SST anomalies in the tropical Pacific; whereas SST variations lying outside of the TOGA forcing domain play a relatively minor role. This inference is substantiated by the results from the MOGA experiment (Figures 6.9g, h), which yielded much weaker atmospheric anomalies in response to SST forcing prescribed in the extratropical North Pacific.

The global extent of the observed SST anomaly pattern associated with ENSO events, as alluded to in the above paragraphs, is illustrated in Figure 6.10a. This figure shows the distribution of temporal correlation coefficients between the SST

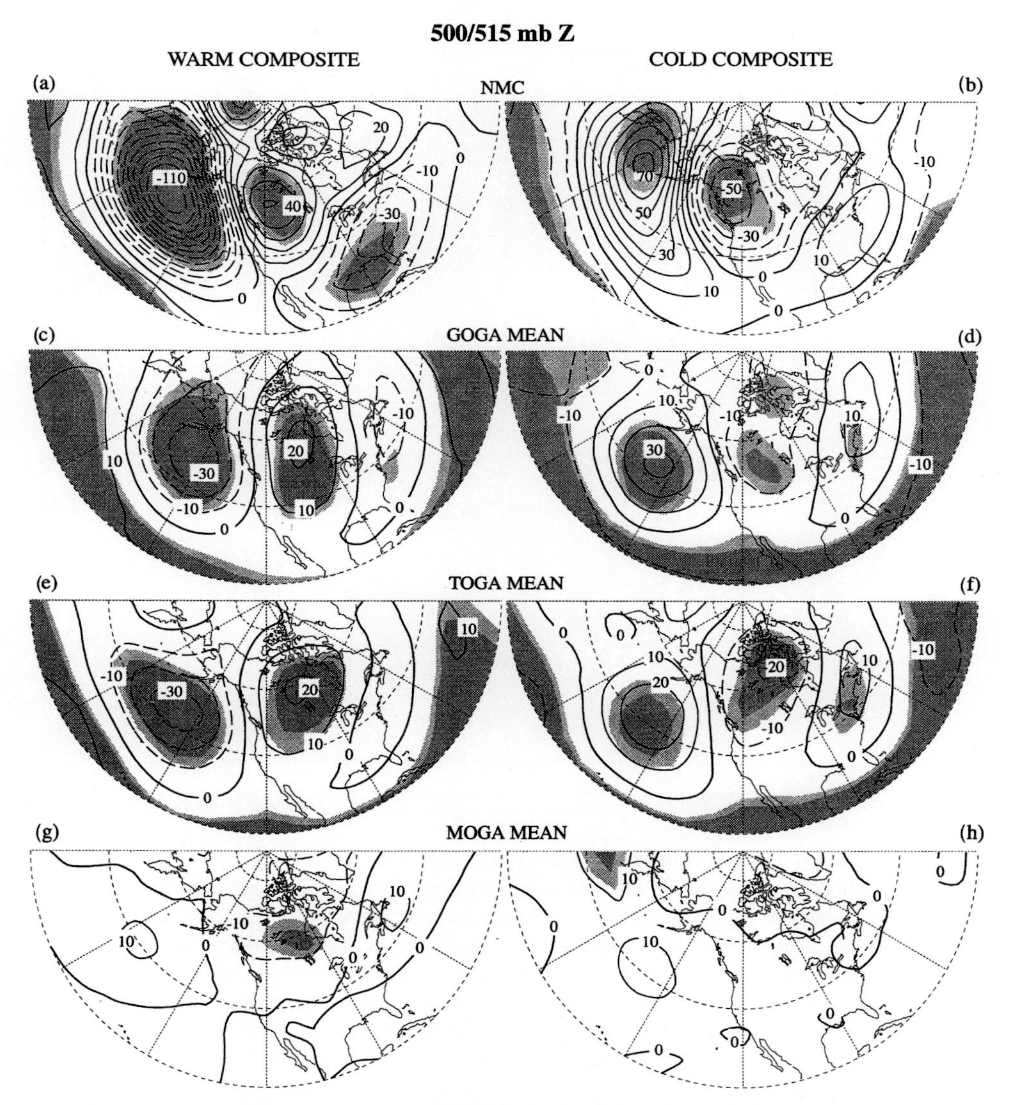

Figure 6.9 Distributions of the composite anomalies of wintertime geopotential height over six selected El Niño events (left panels) and six selected La Niña events (right panels) occurring in the 1946–1988 period (contour: interval: 10 m), as obtained from (a, b) observational NMC analyses at 500 mb, and model output at 515 mb from the (c, d) GOGA, (e, f) TOGA, and (g, h) MOGA experiments (see text for further details). These GCM experiments were conducted using the same GFDL model examined in Figures 6.5, 6.6, and 6.8. Model results correspond to ensemble averages over four parallel runs performed for each of the GOGA, TOGA, and MOGA forcing scenarios. Light and dark stippling represent significant height anomalies at the 90 and 95% levels, respectively. From Lau and Nath (1994).

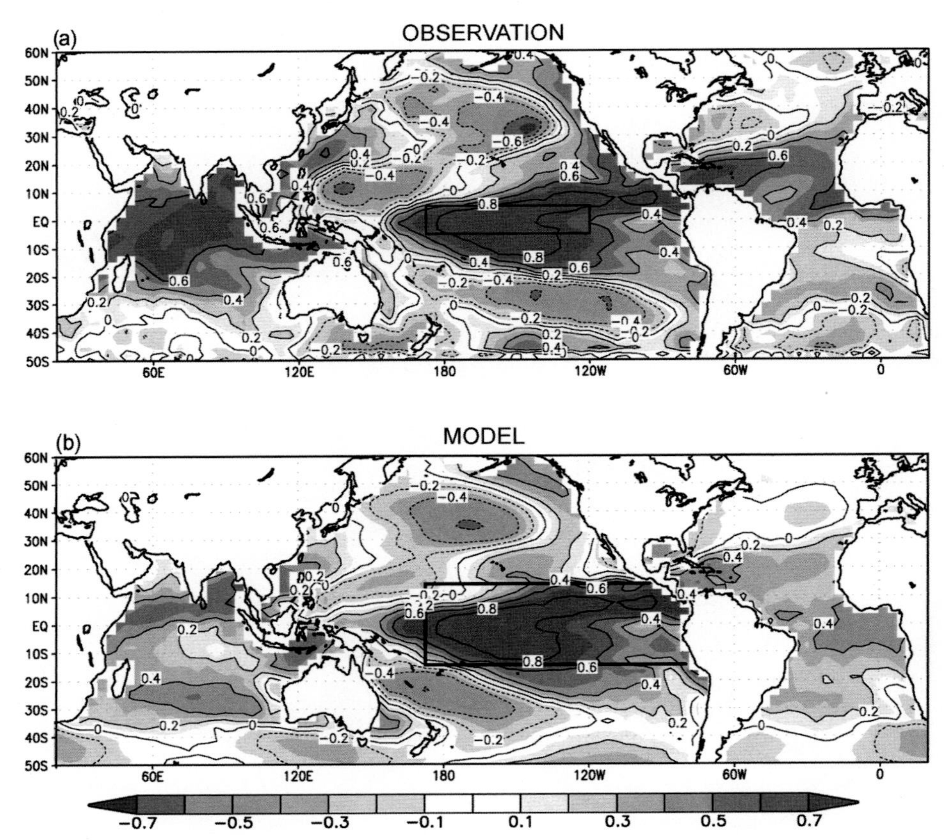

Figure 6.10 Distributions of the temporal correlation coefficients between SST anomalies averaged over the central equatorial Pacific (see rectangular box in upper panel) in the November–December–January season and the SST anomalies in the subsequent February–March–April season at all other gridpoints in the World Oceans. Contour interval: 0.2. Results are shown for (a) observational data compiled by Smith *et al.* (1996) and Reynolds and Smith (1994), and (b) output from an experiment using a 9-level GFDL spectral GCM with rhomboidal truncation at 30 wavenumbers (which corresponds to a spatial resolution of 3.75° in longitude and 2.25° in latitude), all for the 1950–1999 period. The model experiment entailed prescription of time-varying SST forcing in the central and eastern tropical Pacific (indicated by black borders in the lower panel), and computation of SST variations elsewhere using a 31-level oceanic mixed layer. Model values were based on ensemble averages over 16 parallel integrations over the 50-year period. From Alexander *et al.* (2002). See also color plate.

variations in the central equatorial Pacific during the boreal winter (November–December–January) and the SST anomalies at individual gridpoints one season later (February–March–April). It is evident from this correlation chart that warm ENSO episodes in the tropical Pacific are followed by positive SST changes in the Indian Ocean, tropical Atlantic, as well as the waters along the western seaboard of North

America and eastern seaboard of Asia; whereas cold anomalies appear in the central portions of the North Pacific and South Pacific, subtropical western North Pacific, and off the eastern and southern coasts of the United States. The polarity of these anomalies is reversed after La Niña events. It was proposed by Alexander (1992), Luksch and von Storch (1992), and Klein *et al.* (1999) that the SST anomalies at various remote sites are primarily forced by changes in the overlying atmospheric flow pattern, that modulates the surface fluxes across the local air–sea interface through alterations in wind speed, air temperature and humidity, and cloud cover. These meteorological changes are in turn manifestations of the global atmospheric response to ENSO-related SST forcing in the tropical Pacific. In this manner, the atmospheric circulation serves as a "bridge" linking SST variations in the core ENSO region to oceanic conditions throughout the World Oceans.

A series of GCM experiments was launched to examine the extent to which the observed global SST pattern as shown in Figure 6.10a can be explained by the "atmospheric bridge." In one of these modeling studies (Alexander *et al.*, 2002), the observed monthly SST variations were prescribed in the central and eastern equatorial Pacific, in a similar fashion as in the TOGA experiment described above. However, instead of fixing the SST pattern outside this forcing domain at climatological conditions, an oceanic mixed-layer model was used to compute the SST anomalies at all gridpoints lying beyond the tropical Pacific. This experimental set-up allowed for simplified two-way interactions between the global atmospheric responses to ENSO changes originating from the tropical Pacific and the ocean surfaces at various remote sites. The same procedure used in constructing the observed lagged correlation chart (Figure 6.10a) was applied to the SST output from this model experiment. The result, shown in Figure 6.10b, compares favorably with its observed counterpart. Particularly noteworthy is the replication of the observed cold anomaly in the central North Pacific, and the warm anomaly in the Gulf of Alaska. This result suggests that the ENSO-related SST pattern in the North Pacific Basin is primarily a response to atmospheric driving, rather than a forcing agent of the atmospheric circulation. Prescription of this anomaly pattern in the lower boundary of an atmospheric GCM, as was done in the MOGA experiment (Lau and Nath, 1994), is therefore not compatible with the actual nature of air–sea coupling in the North Pacific during ENSO. The generally weak atmospheric signals produced in the MOGA runs (Figures 6.9g, h) attest to the physical inconsistency in that particular experiment.

GCM experiments were also utilized to identify the causes of atmospheric signals observed in specific ENSO events. An example of such attribution studies is the investigation by Hoerling and Kumar (2003) on the role of SST forcing at different oceanic sites in generating widespread and prolonged high temperature and low precipitation anomalies over much of the northern subtropics and extratropics in the

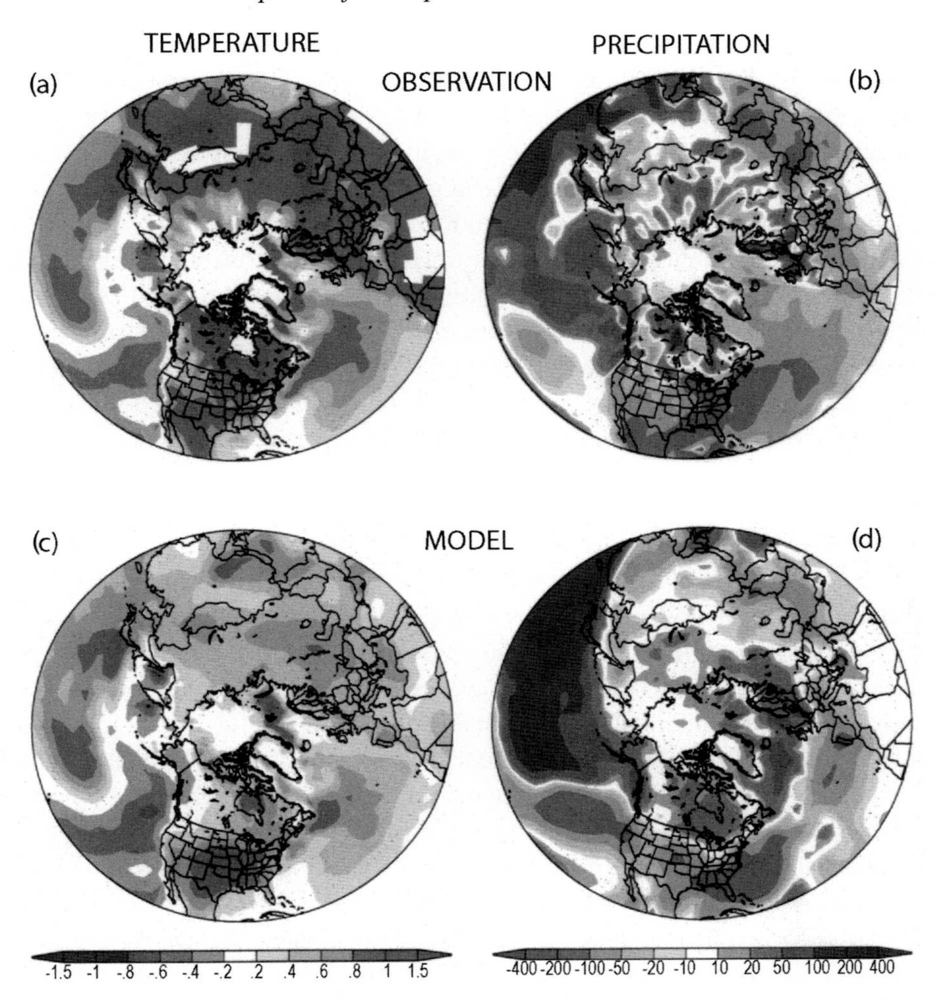

Figure 6.11 Distributions of the departures of (a, c) surface temperature (units: °C) and (b, d) precipitation (units: mm y^{-1}) in the June 1998–May 2002 period from climatology. Results were based on observational data (upper panels) and a grand 50-member ensemble average of simulations using the European Centre-Hamburg, National Centers for Environmental Prediction, and National Aeronautics and Space Administration models (lower panels). From Hoerling, M. and Kumar, A. (2003). Reprinted with permission from AAAS. See also color plate.

1998–2002 period (see observational patterns in the upper panels of Figure 6.11). These authors noted that such atmospheric signals were coincident with a long-lasting La Niña episode in the tropical Pacific, as well as a persistent warm SST anomaly in the western tropical Pacific–Indian Ocean sector. They proceeded to examine the model responses to this SST pattern, as generated in multi-member ensemble runs using three different GCMs. The grand average of these simulations based on observed global SST forcing in the 1998–2002 era is displayed in the lower panels of Figure 6.11. It is evident that the anomalously warm and dry conditions observed over the United States, southern Europe, and many parts of southern Asia were reproduced in these model integrations. Further experiments were performed by considering the cold SST forcing in the tropical central Pacific and the warm SST forcing in the Indo-western Pacific sectors separately. These additional simulations indicate that the oceanic conditions in both sectors act in a synergistic fashion to produce the extended hot and arid spells. Hoerling and Kumar (2003) hence concluded that La Niña in the central-eastern Pacific and warmth in the Indo-western Pacific constitute the "perfect" oceanic scenario for inducing circum-global drought conditions in the mid-latitude zone. The authors noted that the positive SST anomaly in the Indo-western Pacific was part of an ongoing warming trend in that region over the past several decades, which could in turn be linked to effects of increasing greenhouse gases. The joint forcing of this warming trend and of interannual ENSO activity in the tropical Pacific would have implications on the differing severity of climate anomalies in future El Niño and La Niña events (e.g. see the recent model diagnoses reported by Lau *et al.*, 2008).

Various GCM experiments have also been launched to study the impacts of boundary forcings other than SST anomalies. Comparison of these model results with observations have provided insights on the roles of sea ice, snow cover, and ground hydrology in the variability of the climate system.

6.6 Discussion and future prospects

The studies cited in this chapter have been selected by virtue of their representativeness of the strong interplay among observational and model diagnoses. These examples illustrate that the histories of observational studies and model investigations are intertwined. Active feedback between atmospheric research based on both observational and model-generated datasets has been beneficial to the development of both approaches. Specifically, the discernment of certain classes of atmospheric phenomena or processes based on appropriate observational datasets has motivated the application of similar analysis techniques to GCM output, and the design of model experiments to delineate various mechanisms and empirical relationships of interest. Comparison of model results against the observational "truth" according to various performance metrics has facilitated objective assessments of the capability of the models in reproducing different facets of the climate system. Observational findings have also offered helpful clues for experimental designs that serve to address specific scientific issues. These experiments take advantage of the controlled setting of the model environment, which often allows for clearer identification of causes and effects than is feasible with observations alone. In many instances, the insights gained from such model investigations have provided impetus for examining the observational data from fresh or alternate perspectives, or for procuring additional datasets to study the pertinent problems more fully. In summary, the verification of model output by using observational datasets, and the myriad applications of model experimentation for understanding various observed phenomena, have contributed significantly to the evolution of GCMs as an effective tool for climate research.

The examples described here are drawn from the observational and model literature on large-scale dynamical processes and air–sea interactions related to ENSO, which are closely aligned with this author's research interests. It should be stressed that the synergy between observational and GCM-based studies is not confined to the scope of this chapter. In addition to the rawinsonde network and reanalysis products, many other observational data sources, such as those associated with meteorological satellites and various remote sensing platforms, intensive field campaigns, laboratory experiments, and proxy climatic records, have similarly played important roles in validating GCM simulations, and in constraining parameters for representing different physical processes in model atmospheres. These diverse datasets have also facilitated many GCM studies of atmospheric features and mechanisms ranging from sub-cloud scale microphysics to multi-millenial variability of the global climate system.

With the rapid advances in computer technology and design of model numerics in recent years, it is now feasible to conduct long-term integrations with global GCMs with horizontal resolution higher than 50 km (e.g. Branković and Gregory, 2001; Duffy *et al.*, 2003; Ohfuchi *et al.*, 2004; Mizuta *et al.*, 2006; Roeckner *et al.*, 2006; Lau and Ploshay, 2009; Zhao *et al.*, 2009). The availability of these powerful model tools offers a unique opportunity to investigate new families of sub-synoptic scale phenomena, as well as the details of the mean state and variability of microclimates throughout the world. Simulations with the high-resolution GCMs will place unprecedented demands on observational information on comparable spatial scales, both for model validation, and for providing guidance and stimulation for experimental design and analysis of model output. The production of global, mesoscale reanalyses of a comprehensive set of atmospheric fields, as well as the compilation of datasets that take fuller advantage of the rich information content of satellite measurements, would be particularly helpful for assessing the performance of high-resolution GCMs, and for identifying new scientific issues that could be addressed with these models.

The increase in computer power allows for routine integrations of medium-resolution GCMs with durations of hundreds and even thousands of years. Such extended model runs are well suited for understanding the nature and causes of climate variability on decadal and centennial time-scales, for which only scanty observational information is currently available. Inferences from the very long-term model experiments could guide future acquisition efforts and analysis strategies of historical records of the observed climate system. Enhanced computing resources also enable us to investigate model behavior under a given forcing scenario, by making multiple runs either with a simple model or with a group of different models, all of which have been subjected to the same set of prescribed conditions. The number of independent realizations generated in these ensemble experiments is typically much larger than that available from observational records. Many empirical relationships deduced from observational datasets of limited temporal coverage may hence be established on a firmer statistical footing by taking advantage of the larger population of the pertinent phenomena that are present in the multi-member, multi-model simulations. The more comprehensive sampling afforded by the ensemble approach could yield more robust signals of the primary physical processes at work.

In the current era, GCMs are employed to delineate the mechanisms contributing to climate trends in recent decades (e.g. Compo and Sardeshmukh, 2009; Deser and Phillips, 2009), and to make projections of climate change in the future. These studies make use of comprehensive Earth system models, which take into account a more diverse set of interactive processes. It is difficult to assess the reliability of long-term model predictions, since the observational data for validating such model predictions do not yet exist. However, the observations still play a substantial role in model-based climate change research. The traditional types of comparisons between observations and model output still need to be performed, so as to ascertain the degree of fidelity with which the present generation of models can simulate the processes operating in an "unforced" environment. The level of natural variability attributed to these internal processes could then be utilized to assess the significance of model projections of climate change due to anthropogenic forcing (e.g. see Manabe *et al.*, 2001). Verification of the model hindcasts of the climate variations that have occurred in the past several decades against the corresponding observations also provides useful measures of the model skill in projecting future climate changes. Model predictions would identify those indicators and geographical sites that are likely to exhibit large climate change signals. Such information could be utilized to optimize our efforts to design observational networks for monitoring climate variability in the future.

Acknowledgments

I thank Mike Alexander, John Lanzante, Suki Manabe, Mike Wallace, and an official reviewer for offering constructive comments on earlier versions of this article. I am indebted to Maurice Blackmon and Mike Wallace for their guidance and encouragements in the early stages of my research career in climate diagnostics using observational and model tools. Mary Jo Nath provided technical assistance in preparing various figures for publication in their present form.

References

Alexander, M. A. (1992). Midlatitude atmosphere–ocean interaction during El Niño. Part I: The North Pacific Ocean. *Journal of Climate*, 5, 944–958.

Alexander, M. A, Bladé, I., Newman, M., Lanzante, J. R., Lau, N.-C., and Scott, J. D. (2002). The atmospheric bridge: The influence of ENSO teleconnections on air–sea interaction over the global oceans. *Journal of Climate*, 15, 2205–2231.

Blackmon, M. L. (1976). A climatological spectral study of the 500 mb geopotential height of the Northern Hemisphere. *Journal of the Atmospheric Sciences*, 33, 1607–1623.

Blackmon, M. L. and Lau, N.-C. (1980). Regional characteristics of the Northern Hemisphere wintertime circulation: A comparison of the simulation of a GFDL general circulation model with observations. *Journal of the Atmospheric Sciences*, 37, 497–514.

Blackmon, M. L, Wallace, J. M., Lau, N.-C., and Mullen, S. L. (1977). An observational study of the Northern Hemisphere wintertime circulation. *Journal of the Atmospheric Sciences*, 34, 1040–1053.

Blackmon, M. L., Lee, Y.-H., and Wallace, J. M. (1984a). Horizontal structure of 500 mb height fluctuations with long, intermediate and short time scales. *Journal of the Atmospheric Sciences*, 41, 961–980.

Blackmon, M. L., Lee, Y.-H., Wallace, J. M., and Hsu, H.-H. (1984b). Time variation of 500 mb height fluctuations with long, intermediate and short time scales as deduced from lag-correlation statistics. *Journal of the Atmospheric Sciences*, 34, 981–991.

Branković, Č. and Gregory, D. (2001). Impact of horizontal resolution on seasonal integrations. *Climate Dynamics*, 18, 123–143.

Cattle, H. (ed.) (1987). *Atmospheric and Oceanic Variability*. Royal Meteorological Society, U.K., 182 pp.

Compo, G. P. and Sardeshmukh, P. D. (2009). Oceanic influences on recent continental warming. *Climate Dynamics*, 32, 333–342.

Crutcher, H. L. (1961). *Meridional Cross-sections of Upper Winds over the Northern Hemisphere*. Tech. Paper No. 41, U.S. Department of Commerce, Washington, DC, 307 pp.

Deser, C. and Phillips, A. S. (2009). Atmospheric circulation trends, 1950–2000: The relative roles of sea surface temperature forcing and direct atmospheric radiative forcing. *Journal of Climate*, 22, 396–413.

Duffy, P. B., Govindasamy, B., Iorio, J. P. *et al*. (2003). High-resolution simulations of global climate, Part 1: Present climate. *Climate Dynamics*, 21, 371–390.

Hoerling, M. and Kumar, A. (2003). The perfect ocean for drought. *Science*, 299, 691–694.

Hoskins, B. J. and Pearce, R. P. (eds.) (1983). *Large-scale Dynamical Processes in the Atmosphere*. Academic Press, 397 pp.

Hoskins, B. J., James, I. N., and White, G. H. (1983). The shape, propagation and mean-flow interaction of large-scale weather systems. *Journal of the Atmospheric Sciences*, 40, 1595–1612.

Hoskins, B. J., Hsu, H. H., James, I. N., Masutani. M., Sardesmukh, P. D., and White, G. H. (1989). *Diagnostics of the Global Atmospheric Circulation Based on ECMWF*

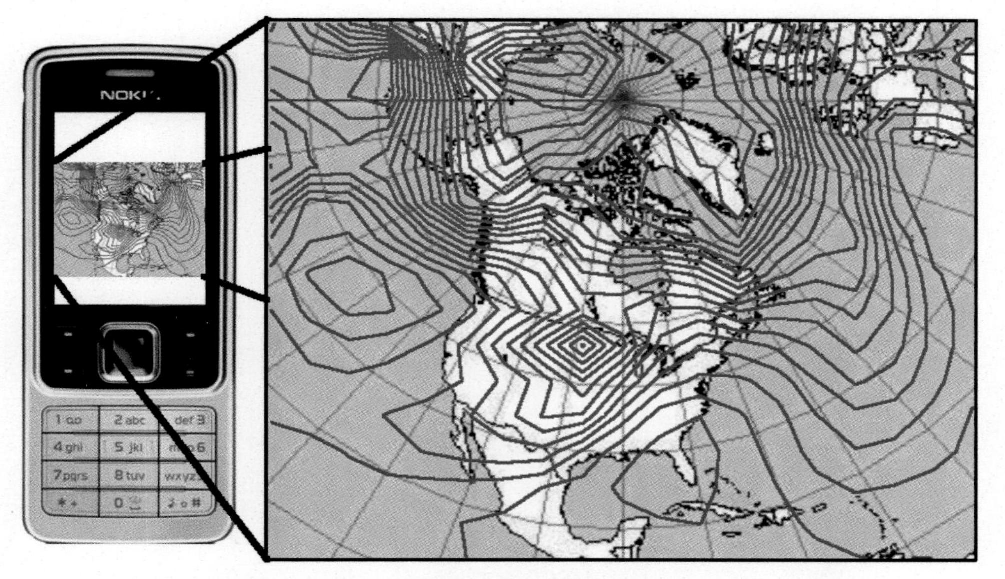

Figure 2.3 The Nokia 6300, dubbed PHONIAC (left) and the forecast for 0300 UTC, January 6, 1949 (right) made with the program phoniac.jar. The contour interval is 50 m (from Lynch and Lynch, 2008).

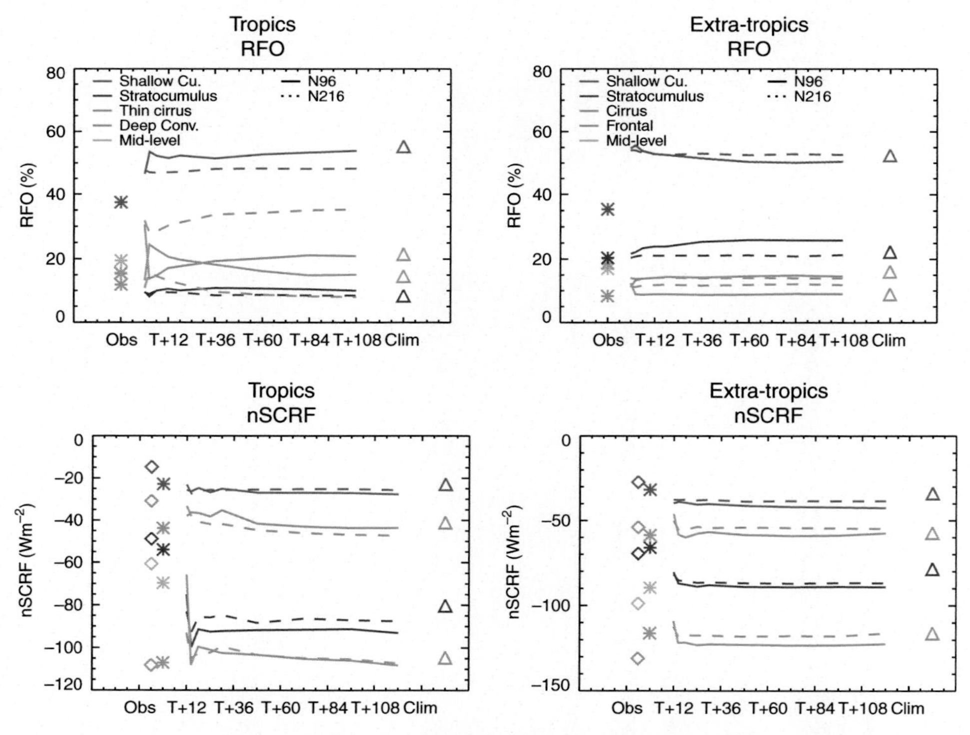

Figure 5.6 Evolution of cloud-regime relative frequency of occurrence (RFO) and normalized shortwave cloud radiative forcing (nSCRF) through the model forecast. N96 is shown solid; N216 is shown dashed. Observed climatology from ISCCP is shown with an asterisk; from Earth radiation budget experiment (ERBE) data is shown with a diamond; climate model climatology is shown with a triangle. Note that the ISCCP tropical stratocumulus and deep convective RFO are identical (from Fig. 3 of Williams and Brooks, 2008).

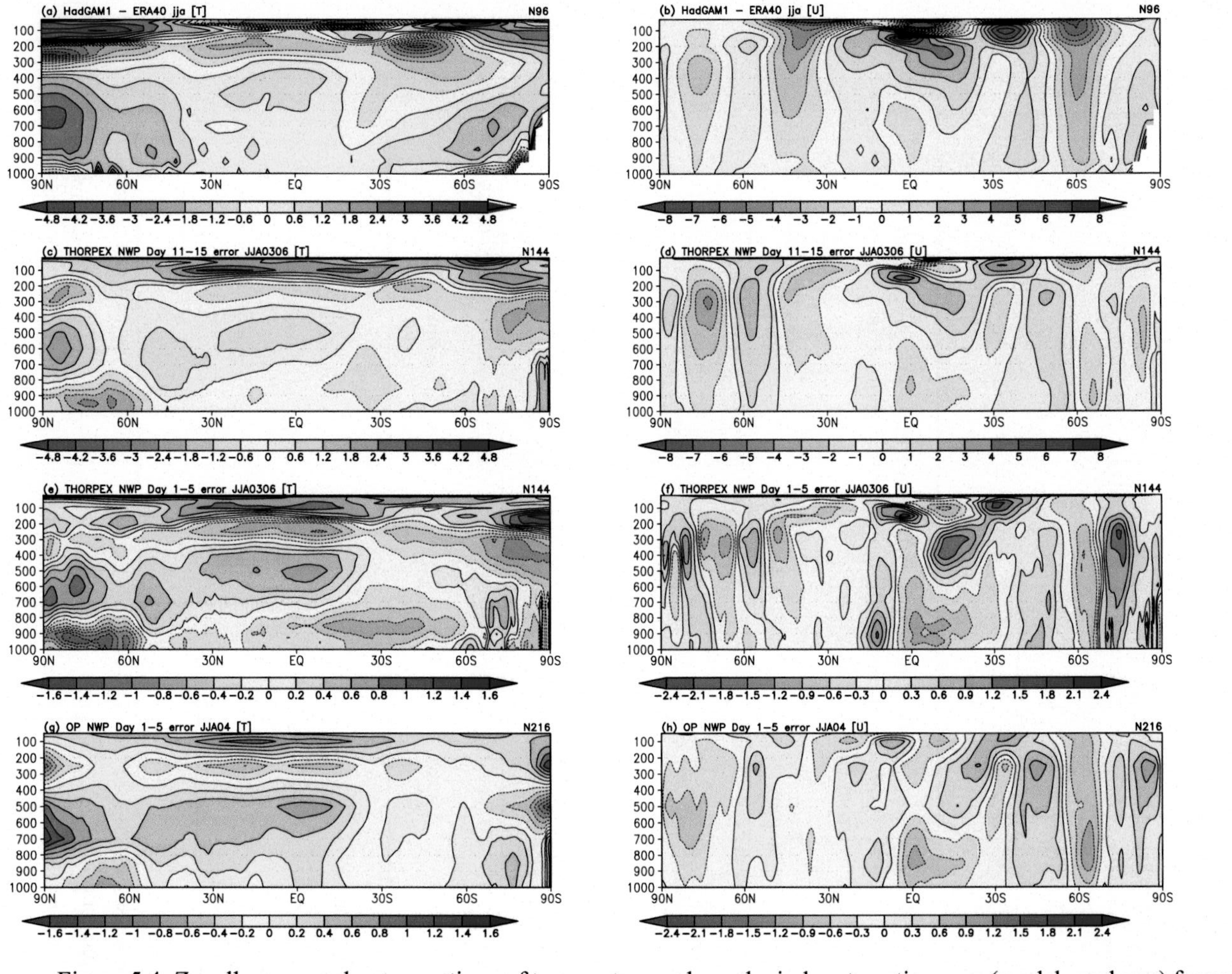

Figure 5.4 Zonally averaged cross-sections of temperature and zonal wind systematic errors (model–analyses) for model forecasts/simulations during June–August performed across a range of temporal scales and at differing horizontal resolutions. (a) & (b) HadGAM1 climate model 20 year AMIP run–ERA40 re-analysis, N96 (135 km) horizontal resolution (c) & (d) MOGREPS-15 (THORPEX) days 11–15 forecast – Met Office analyses for JJA 2003 and 2006, N144 (90 km) horizontal resolution (e) & (f) as (c) & (d) but for days 1–5, (g) & (h) days 1–5 from the operational deterministic global NWP model (circa 2004) – Met Office analyses for JJA 2004, N216 (60 km) resolution.

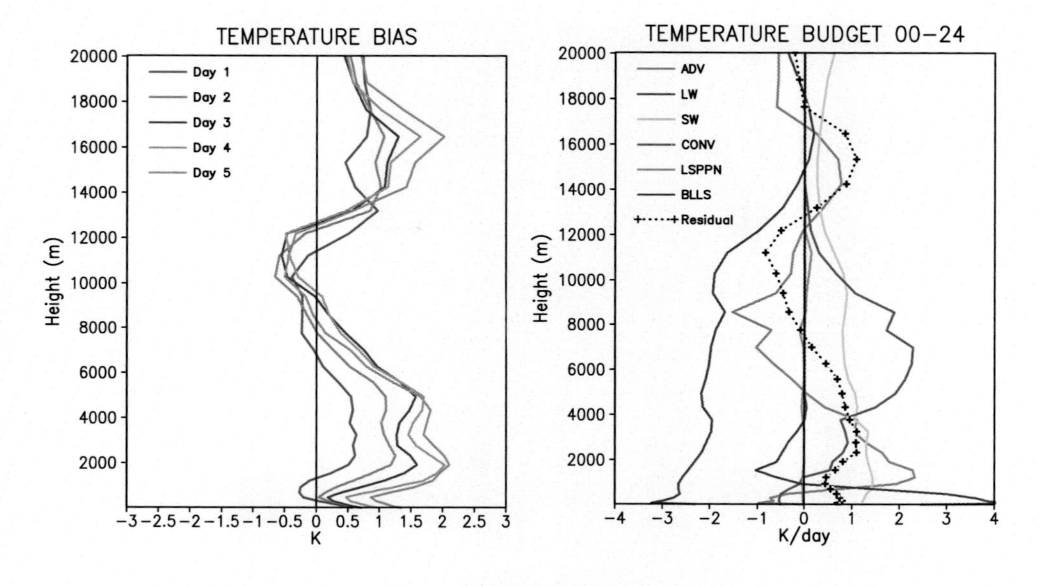

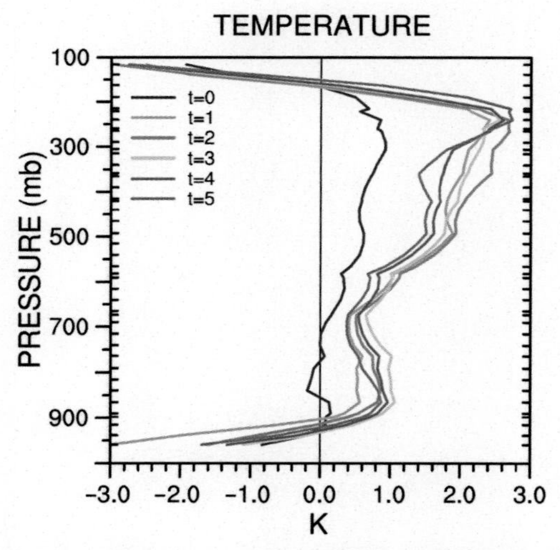

Figure 5.5 (cont'd overleaf) Diagnosing MetUM and NCAR-CAM2 systematic temperature biases at ARM-SGP during IOP June–July 1997 as part of Transpose-AMIP intercomparisons (see text). Results from the MetUM are shown overleaf and in the upper panels above. Results from NCAR-CAM2 are shown in the lower panel above. The figure overleaf shows the evolution of MetUM vertical temperature biases (model-sonde) as a function of forecast range (0 to 5 days, showing 5 diurnal cycles). Results are averaged over all forecasts during the IOP. The upper left panel shows the evolution of MetUM temperature bias integrated over a diurnal cycle (and over all forecasts), and the upper right panel shows MetUM mean 00–24 hour tendencies from the individual physics routines. Note the strong similarity between evolving temperature bias and the residual amongst the tendencies and also how the contributions to the temperature bias come from different parts of the diurnal cycle and point to specific errors in physical processes (see text for more details). The lower panel above shows the NCAR-CAM2 (from Williamson et al, 2005) temperature biases as a function of forecast range, which also shows an evolving warm bias but differs from MetUM in that larger errors are seen in the upper troposphere.

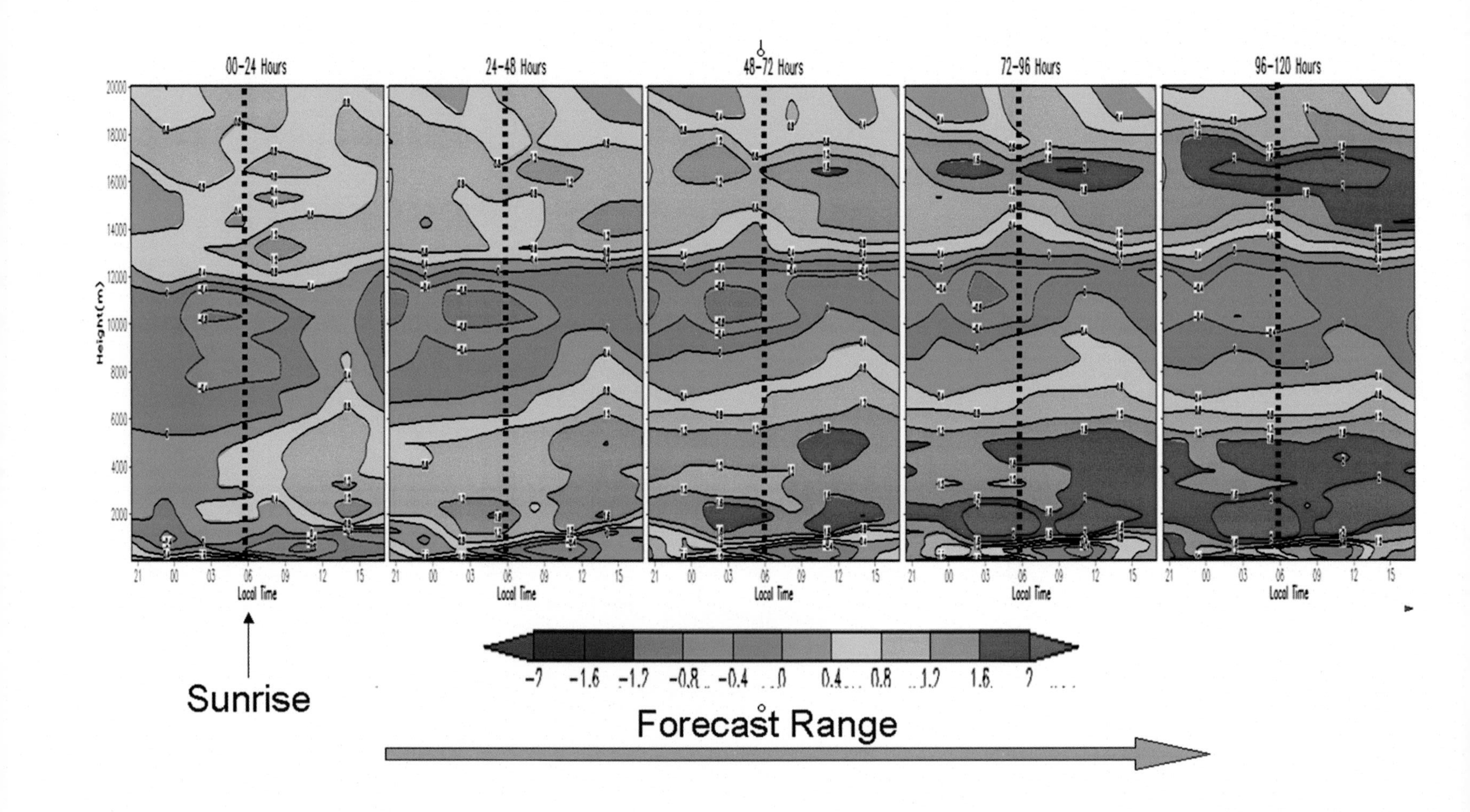
00-24 Hours
24-48 Hours
48-72 Hours
72-96 Hours
96-120 Hours
Height(m)
20000
18000
16000
14000
12000
10000
8000
6000
4000
2000
21 00 03 06 09 12 15
Local Time
Sunrise
-2 -1.6 -1.2 -0.8 -0.4 0 0.4 0.8 1.2 1.6 2
Forecast Range

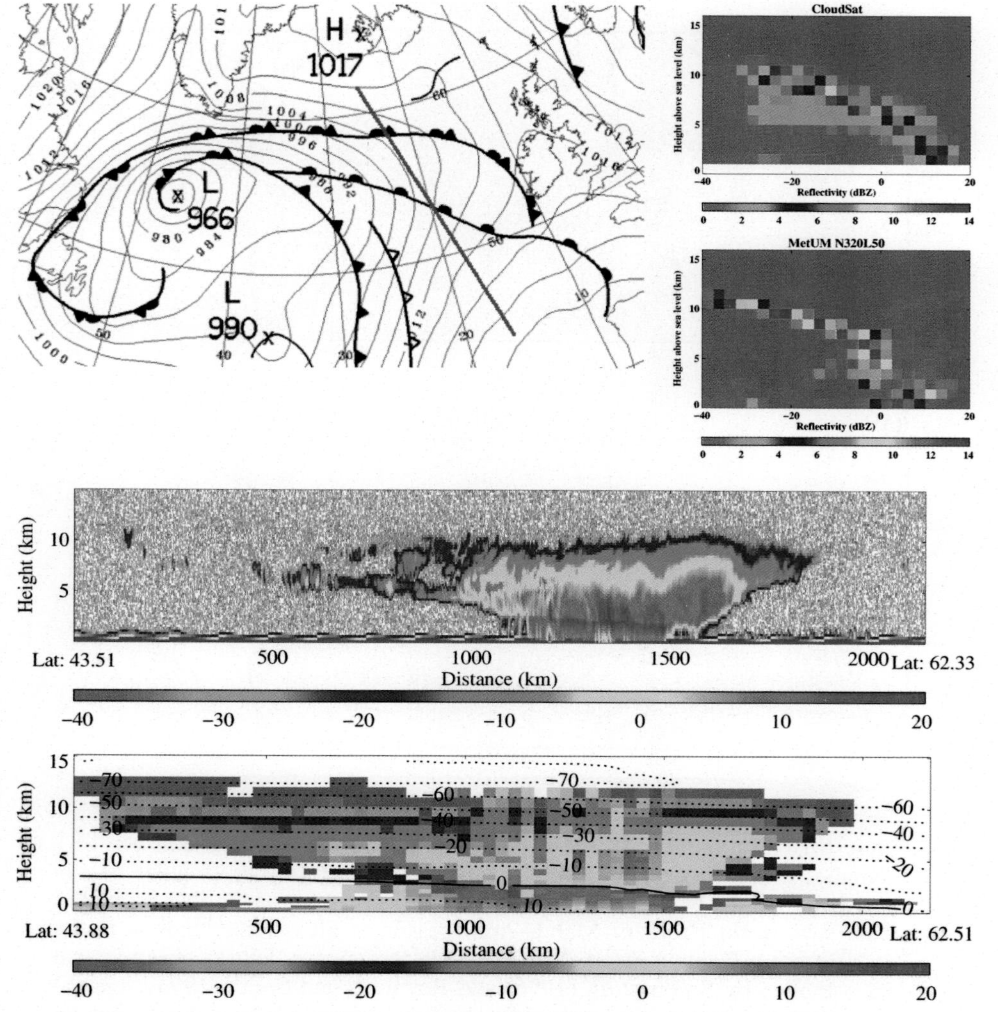

Figure 5.7 Top left: Met Office analysis chart for North Atlantic at 1200 UTC on 26 February 2007. The CloudSat track is shown in red. Bottom: radar reflectivity (dBZ) observed by CloudSat (upper) and simulated by the MetUM global forecast model (lower) along this track. Top right: joint height–reflectivity hydrometeor frequency of occurrence (%) observed by CloudSat (upper) and simulated by the MetUM global forecast model (lower) for this case study. Taken from Bodas-Salcedo *et al.* (2008).

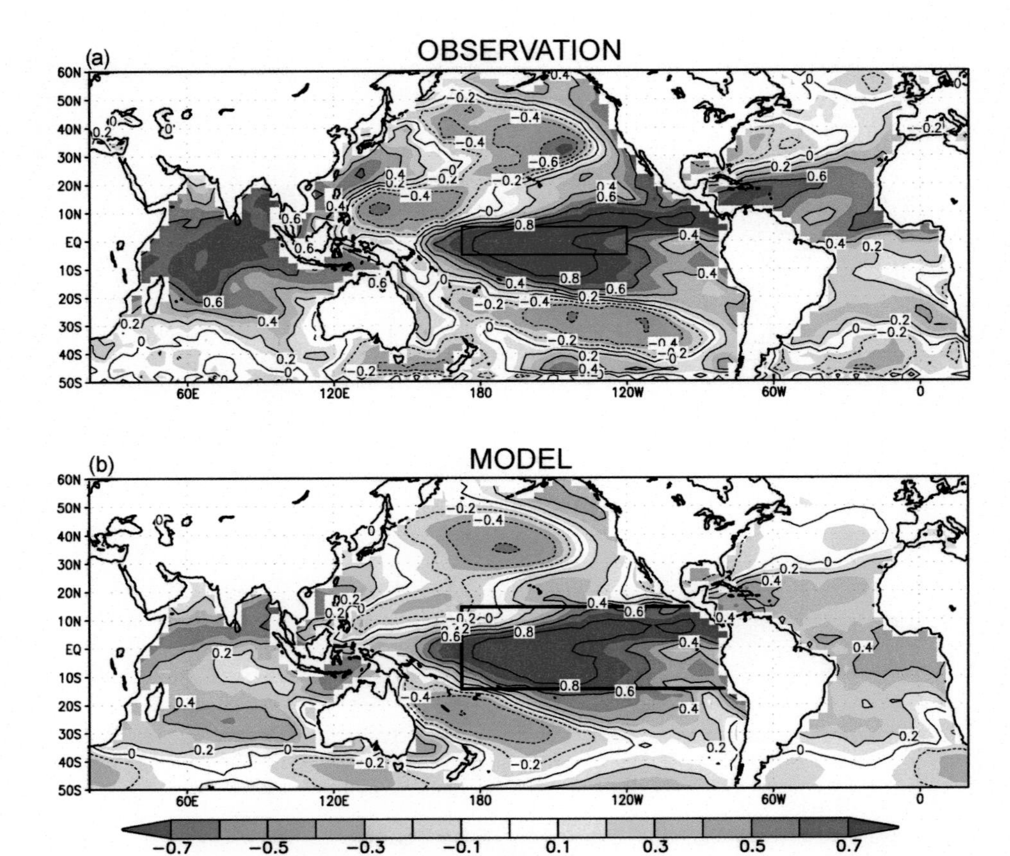

Figure 6.10 Distributions of the temporal correlation coefficients between SST anomalies averaged over the central equatorial Pacific (see rectangular box in upper panel) in the November–December–January season and the SST anomalies in the subsequent February–March–April season at all other gridpoints in the World Oceans. Contour interval: 0.2. Results are shown for (a) observational data compiled by Smith *et al.* (1996) and Reynolds and Smith (1994), and (b) output from an experiment using a 9-level GFDL spectral GCM with rhomboidal truncation at 30 wavenumbers (which corresponds to a spatial resolution of 3.75° in longitude and 2.25° in latitude), all for the 1950–1999 period. The model experiment entailed prescription of time-varying SST forcing in the central and eastern tropical Pacific (indicated by black borders in the lower panel), and computation of SST variations elsewhere using a 31-level oceanic mixed layer. Model values were based on ensemble averages over 16 parallel integrations over the 50-year period. From Alexander *et al.* (2002).

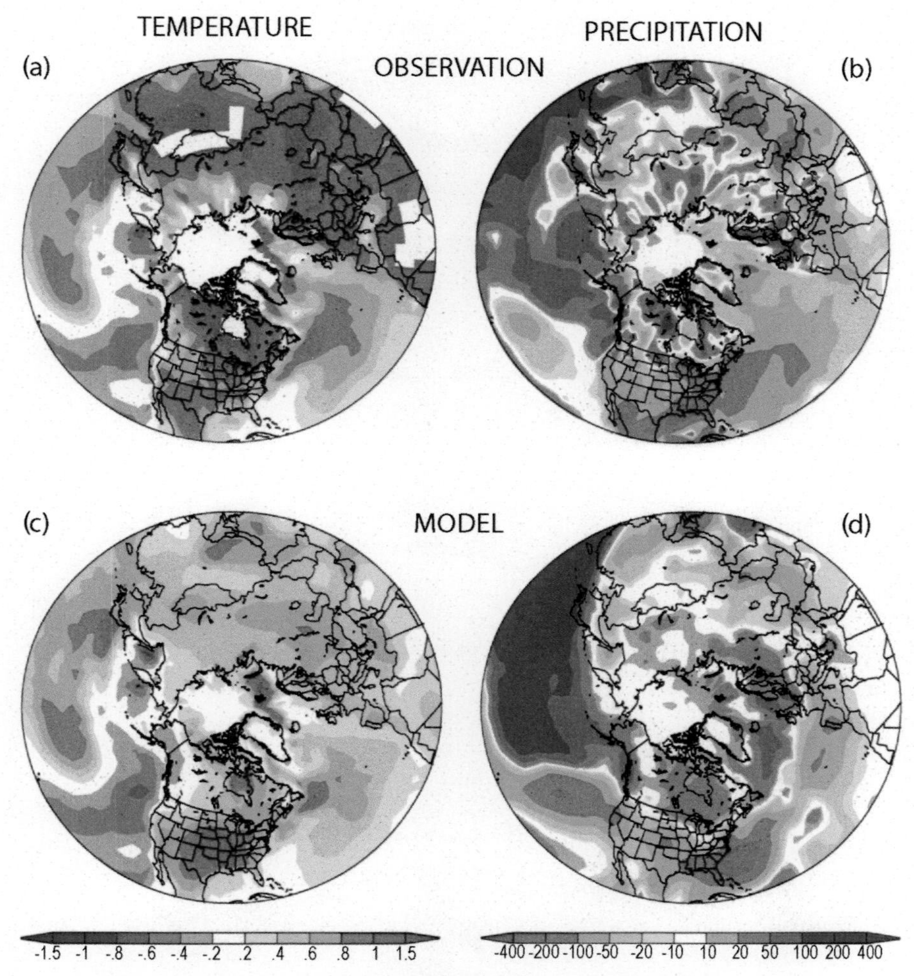

Figure 6.11 Distributions of the departures of (a, c) surface temperature (units: °C) and (b, d) precipitation (units: mm y^{-1}) in the June 1998–May 2002 period from climatology. Results were based on observational data (upper panels) and a grand 50-member ensemble average of simulations using the European Centre-Hamburg, National Centers for Environmental Prediction, and National Aeronautics and Space Administration models (lower panels). From Hoerling, M. and Kumar, A. (2003). Reprinted with permission from AAAS.

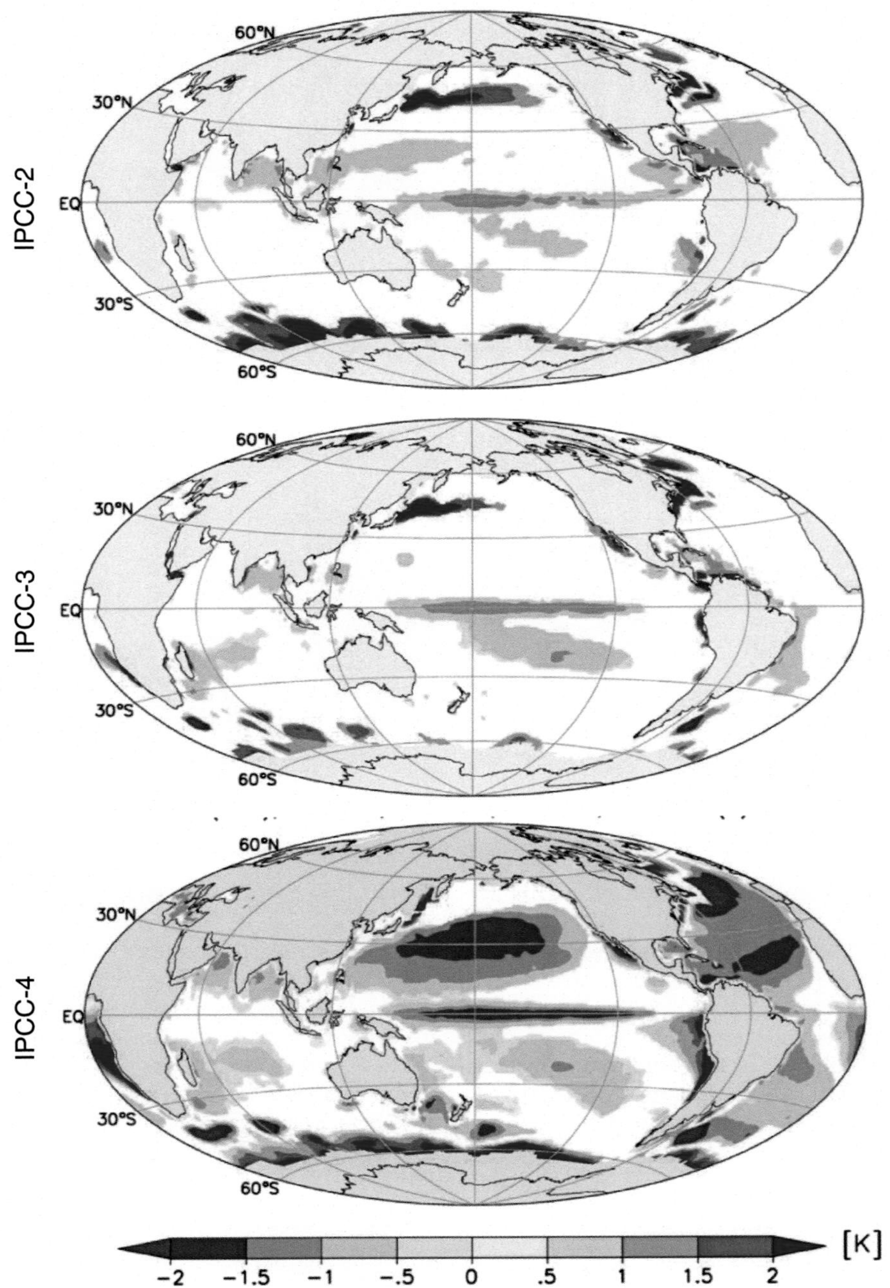

Figure 7.1 Biases in the annual mean climatological sea surface temperature averaged over all IPCC models. (From Reichler and Kim, 2008). Note that without flux adjustment in most models, the average error in SST is larger in IPCC-4.

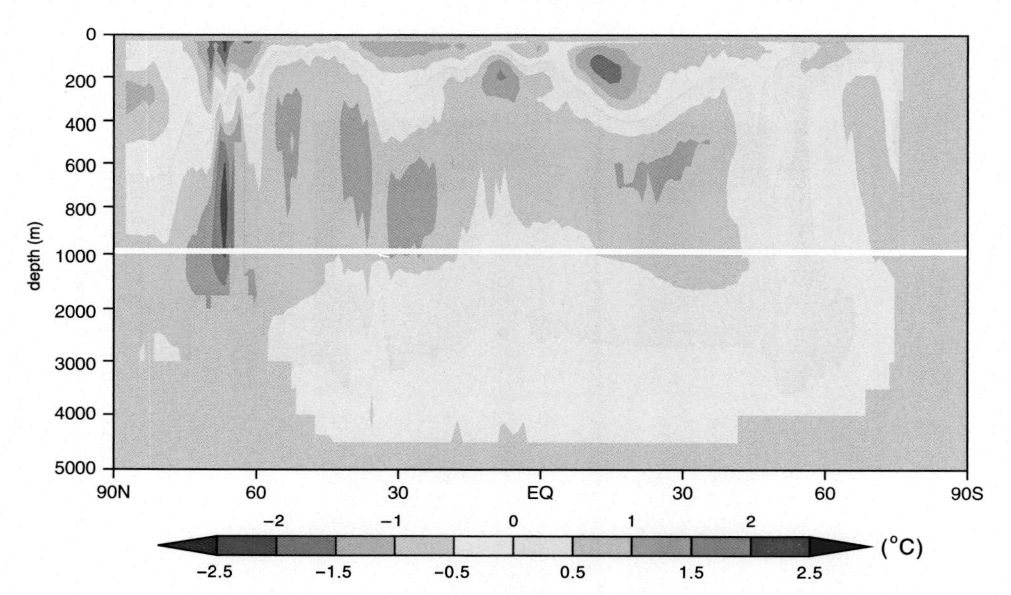

Figure 7.3 (Modified and based on Randall *et al.* 2007) Background shading represents the multi-model zonally averaged temperature bias of the climate models submitted to the IPCC-4 assessment. The observations are from the 2005 World Ocean Atlas, Vol. 1 (Levitus, 2005).

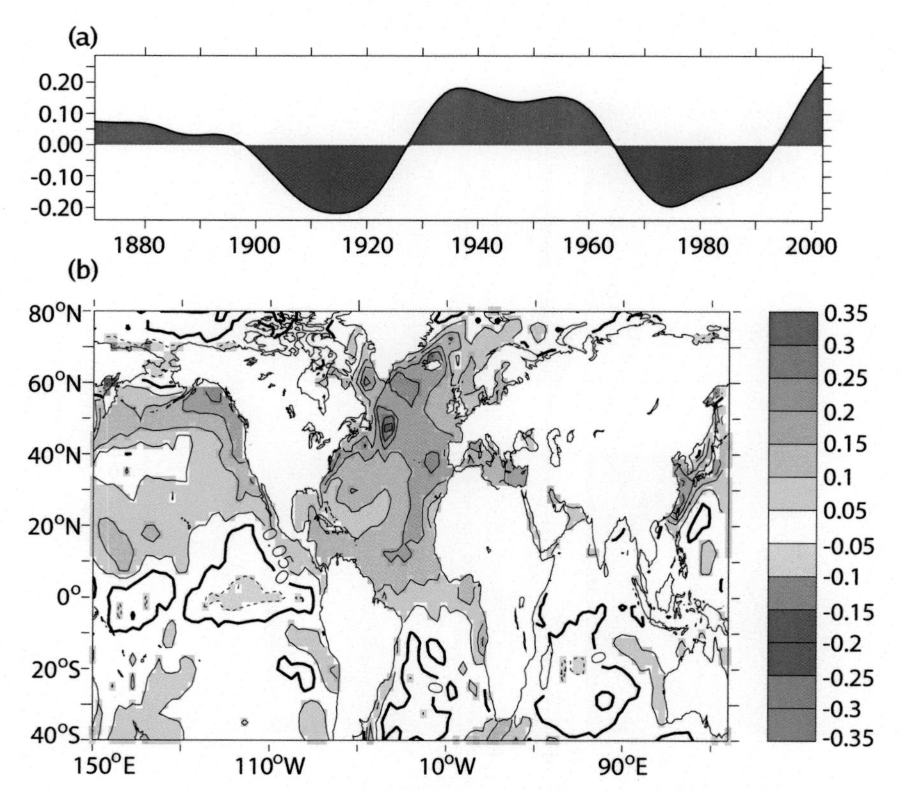

Figure 7.5 (From Sutton and Hodson, 2005. Reprinted with permission from AAAS) (a) Observed AMO temperature index, which is averaged SST over the North Atlantic from the equator to 60 N, and 75 W to 7.5 W. (b) Regression of SST to the AMO time series normalized by one unit of variance. The regression is calculated over the same time period shown in (a).

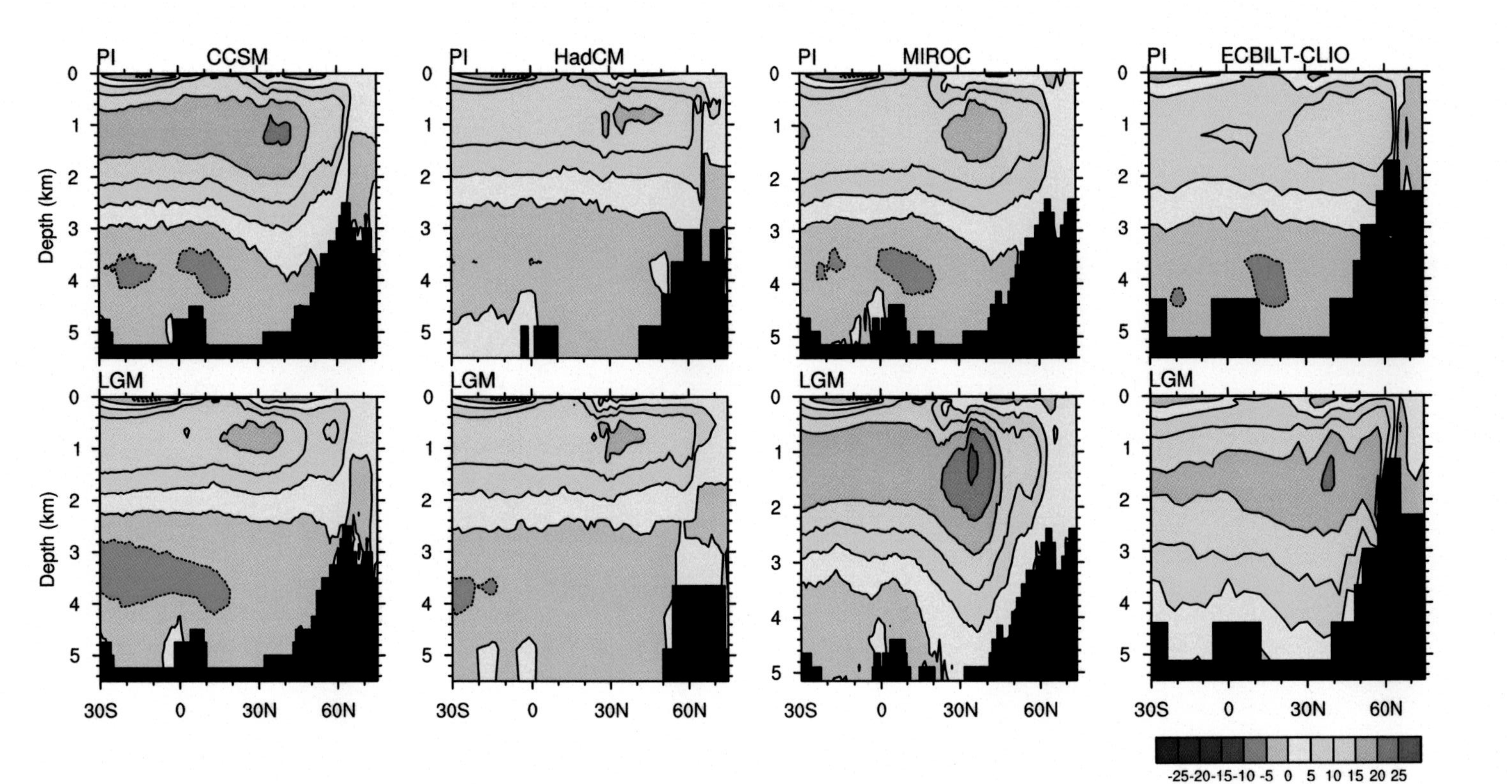

Figure 7.4 (From Otto-Bliesner *et al.*, 2007) Meridional overturning of the Atlantic Ocean sector of the PMIP coupled climate models, which are of lower resolution than the most recent IPCC climate models by approximately a factor of 2. Units are Sverdrups, defined as one million cubic meters/sec. Flow is clockwise around maxima. The upper panels represent a simulation of the present climate as control. The lower panels represent simulations of meridional overturning during the Last Glacial Maximum (LGM). CCSM (NCAR), HadCM (Hadley Centre), MIROC (Combined Japanese), ECBILT-CLIO (Netherlands).

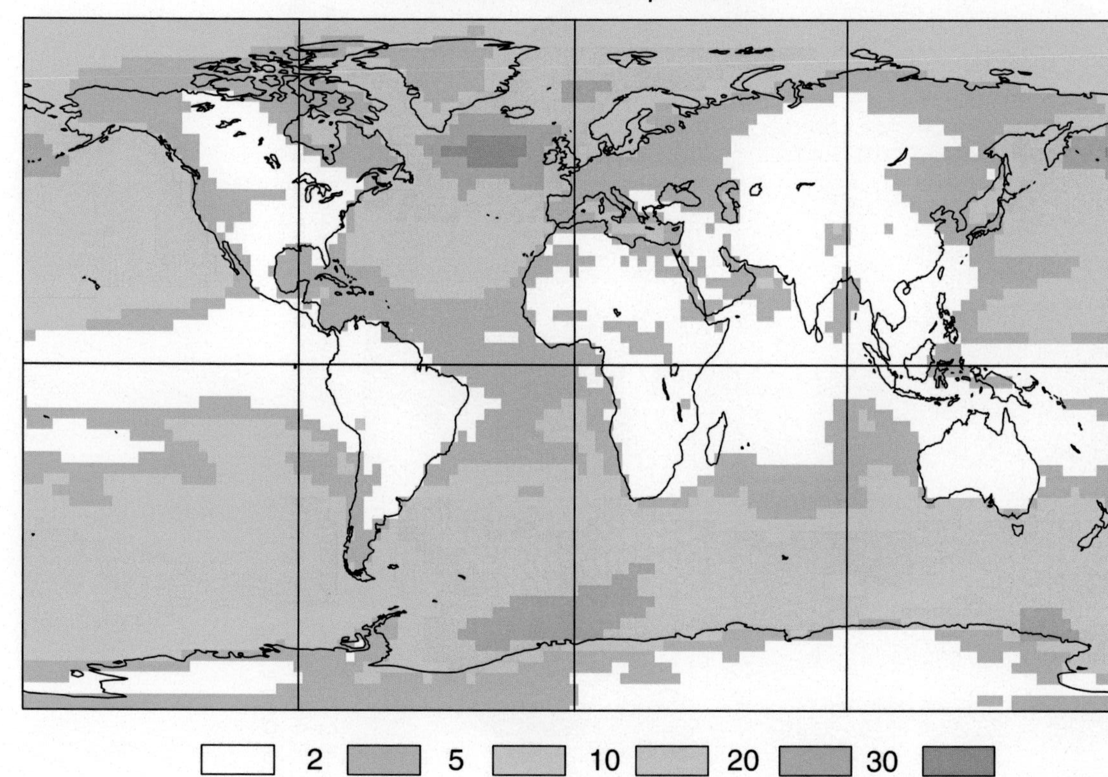

Figure 7.8 The potential predictability of decadal time-scales calculated from the output of several coupled models. Color shading indicates the ratio of variance of decadal means to the total variance, which is shown as a percentage (from Boer and Lambert, 2008). Note that the highest potential predictability is in the polar seas of both hemispheres.

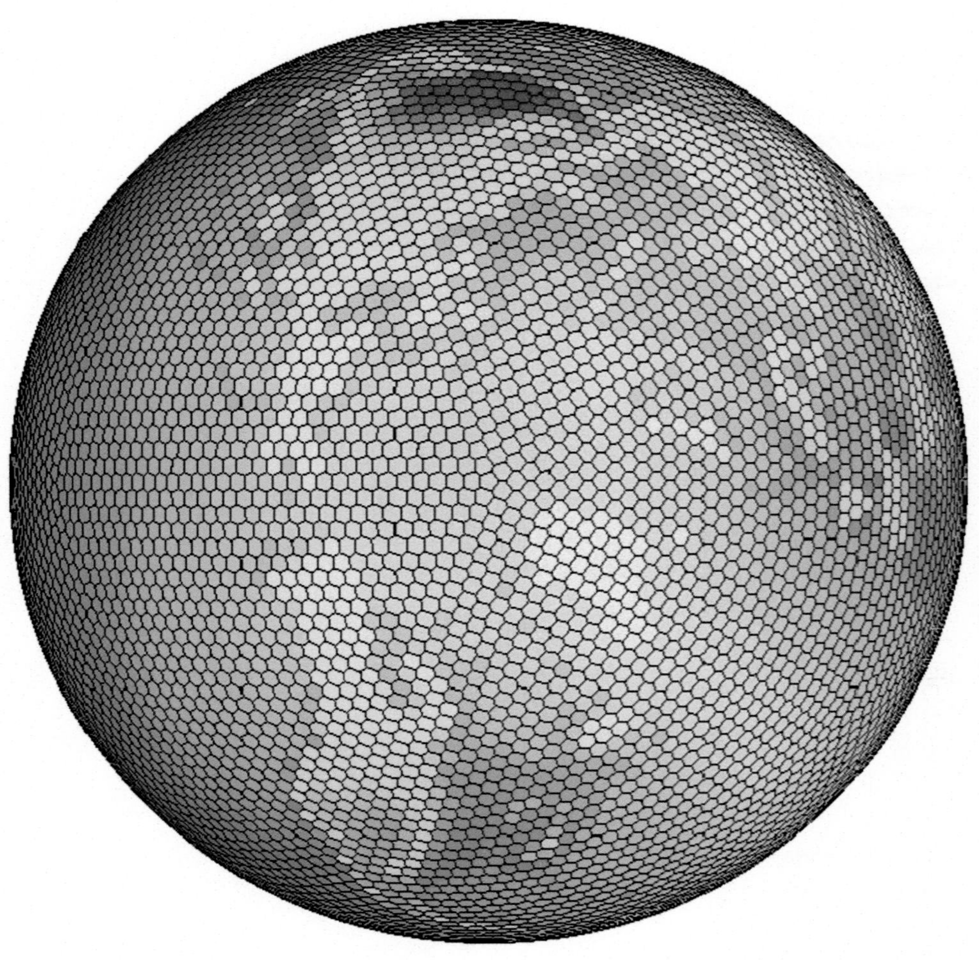

Figure 9.2 The distribution of surface elevation, plotted on a geodesic grid with about 10,000 cells. The figure has been made in such a way that the individual grid cells are visible.

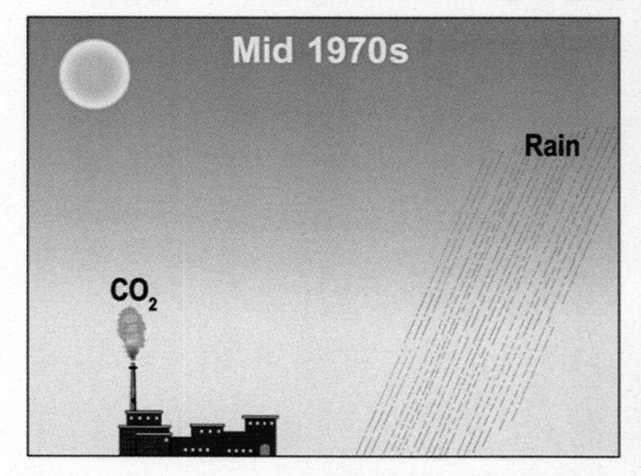

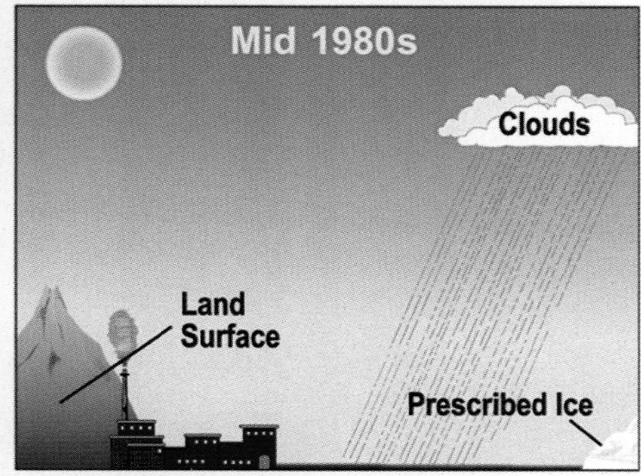

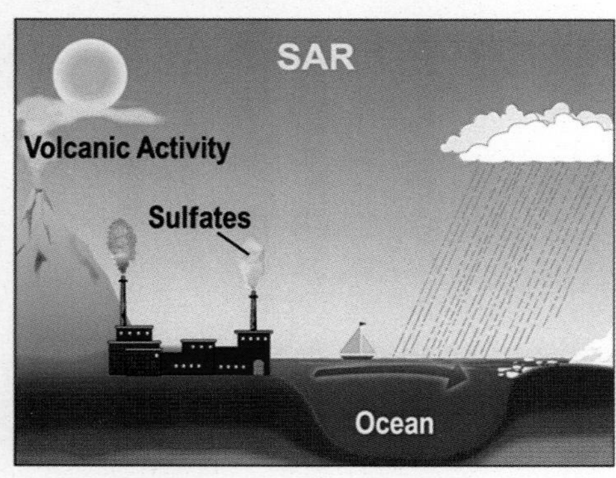

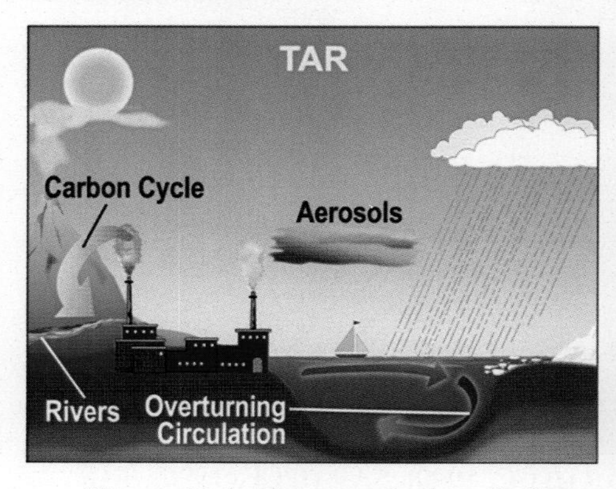

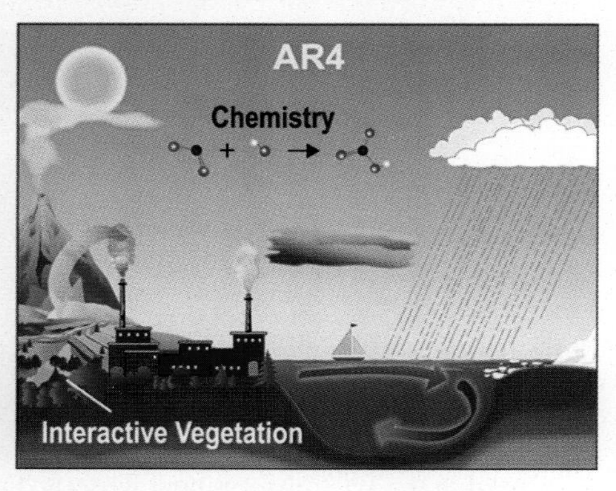

Figure 10.1 The complexity of climate models has increased over the last few decades. The additional physics incorporated in the models are shown pictorially by the different features of the modeled world. This is Figure 1.2 in Le Treut *et al.* (2007).

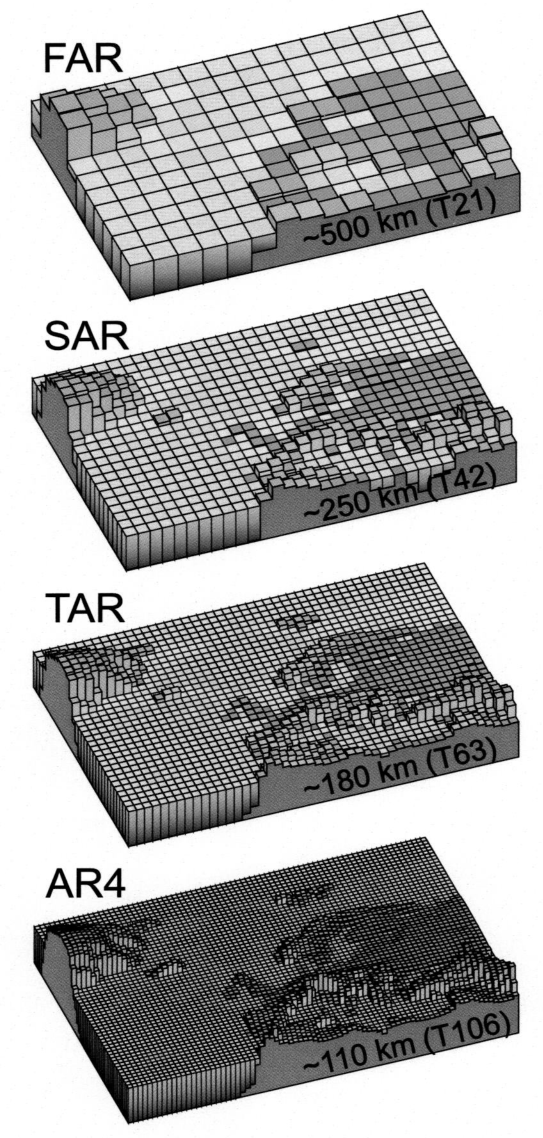

Figure 10.2 Geographic resolution characteristic of the generations of climate models used in the IPCC Assessment Reports: FAR (IPCC, 1990), SAR (IPCC, 1996), TAR (IPCC, 2001), and AR4 (IPCC, 2007). The figure shows how successive generations of these global models increasingly resolve northern Europe. These illustrations are representative of the most detailed horizontal resolution used for short-term climate simulations. The century-long simulations cited in IPCC Assessment Reports after the FAR were typically run with the previous generation's resolution. Vertical resolution in both atmosphere and ocean models is not shown, but it has increased comparably with the horizontal resolution, beginning typically with a single-layer slab ocean and ten atmospheric layers in the FAR and progressing to about thirty levels in both atmosphere and ocean. This is Figure 1.4 in Le Treut *et al.* (2007).

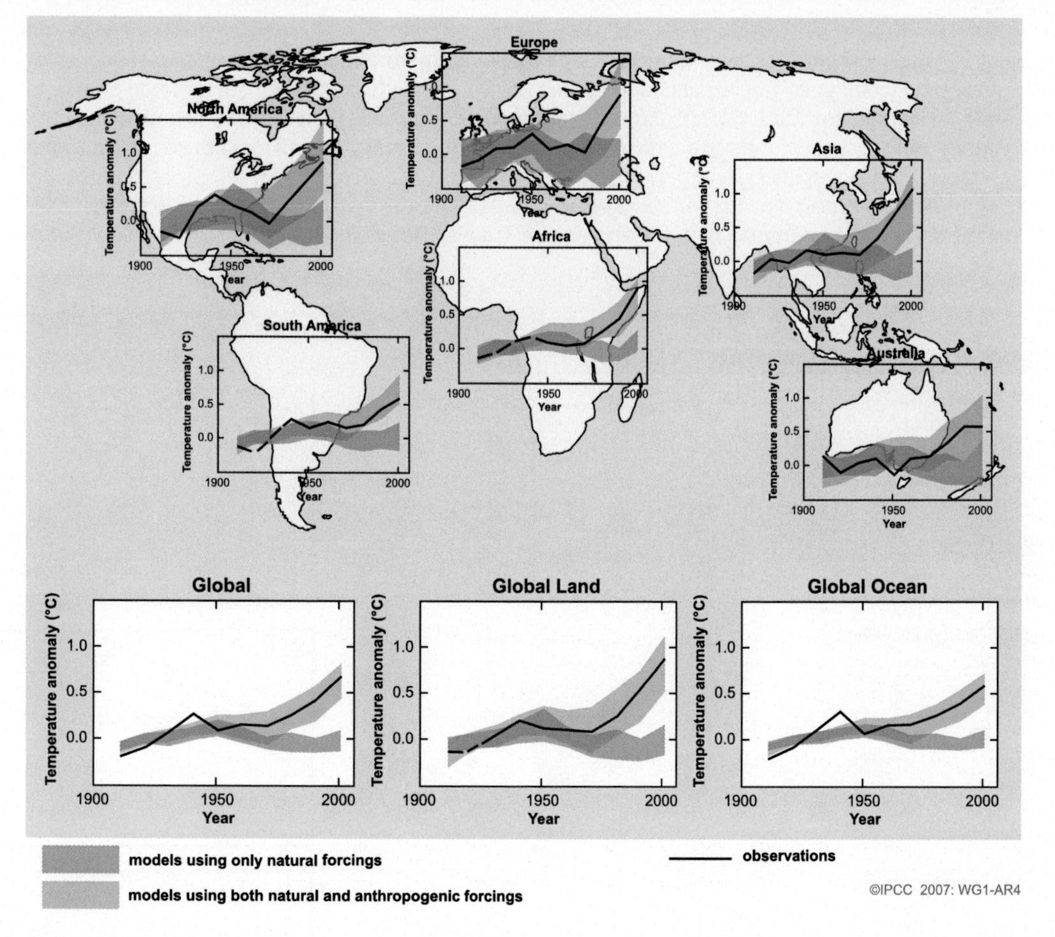

Figure 10.3 Temperature changes relative to the corresponding average for 1901–1950 (°C) from decade to decade from 1906 to 2005 over the Earth's continents, as well as the entire globe, global land area, and the global ocean (lower graphs). The black line indicates observed temperature change, while the shaded bands show the combined range covered by 90% of recent model simulations. The upper shaded band indicates simulations that include natural and human factors, while the lower shaded band indicates simulations that include only natural factors. Dashed black lines indicate decades and continental regions for which there are substantially fewer observations. This is FAQ 9.2, Figure 1 in Hegerl *et al.* (2007).

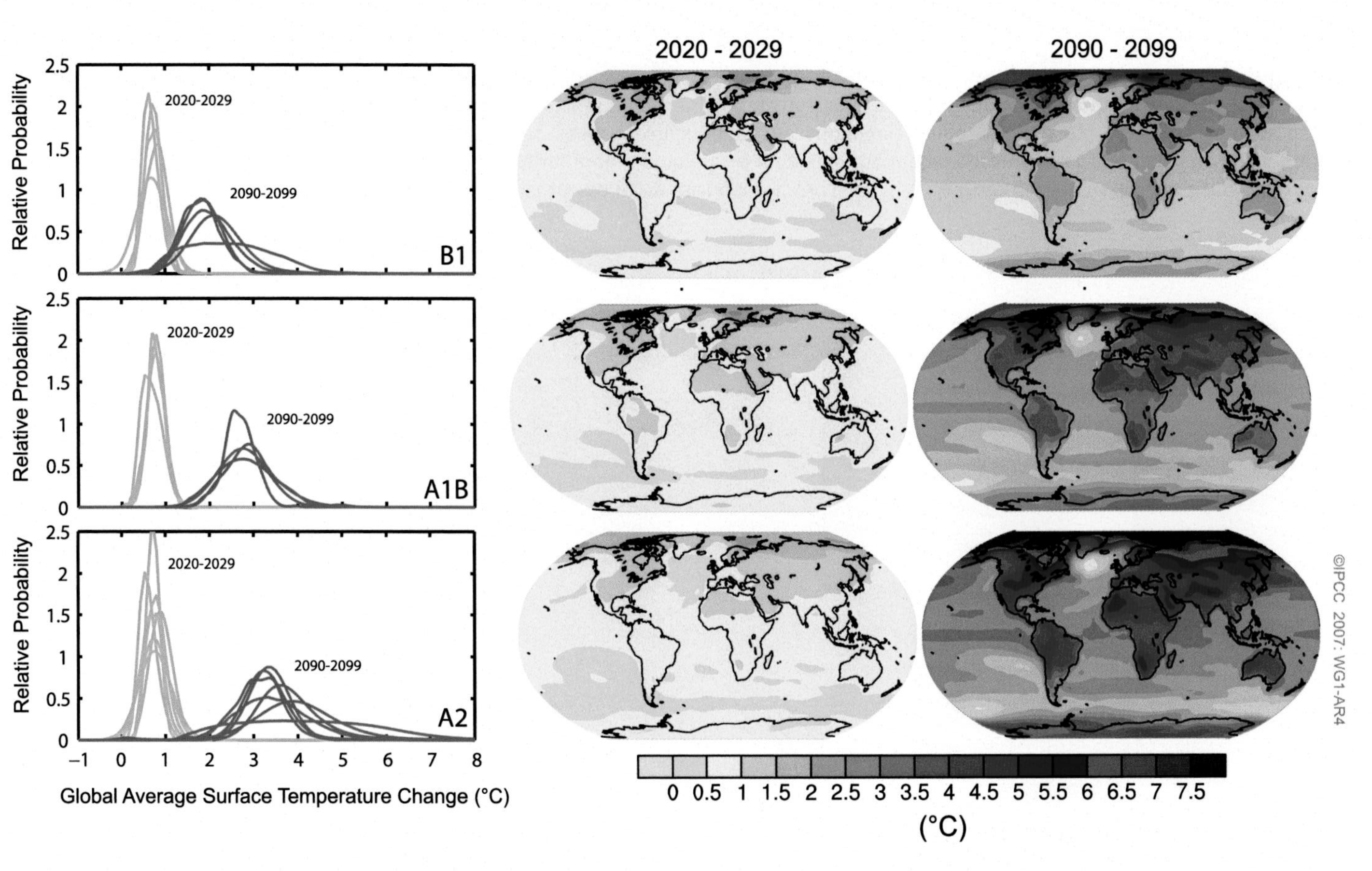

Figure 10.4 Projected surface temperature changes for the early and late twenty-first century relative to the period 1980–1999. The central and right panels show the AOGCM multi-model average projections for the B1 (top), A1B (middle) and A2 (bottom) scenarios averaged over the decades 2020–2029 (centre) and 2090–2099 (right). The left panels show corresponding uncertainties as the relative probabilities of estimated global average warming from several different AOGCM and Earth System Model of Intermediate Complexity studies for the same periods. Some studies present results only for a subset of the scenarios, or for various model versions. Therefore the difference in the number of curves shown in the left-hand panels is due only to differences in the availability of results. This is adapted from Figures 10.8 and 10.28 of Meehl *et al*. (2007) and is Figure TS.28 from the Technical Summary of IPCC (2007).

Analyses 1979-1989. WMO/TD No. 326. World Meteorological Organization, Geneva, Switzerland, 217 pp.
Kalnay, E., Kanamitsu, M., Kistler, R. *et al.* (1996). The NCEP/NCAR 40-year reanalysis project. *Bulletin of the American Meteorological Society*, 77, 437–471.
Kasahara, A. and Washington, W. M. (1971). General circulation experiments with a six-layer NCAR model, including orography, cloudiness and surface temperature calculations. *Journal of the Atmospheric Sciences*, 28, 657–701.
Klein, S. A., Soden, B. J., and Lau, N.-C. (1999). Remote sea surface temperature variations during ENSO: Evidence for a tropical atmospheric bridge. *Journal of Climate*, 12, 917–932.
Kutzbach, J. E. (1967). Empirical eigenvectors of sea-level pressure, surface temperature and precipitation complexes over North America. *Journal of Applied Meteorology*, 6, 791–802.
Lau, N.-C. (1981). A diagnostic study of recurrent meteorological anomalies appearing in a 15-year simulation with a GFDL general circulation model. *Monthly Weather Review*, 109, 2287–2311.
Lau, N.-C. (1984). *Circulation Statistics Based on FGGE Level III-B Analyses Produced by GFDL*. NOAA Data Report ERL GFDL-5. Geophysical Fluid Dynamics Laboratory, Princeton, NJ, 427 pp.
Lau, N.-C. and Nath, M. J. (1987). Frequency dependence of the structure and temporal development of wintertime tropospheric fluctuations – Comparison of a GCM simulation with observations. *Monthly Weather Review*, 115, 251–271.
Lau, N.-C. and Nath, M. J. (1994). A modeling study of the relative roles of tropical and extratropical SST anomalies in the variability of the global atmosphere-ocean system. *Journal of Climate*, 7, 1184–1207.
Lau, N.-C. and Nath, M. J. (2001). Impact of ENSO on SST variability in the North Pacific and North Atlantic: Seasonal dependence and role of extratropical sea-air coupling. *Journal of Climate*, 14, 2846–2866.
Lau, N.-C. and Oort, A. H. (1981). A comparative study of observed Northern Hemisphere circulation statistics based on GFDL and NMC analyses. Part I: The time-mean fields. *Monthly Weather Review*, 109, 1380–1403.
Lau, N.-C. and Oort, A. H. (1982). A comparative study of observed Northern Hemisphere circulation statistics based on GFDL and NMC analyses. Part II: Transient eddy statistics and the energy cycle. *Monthly Weather Review*, 110, 889–906.
Lau, N.-C. and Ploshay, J. J. (2009). Simulation of synoptic- and subsynoptic-scale phenomena associated with the East Asian summer monsoon using a high-resolution GCM. *Monthly Weather Review*, 137, 137–160.
Lau, N.-C., Leetmaa, A., and Nath, M. J. (2008). Interactions between the responses of North American climate to El Niño–La Niña and to the secular warming trend in the Indian–Western Pacific Oceans. *Journal of Climate*, 21, 476–494.
Lau, N.-C. White, G. H., and Jenne, R. L. (1981). *Circulation Statistics for the Extratropical Northern Hemisphere Based on NMC Analyses*. NCAR Tech. Note 171+STR. National Center for Atmospheric Research, Boulder, CO, 138 pp.
Lorenz, E. N. (1955). Available potential energy and the maintenance of the general circulation. *Tellus*, 7, 157–167.
Lorenz, E. N. (1967). *The Nature and Theory of the General Circulation of the Atmosphere*. WMO Publication No. 218, T.P. 115. World Meteorological Organization, Geneva, Switzerland, 161 pp.

Luksch, U. and von Storch, H. (1992). Modeling the low-frequency sea surface temperature variability in the North Pacific. *Journal of Climate*, 5, 893–906.

Manabe, S. and Hahn, D. G. (1981). Simulation of atmospheric variability. *Monthly Weather Review*, 109, 2260–2286.

Manabe, S. and Smagorinsky, J. (1967). Simulated climatology of a general circulation model with a hydrologic cycle. II. Analysis of the tropical atmosphere. *Monthly Weather Review*, 95, 155–169.

Manabe, S., Hahn, D. G., and Holloway, J. L. (1974). The seasonal variation of the tropical circulation as simulated by a global model of the atmosphere. *Journal of the Atmospheric Sciences*, 31, 43–83.

Manabe, S., Hahn, D. G., Holloway, J. L., and Stone, H. M. (1970) Simulated climatology of a general circulation model with a hydrologic cycle. III. Effects of increased horizontal computational resolution. *Monthly Weather Review*, 98, 175–212.

Manabe, S., Knutson, T. R., Stouffer, R. J., and Delworth, T. L. (2001). Exploring natural and anthropogenic variation of climate. *Quarterly Journal of the Royal Meteorological Society*, 127, 1–24.

Mintz, Y. (1965). Very long term global integration of the primitive equation of atmospheric motion. *Proc. WMO-IUGG Symp. Research and Development Aspects of Long Range Forecasting*. WMO Tech. Note 66, 141–161.

Mizuta, R., Oouchi, K., Yoshimura, H. *et al.* (2006). 20-km-Mesh global climate simulations using JMA-GSM model – Mean climate states. *Journal of the Meteorological Society of Japan*, 84, 165–185.

Newell, R. E., Kidson, J. W., Vincent, D. G., and Boer, G. J. (1972 and 1974). *The General Circulation of the Tropical Atmosphere, Vols. 1 and 2*, MIT, Cambridge, MA, 258 pp and 371 pp.

O'Conner, J. F. (1961). Mean circulation patterns based on 12 years of recent Northern Hemispheric data. *Monthly Weather Review*, 89, 211–227.

Ohfuchi, W., Nakamura, H., Yoshioka, M. K. *et al.* (2004) 10-km mesh meso-scale resolving simulations of the global atmosphere on the Earth Simulation-preliminary outcomes of AFES (AGCM for the Earth Simulator). *Journal of the Earth Simulator*, 1, 8–34.

Oort, A. H. (1964). On estimates of the atmospheric energy cycle. *Monthly Weather Review*, 92, 483–493.

Oort, A. H. (1983). *Global Atmospheric Circulation Statistics, 1958–1973*. NOAA Professional Paper No. 14. U.S. Government Printing Office, Washington DC, 180 pp.

Oort, A. H. and Rasmusson, E. M. (1971). *Atmospheric Circulation Statistics*. NOAA Professional Paper No. 5. U.S. Government Printing Office, Washington DC, 323 pp.

Palmén, E. H. (1964). General circulation of the tropics. *Proc. Symp. Tropical Meteor.*, New Zealand Meteorological Service, Wellington, 3–30.

Reynolds, R. W. and Smith, T. M. (1994). Improved global sea surface temperature analyses using optimum interpolation. *Journal of Climate*, 7, 929–948.

Riehl, H. (1954). *Tropical Meteorology*. McGraw-Hill, New York, 392 pp.

Roeckner, E., Brokopf, R., Esch, M. *et al.* (2006). Sensitivity of simulated climate to horizontal and vertical resolution in the ECHAM5 atmosphere model. *Journal of Climate*, 19, 3771–3791.

Schubert, S., Park, C.-K., Higgins, W., Moorthi, S., and Suarez, M. (1990a). *An Atlas of ECMWF Analyses (1980–87). Part I – First Moment Quantities*. NASA Tech. Memo. 100747. NASA Goddard Space Flight Center, Greenbelt, MD, 258 pp.

Schubert, S., Higgins, W., Park, C.-K., Moorthi, S., and Suarez, M. (1990b). *An Atlas of ECMWF Analyses (1980–87). Part II – Second Moment Quantities*. NASA Tech. Memo. 100762. NASA Goddard Space Flight Center, Greenbelt, MD, 262 pp.

Simmons, A. J. and Hoskins, B. J. (1978). The life cycles of some nonlinear baroclinic waves. *Journal of the Atmospheric Sciences*, 35, 414–432.

Smagorinsky, J. (1963). General circulation experiments with the primitive equations. I. The basic experiment. *Monthly Weather Review*, 91, 99–164.

Smagorinsky, J., Manabe, S., and Holloway, J. L. (1965). Numerical results from a nine-level general circulation model of the atmosphere. *Monthly Weather Review*, 93, 727–768.

Smith, T. M., Reynolds, R. W., Livezey, R. E., and Stokes, D. C. (1996). Reconstruction of historical sea surface temperatures using empirical orthogonal functions. *Journal of Climate*, 9, 1403–1420.

Starr, V. P., Peixóto, J. P., and Sims, J. E. (1970). A method for the study of the zonal kinetic energy balance in the atmosphere. *Pure and Applied Geophysics*, 80, 346–358.

Trenberth, K. E. (1992). *Global Analyses from ECMWF and Atlas of 1000 to 10 mb Circulation Statistics*. NCAR Tech. Note TN-373+STR. National Center for Atmospheric Research, Boulder, CO, 191 pp.

Uppala, S. M., Kallberg, P. W., Simmons, A. J. *et al.* (2005). The ERA-40 re-analysis. *Quarterly Journal of the Royal Meteorological Society*, 131, 2961–3012.

van Loon, H. (1961). Charts of average 500 mb absolute topography and sea-level pressure in the Southern Hemisphere in January, April, July and October. *Notos*, 10, 105–112.

van Loon, H., Taljaard, J. J., Jenne, R. L., and Crutcher, H. L. (1971). *Climate of the Upper Air: Part 1. Southern Hemisphere. Zonal and Meridional Geostrophic Winds, Vol. 2*. NCAR TN/STR-57, NAVAIR 50–1C-56. National Center for Atmospheric Research, Boulder, CO, 40 pp.

Vincent, D. G. (1969). Seasonal changes in the global atmospheric energy balance and results for restricted regions. Ph.D. thesis, Dept. of Meteorology, Massachusetts Institute of Technology, 174 pp.

Wallace, J. M. and Gutzler, D. S. (1981). Teleconnections in the geopotential height field during the Northern Hemisphere winter. *Monthly Weather Review*, 109, 784–812.

Wallace, J. M. and Lau, N.-C. (1985). On the role of barotropic energy conversions in the general circulation. *Advances in Geophysics*, 28A, 33–74.

Wallace, J. M., Lim, G.-H., and Blackmon, M. L. (1988). Relationship between cyclone tracks, anticyclone tracks and baroclinic waveguides. *Journal of the Atmospheric Sciences*, 45, 439–462.

Zhao, M., Held, I. M., Lin, S. J., and Vecchi, G. A. (2009). Simulations of global hurricane climatology, interannual variability, and response to global warming using a 50 km resolution GCM. *Journal of Climate*, 22, 6653–6678.

7

Coupling atmospheric general circulation to oceans

KIRK BRYAN

7.1 Introduction

One can make a reasonable case that the development of comprehensive climate system models is one of the great scientific and technical achievements of the "computer age" (Ruttiman, 2006). Although the original concept of a global climate model must be credited to Richardson (1922), the origins of climate models as we know them today go back to the decade after the Second World War. The early success of numerical weather prediction naturally suggested numerical models of the ocean circulation. A continual concern in the early days was the relative paucity of oceanographic data to verify ocean models. Another formidable difficulty in early ocean modeling was the lack of computer power to resolve the relatively small synoptic scales of the ocean, such as the eddies in the Gulf Stream or the Antarctic Circumpolar Current, as well as very small scales in the time-mean flow itself. These two stumbling blocks, lack of data and limitations on computer resources, are now being addressed. On the computing side, great advances in technology have allowed ocean models of higher and higher resolution. On the observational side, a revolution in ocean measurements has taken place, thanks to WOCE (World Ocean Circulation Experiment) and follow up programs such as CLIVAR (Climate Variability and Predictability Programme). Ocean models in turn led to coupling atmospheric and ocean models together. The potential of coupling ocean models to atmospheric models was first demonstrated in prototype numerical experiments carried out in the 1960s and 1970s (Manabe and Bryan, 1969). The goal of this chapter is to briefly describe ocean models, some of the technical challenges to coupling, and briefly survey the highlights of scientific goals achieved. The emphasis will be on the ocean component of the coupled models, since atmospheric models are the main topic of other chapters in this volume.

A complete review of all the many applications of coupled models is naturally beyond the scope of this chapter. Our goal is simply to provide some background and a sampling of the real progress that is being made to use coupled models to understand climate as an integrated system, particularly those areas that involve large-scale air–sea interaction.

7.2 Ocean models

The atmosphere and the ocean have much in common. Both are governed by the same laws of geophysical fluid dynamics and have a range of time and space scales of motion from planetary waves to small-scale turbulence. The Antarctic Circumpolar Current can be considered the analog of the jet stream in the atmosphere. The mesoscale eddies, which are associated with major mid-latitude currents in the ocean, are physical analogs of atmospheric synoptic disturbances, but are much smaller in scale. There are very significant physical differences between the ocean and atmosphere, however, which make it impractical to apply exactly the same numerical methods to the different media. The most significant differences are that density contrasts within the ocean are small relative to the density itself, and that the ocean is opaque to longwave radiation and nearly opaque to shortwave radiation.

One of the greatest challenges of ocean modeling is the great range of scales that are important for heat transport and tracer transport in the ocean. To get some insight into this, consider the Rossby radius of deformation, L_d, the length at which rotational effects become as important as buoyancy effects (Vallis, 2006):

$$L_d = \frac{NH}{f_o} \tag{7.1}$$

where

$$N^2 = \frac{g\delta\rho}{\rho_o H}.$$

N is the Brunt–Vaisalla frequency, g is the acceleration of gravity, $\delta\rho/\rho_o$ is the fractional change in density over the depth H, and f_o is the Coriolis parameter. The Rossby radius of deformation determines the scale of synoptic disturbances in the atmosphere and mesoscale eddies in the ocean. Equation 7.1 shows that the Rossby radius is proportional to N, but inversely proportional to rotation. Since N^2 is so much smaller in the ocean than in the atmosphere, synoptic disturbances in the ocean are at least an order of magnitude shorter. The largest synoptic scales in the ocean will be close to the equator where the vertical stratification is greatest, and the Coriolis parameter is smallest. The smallest scales will be found in the polar oceans where the opposite is true. The range in the ocean's Rossby radius is large compared to the range of the Rossby radius in the atmosphere. To provide a realistic simulation an ocean model should resolve the span between the planetary scale, 10^4 km, and the ocean Rossby radius scale at higher latitudes, which is less than 10^2 km.

In contrast, the synoptic scale of the atmosphere is closer to 10^3 km. Thus the scale spread from planetary to synoptic is approximately two orders of magnitude for the ocean and approximately only one order of magnitude for the atmosphere. In the jargon of numerical analysis, the ocean represents a "stiff" system. It is only recently that the technology of computers allowed the construction of World Ocean general circulation models that had enough horizontal resolution to span this two orders of magnitude range. Ocean modeling until recently has not had the resources which have been available for the development of numerical weather forecasting. The critical differences in the physical properties of the atmosphere and ocean have also been a major reason that numerical weather forecasting is a relatively mature field, while ocean modeling is still in its infancy. Coupled models with enough resolution to resolve the global circulation and mesoscale eddies, which are the synoptic features of the ocean, are just beginning to appear in the literature.

The geostrophic "thermal wind" scale in the ocean is

$$V = \frac{g\delta\rho H}{Lf_o\rho_o} = \frac{(NH)^2}{Lf_o}. \tag{7.2}$$

Using the radius of deformation defined in Eq. 7.1,

$$V = f_o \frac{L_d^2}{L} \qquad (7.3)^1$$

Since the radius of deformation in the ocean is only 100 km at mid latitudes compared to a radius of deformation in the atmosphere of 1000 km, the ocean scale velocity given by Eq. 7.3 is almost two orders of magnitude smaller than the equivalent atmospheric "thermal wind" scale. This weak ocean "thermal wind" explains the relatively sluggish circulation of the ocean. With respect to large-scale heat transfer, the ocean's slower currents are compensated by a much greater heat capacity per unit volume. If this compensation did not occur, the atmosphere would completely dominate the poleward transport of heat in the Earth's fluid envelope. As it is, the poleward heat transport of the atmosphere and ocean are comparable at low latitudes. The atmospheric poleward heat transport only dominates at higher latitudes.

External gravity waves move at a speed proportional to $g^{1/2}$, but internal gravity waves move at a speed proportional to $(g\Delta\rho/\rho o)^{1/2}$. Thus there is a large spread between the speed of external and internal gravity waves. This is an important factor in the choice of a numerical method for ocean models. The limitation on the time

[1] The author is indebted to an anonymous reviewer for suggesting Eq. 7.3.

step for numerical integrations is the classical CFL (Courant–Friedrichs–Lewy, 1928) condition,

$$\Delta t < \frac{\Delta x}{V}, \tag{7.4}$$

where Δt is the timestep of the numerical integration, and Δx is the grid spacing. V in the denominator of Eq. 7.4 may be either the current speed or the wave speed. Thus, except for the fast-moving external planetary gravity waves, relatively slow internal waves and relatively slow currents make it possible to carry out numerical integrations of ocean models with a much longer timestep than that used in atmospheric models with the same lateral grid spacing. In the very early three-dimensional ocean models the fast-moving external gravity waves were simply filtered out (Bryan, 1969, 1997) by forcing the vertically integrated flow to be non-divergent. In ocean models that are currently being used at the major climate centers external gravity waves are treated implicitly (Dukowicz and Smith, 1994) or included with a "split time step" (Marchuk, 1974, see also Griffies, 2004 for details). Split timesteps are one way to deal efficiently with two time-scales. The fast-moving external waves are integrated explicitly with short timesteps, which are then linked at evenly spaced time intervals with the slower interior dynamics.

Another important difference between the numerics of atmospheric and ocean models involves the vertical coordinate system. Since the ocean is nearly opaque to radiation, the stratified interior of the ocean can be considered a region of nearly "ideal" flow in which water mass properties are conserved to a remarkable extent over long trajectories and for long periods of time. That is why the study of water masses in the ocean is so much more fruitful than the study of air masses in the atmosphere, where diabatic processes are important. The ocean interior is bounded by surface and bottom boundary layers. In the surface boundary layer turbulent diabatic processes are dominant and determine the coupling of the ocean to the atmosphere. Since the advent of ocean modeling there has been an ongoing debate on the optimal choice of vertical coordinate to represent the nearly "ideal" interior flow as well as the top and bottom boundary layers. What has emerged are three categories of vertical coordinates. The most commonly used vertical coordinate is pressure or a z-coordinate. The advantage of a pressure or a z-coordinate is that it can be adapted easily to simulate the surface boundary of the ocean and it can handle the complicated equation of state of sea water in a straightforward way. The disadvantage of a pressure or a z-vertical coordinate, is that it is inconvenient to simulate the "ideal" water mass conserving flows in the ocean's thermocline and bottom boundary layers. The natural trajectories of water parcels along density surfaces are difficult to represent accurately in a numerical model using horizontal or nearly horizontal pressure surfaces. This issue is important in ocean models, because interior non-adiabatic mixing is weak and direct radiative heating is only important near the surface. An exception is the case of very clean waters with low biologic activity (Anderson *et al.*, 2007). A pressure coordinate system has the advantage over a z-coordinate in that it conserves mass rather than volume. This is an important advantage when considering sea-level changes.

Another approach for ocean modeling is to use a hybrid isopycnic–Eulerian vertical coordinate (Bleck, 2002), where the coordinates in the main thermocline are aligned along density surfaces and move up and down as the density field changes. This coordinate system uses a Eulerian coordinate system near the surface, and a quasi-isopycnal, semi-Lagrangian coordinate in the interior. The Eulerian coordinate is ideal to represent the flat upper boundary of the ocean and the isopycnal coordinate system has obvious advantages to represent nearly ideal, adiabatic flow in the interior. Unfortunately the equation of state of the ocean is very complicated. It is not possible to find a global coordinate which will exactly correspond to local density surfaces. Compromises exist, which provide vertical coordinates that are reasonably parallel to local "neutral" surfaces. In this case advection and lateral mixing tend to be density conserving and provide a much better simulation of the physics of the real ocean as far as can be determined by observational data. The disadvantage of this coordinate is in dealing with polar oceans, in which the density over most of the water column is nearly uniform. The same choice of density layers, which may be more than adequate at lower latitudes, will provide minimal resolution of the water column in the polar seas.

The "sigma" coordinate system, which is applied widely in atmospheric models, has been used for ocean circulation models as well. Basically the depth of the ocean is remapped so that it is uniform in a new "sigma" coordinate. Thus there is a uniform number of coordinate levels at each point between the surface and the ocean bottom. The vertical spacing between coordinate surfaces is proportional to the total depth. Most of the ocean is 3 to 5 km in depth, but there are also extended areas of shelf that are only a few hundred meters deep, implying a global frequency distribution of depths with two maxima. The shelf area and the deep ocean are separated by narrow, very steep slopes. Numerical formulations of the equations of motion require the accurate calculation of horizontal pressure gradients in these shelf regions, which are often the locus of strong ocean currents. Approximating horizontal pressure gradients is very difficult along the steeply sloping coordinates of a sigma coordinate. For this reason most ocean circulation models, particularly those used in coupled climate models, use pressure coordinates (Griffies, 2004) or isopycnal (Bleck, 2002) coordinate systems.

In an Eulerian coordinate system a significant dispersion error is due to the representation of vertical advection of tracers and momentum. This difficulty is often associated with the sharp vertical gradients at the base of the mixed layer. In recent years some success in atmospheric models has been achieved by using a floating semi-Lagrangian vertical coordinate. Unlike isopycnal coordinates, "finite volume" models use Eulerian coordinates as a reference coordinate, but allow coordinate surfaces to float up and down with the flow. Variables are interpolated in the vertical direction back to their reference positions at suitable intervals. This scheme is currently being used in hybrid ocean models, but could be applied to advantage in z- and p-coordinate models as well.

Most ocean models in climate applications use relatively simple, second-order numerics. Higher-order differencing schemes are in most cases not justified at this stage because the ocean fields and boundary conditions are not "smooth" due to the great range of geostrophic motions discussed earlier. The first ocean models could not be used for extended numerical integrations because they accumulated spurious energy through nonlinear terms. This was avoided using the energy conservation principles set out by Arakawa (1966). Griffies (2004) shows that the energy conservation idea can be generalized by deriving numerical differencing formulas from a variational principle.

Eddy-resolving ocean models of the World Ocean exist, but only in recent years has it become possible to integrate them with respect to time for more than a few decades. The practical requirements for a useful climate model require an ability to be numerically integrated repeatedly over many years in ensemble experiments. In the early ocean models the effect of eddies was represented simply as horizontal diffusion of heat, salt, and momentum. It was pointed out (Veronis, 1975) that this was a very poor representation of the ocean, in which lateral mixing in the main thermocline takes place almost entirely parallel to local density surfaces, so that density is conserved. There has been an effort to parameterize density-conserving mesoscale eddy mixing in climate models. For ocean models based on Eulerian coordinates, unresolved mesoscale eddies are usually represented by the Gent–McWilliams (1990) scheme. This parameterization accomplishes two things. First, heat and water mass properties are mixed along local density surfaces, and not along nearly horizontal coordinate surfaces. Second, the thickness between isopycnal surfaces is smoothed laterally as an approximation of the smoothing of potential vorticity by mesoscale eddies.

7.3 Coupled models compared to observations

The summaries prepared by the periodic reports of the IPCC provide an excellent opportunity to trace progress in the development of coupled models. In addition, model datasets have been compiled allowing in-depth analysis and comparison of different models. An example is a summary of SST average departures from observations of the control solutions submitted to IPCC-2 (1995), IPCC-3 (2001) and IPCC-4 (2007) shown in Figure 7.1. The analysis of IPCC models was made by Reichler and Kim (2008). In the years between 1995 and 2007 coupled models had increased in horizontal resolution, allowing a much better representation of atmospheric and ocean currents. Surprisingly, the overall SST average bias shown in IPCC-4 models is larger than that of earlier models. This is due to the fact that earlier models used a semi-empirical correction to the sea surface heat and water flux termed "flux correction" (Sausen *et al.*, 1988; Manabe *et al.*, 1991). This method added a fixed flux increment to the calculated flux which kept the model SST and surface salinity within the observed range. The flux increment was normalized to avoid global energy or water imbalances. Over time this procedure has been used less and less. In the IPCC-4 assessment only 5 of the 23 models submitted used "flux correction". The motive for dropping this procedure is that "flux correction" can mask fundamental problems of the coupled models, and as a partial linearization, which is fitted to the present climate, limits the model's application to future climates.

Concentrating on the lower panel of Figure 7.1, corresponding to IPCC-4, we see that the tropical and mid-latitude oceans are generally too cold. In particular the equatorial Pacific is 1.5 to 2.0 °C colder than observed. This is partly due to the fact that many models have easterly trades that are too strong (see Figure 7.2). Upwelling areas along the eastern boundaries of the tropical oceans, on the other hand, are too warm. This is a typical bias in coupled models and is associated with the lack of lateral resolution in both the atmospheric and ocean component of the models to fully define the narrow meridional winds and ocean currents in these

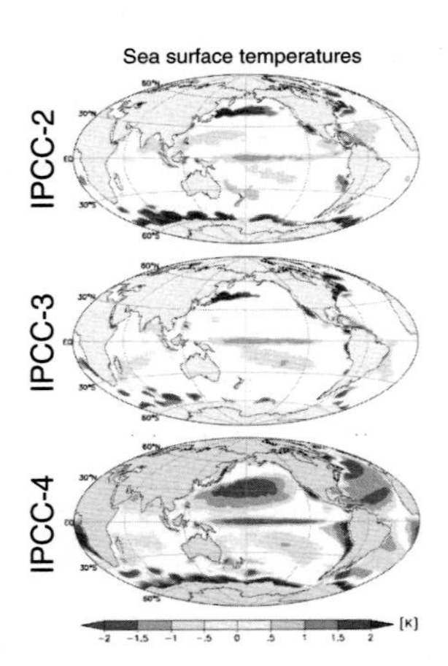

Figure 7.1 Biases in the annual mean climatological sea surface temperature averaged over all IPCC models. (From Reichler and Kim, 2008). Note that without flux adjustment in most models, the average error in SST is larger in IPCC-4. See also color plate.

coastal upwelling areas. Higher resolution should bring a big improvement in these climatically important coastal upwelling areas, where relatively low surface temperatures allow high heat uptake by the ocean. It should also be noted that many atmospheric GCMs have difficulty simulating marine stratocumulus, and the resulting positive bias in surface shortwave radiation also contributes to unrealistic ocean warmth in these areas.

The boundary between the subtropical and subarctic ocean gyres is characterized by a dipole anomaly in the IPCC-4 SST shown in Figure 7.1. The dipole consists of a cold anomaly towards the equator flanked by a poleward warm anomaly. The models on average appear to be too diffusive at higher latitudes and do not show the strong poleward SST gradients across the Gulf Stream, Kuroshio, and Antarctic Circumpolar Current that are actually observed. This does not mean that the lack of strong currents is caused directly by diffusion in the models, but is more likely to be due to more complex factors associated with inadequate lateral resolution. As in the case of the coastal upwelling, higher lateral model resolution should allow tighter, more realistic SST gradients along major ocean currents.

The IPCC 4th Assessment report did not attempt to diagnose individual biases of the models, but through the comparison of such a large number of models it is clear which errors were unique to one particular model, and which errors were prevalent in many models. The combined zonally averaged zonal wind stress over the ocean from all the climate models is shown in Figure 7.2. The zonally averaged wind stress

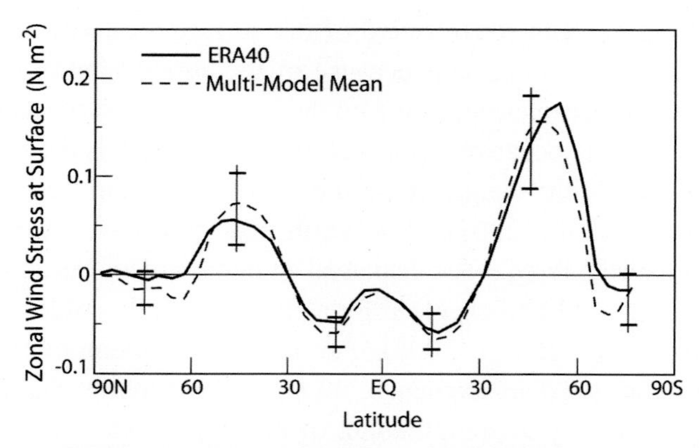

Figure 7.2 (Modified and based on Randall *et al.* 2007) Annual mean east–west component of atmospheric stress averaged over the oceans. The spread of different models in IPCC-4 model assessments shown by the vertical bars. The dashed black line is the average over all models and the solid black line is the wind stress inferred from the European Center operational atmospheric model (Trenberth and Carron, 2001). Note the average peak westerly wind stress in the models is shifted equatorward compared to the European Center in both hemispheres. The stress in the trade-wind regions is also slightly stronger than observed in both hemispheres.

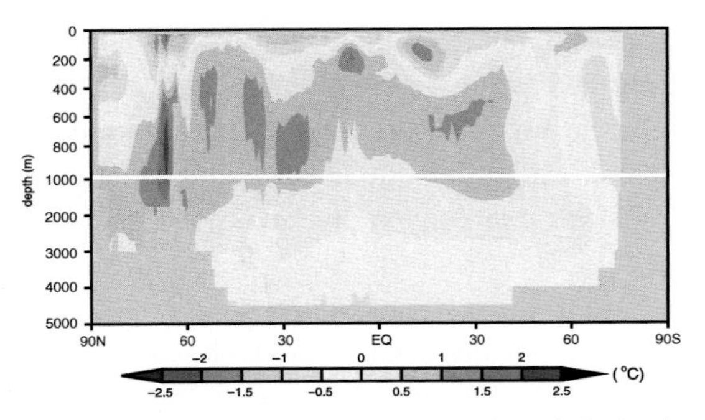

Figure 7.3 (Modified and based on Randall *et al.* 2007) Background shading represents the multi-model zonally averaged temperature bias of the climate models submitted to the IPCC-4 assessment. The observations are from the 2005 World Ocean Atlas, Vol. 1 (Levitus, 2005). See also color plate.

over the ocean in the coupled models was compared with the output of the European Center operational atmosphere-only model that was constrained by daily assimilation of observed data. Vertical bars show the spread of individual model results. General agreement is shown with ECMWF, but there are significant differences that may be important in determining the response of the ocean component of the model. Relative to the European Center analysis, the extratropical surface winds of the climate models in both hemispheres appear to be displaced equatorward. In the Southern Hemisphere the climate models indicate surface westerlies that are displaced almost ten degrees equatorward of the surface westerlies in the European Center analysis. This is important because of the strong control of surface wind stress on the path of ocean currents and on the water mass distribution. Any scoring of the simulation of the ocean components of coupled models based on water mass properties will be highly weighted by the Southern Ocean because of its huge volume (Schneider *et al.*, 2007). It is worth noting that the average bias in the IPCC-4 models, surface westerlies displaced equatorward and easterlies slightly too strong, is just what would be expected for a climate slightly colder than the present climate (Williams and Bryan, 2006). For angular momentum to be conserved globally an equatorward displacement of the westerlies without change in amplitude must be compensated by a strengthening of the tropical easterlies. The cold SST anomaly shown in Figure 7.1 for the IPCC-4 coupled models along the equator in the Pacific is consistent with too strong trades.

The zonally and time-averaged temperature difference from observations of all the IPCC-4 assessment coupled models is shown in the cross-section of Figure 7.3. In general, the models are too cold in the upper ocean and too warm in the main thermocline. The models appear to be too diffusive in the vertical in spite of the fact that the most recent versions have much more vertical resolution than the earlier models. The cold bias in the upper thermocline is consistent with the cold SST anomalies shown for IPCC-4 models in Figure 7.1. There is a great difference between individual ocean climate models, but it appears that the simulation of the thermocline is a challenge for future coupled models.

7.4 Paleoclimate

Geological evidence from the last glacial maximum (LGM) indicates that less than 21,000 years ago the Earth had a very different climate from that of today. A major goal of the IPCC effort is to foster the development of climate models that can project climates in the future, which (in the absence of drastic curbs in the use of fossil fuels) will also be very different from the climate of today. It has long been suggested that geological evidence of past climates should be used as a test of the robustness of coupled climate models. Since it is very hard to assess the relative performance of different models using a wide variety of boundary conditions, there has been an effort to foster cooperative experiments under an umbrella organization called PMIP (Paleoclimate Model Intercomparison Project, Otto-Bliesner *et al.*, 2007). In this project modeling groups are encouraged to run coupled models using a uniform geometry of the LGM World Ocean, a uniform geometry of the ice sheets, and uniform radiative forcing. The advantage in focusing on the LGM climate as opposed to investigating more remote geological periods is the much greater amount of data to constrain the coupled models. Geochemical evidence taken from bottom cores in the North Atlantic and Southern Ocean indicate that bottom waters were 1–2 °C colder in the LGM than at present. There is also evidence that deep water of southern origin penetrated much further into the North Atlantic than at present. Proxy data suggest that the present cell transporting North Atlantic Deep Water was shallower than now (Duplessy *et al.*, 1988), but there is ambiguity concerning its intensity. A recent advance has been evidence for the salinity structure of the deep ocean during the LGM (Adkins *et al.*, 2002). It has long been known that the oceans as a whole would be more salty in glacial times because of the amount of fresh water sequestered in the enormous ice caps over land. However, the new data indicate a very different distribution of salinity in the LGM deep sea as well. The LGM deep sea was extremely salty with the greatest salinity in the Southern Ocean. This new evidence is consistent with greatly augmented sea-ice formation in the Southern Ocean, which produced brine flowing into the deep ocean.

A very great technical obstacle to using coupled climate models to study the LGM is the long numerical integrations required to search for near-equilibrium regimes. The natural time-scale of deep waters of the World Ocean in the Central Pacific is of the order of thousands of years. Departures from equilibrium in the deep water

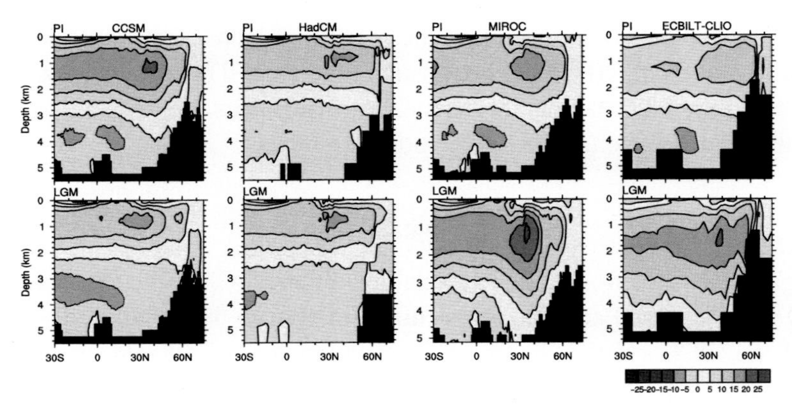

Figure 7.4 (From Otto-Bliesner *et al.*, 2007) Meridional overturning of the Atlantic Ocean sector of the PMIP coupled climate models, which are of lower resolution than the most recent IPCC climate models by approximately a factor of 2. Units are Sverdrups, defined as one million cubic meters/sec. Flow is clockwise around maxima. The upper panels represent a simulation of the present climate as control. The lower panels represent simulations of meridional overturning during the Last Glacial Maximum (LGM). CCSM (NCAR), HadCM (Hadley Centre), MIROC (Combined Japanese), ECBILT-CLIO (Netherlands). See also color plate.

imply an artificial source or sink of heat for the climate system. This means that the latest, high-resolution climate models are difficult to use. Shin *et al.* (2003) use a coarse resolution version of the NCAR climate model to simulate the LGM. To accelerate the approach to equilibrium in deep water, the authors use longer time-steps in the deep ocean relative to the upper ocean using a method introduced by Bryan (1984). It is not clear whether this multi-timestepping method will be useful in models of higher spatial resolution.

The PMIP2 model results (Figure 7.4) show no consensus on the LGM over-turning circulation. The CCSM and HadCM indicate a shallower meridional cell with about the same amplitude as the control solutions, while the MIROC and ECBILT-CLIO show deeper overturning cells with a slightly greater amplitude than the control. The scanty geochemical evidence (Duplessy *et al.*, 1988) suggests that the inflow of Antarctic Bottom water was stronger in the LGM and the outflow of North Atlantic Deep water more shallow. On this basis, the NCAR CCSM results shown in Figure 7.4 look the most reasonable. PMIP plans to redo these experiments as a new generation of climate models of higher resolution becomes available.

7.5 El Niño

No single application of coupled models has been so successful as modeling El Niño. There are two main reasons for this. First, El Niños exist near the equator, where warm tropical oceans allow a very high rate of evaporation, and thus a very large energy exchange between the ocean and atmosphere. This high energy exchange permits a very strong connection between sea surface temperature and the state of the lower atmosphere, providing a basis for a much stronger coupling than is possible at higher latitudes. Second, the time and space scales of El Niño are favorable to coupled modeling. The time-scale of El Niños is 3–5 years, so a fair number of events exist in the climate record. The Rossby radius in the upper ocean (see Eq. 7.1) is much larger in the tropics than at higher latitudes. Thus the main features of El Niño were much easier to resolve with the computers available a few decades ago than the ocean circulation at higher latitudes, where important features like the Gulf Stream and Kuroshio have scales of less than 100 kilometers. Initial studies of El Niño with three-dimensional coupled models were first carried out almost two decades ago (Philander *et al.*, 1989).

Progress in understanding coupled systems depends on "a hierarchy of models of increasing complexity" in the words of Jule Charney. While the three-dimensional coupled models formed the high end of the hierarchy, understanding of El Niño was greatly aided by simpler layered models (O'Brien and Hurlburt, 1974). Later, more complex coupled layered models were developed (Zebiac and Cane, 1987), which proved to be very efficient and extremely useful for exploring many aspects of El Niño, and provided the basis for the first attempts to forecast El Niño when data became available for the tropical Pacific. A basis for three-dimensional models of El Niño, like those used in actual forecasting today, were pioneering ocean-only calculations. Two of these were simulations of the seasonal changes in the Indian Ocean (Cox, 1970) and the Pacific Ocean (Cox, 1980). Since seasonal changes in the tropical oceans involve many of the elements of planetary wave dynamics of El Niño, these studies established the feasibility of three-dimensional ocean models, which could be coupled to atmospheric models to simulate the longer time-scales of El Niño.

In a seminal paper Bjerknes (1969) proposed the basic feedback which initiates El Niño (see McPhaden *et al.*, 2006 for a concise, but excellent review). This feedback depends on the close relationship of the east–west gradient of SST along the Pacific Equator and the strength of the trade winds. Bjerknes conjectured that the initiation of El Niño was caused by a relaxation of the trade winds, reducing the upwelling of colder deep water in the eastern equatorial Pacific. Such an event would reduce the east–west gradient of temperature along the equator in the Pacific and would further weaken the trades, providing a positive feedback. This simple and elegant conjecture has been amply verified by subsequent analysis of data and models (Chang *et al.*, 2006). The response of the oceans to a weakening of the trades is a relatively fast moving, equatorially trapped Kelvin wave, which is confined to within a few hundred kilometers of the equator. It was very exciting that this wave was predicted in theoretical studies, and simulated in the early models, before it was fully verified in observational data (Moore, 2006). As a response to the eastward-moving planetary wave impinging on the eastern boundary, slower moving, "Rossby-like" planetary waves slosh back to the west, gradually restoring the ocean to its original state. The description "Rossby-like" is used because these westward-propagating planetary waves are not observed to move at precisely the predicted speed of Rossby waves. Since the eastward-propagating equatorial Kelvin waves and westward-propagating Rossby waves in the upper ocean can be observed in both subsurface measurements and surface data obtained by satellites, a basis exists for prediction on time-scales of seasons, which is much longer than ordinarily feasible in numerical weather prediction.

In spite of two decades of active research and measurement, there are still many aspects of El Niño that are poorly understood. For example, why is the duration of El Niño 3–5 years, and what causes the variations in the length and amplitude of the cycle? One important concept is the "delayed oscillator theory" (Schopf and Saurez, 1988, 1990; Battisti, 1988). This idea, which was suggested by coupled models, relates the recovery phase of the El Niño to a change in the density structure of the central equatorial Pacific, which provides the seed for the birth of another El Niño. A competing concept is that the initiation of an El Niño is due to chaotic, extratropical forcing from higher latitudes (see Chang *et al.*, 2006 for a review).

The simulation of El Niño in the coupled climate models submitted to IPCC-4 (Randall *et al.*, 2007) show both impressive agreement and significant biases relative to observed data. In most cases the overall frequency of El Niño is more or less correct, but the amplitude is much less than observed (van Oldenborgh, *et al.*, 2005). A precondition for modeling El Niño is a good simulation of the background climatology. A systematic effort has been made to study all the coupled climate models submitted to the IPCC-4 to determine the quality of the simulations of tropical climatology in the Indian, Pacific, and Atlantic Oceans (Saji *et al.*, 2006 (Indian) de Szoeke and Xie, 2008 (Pacific) and Richter and Xie, 2008, (Atlantic). The simulations in each ocean have distinctive strengths and weaknesses that are common to many of the models. These studies are an excellent foundation for model improvements in the next IPCC, and focus attention on the tropical Atlantic and Indian Oceans, which are often somewhat overlooked because of the strong interest in El Niño in the Pacific.

A recent paper by Luo *et al.* (2005), analyzing the results of a relatively high-resolution coupled model, suggested that some of the common biases in rainfall patterns of the models might be partially eliminated by a more careful treatment of air–sea interaction physics. Luo *et al.* (2005) pointed out that many models neglect the speed of ocean currents in calculating momentum transfer and evaporation at the ocean surface. Since ocean currents can be quite large in the equatorial Pacific and are often in the same direction as the surface wind, neglect of this effect causes an overestimate of both evaporation and momentum transfer, as originally pointed out by Pacanowski (1987). Luo *et al.* (2005) found that the elimination of this error reduced some of the common biases in SST in the IPCC-4 models. However, it turns out that this correction for ocean currents has already been included in over half of the IPCC-4 climate models. It may have been that many of the models do not have high enough spatial resolution to fully benefit from a more accurate formulation of the air–sea coupling.

A recent study by Chang *et al.* (2007) used a combination of observations and coupled model results to throw new light on the initiation of El Niño. The authors find that 12 of 17 El Niño events in the historic record have been preceded by a marked weakening of the northeast trades in a broad band across the Pacific north of the equator. A conjectured mechanism is called WES (wind–evaporation–sea surface temperature, Xie and Philander, 1994). The WES mechanism works as follows; a dipole SST anomaly with slightly higher temperatures north of the equator and slightly lower temperatures south of the equator, produces a northward surface wind anomaly. Conservation of vorticity causes the wind anomaly to veer eastward north of the equator, causing a local reduction of the Northeast trades. This in turn decreases surface evaporation locally, allowing surface temperature to increase. The increase of SST results in increased convection, which draws more surface air inward, creating a positive feedback. The effectiveness of the WES mechanism has been tested in coupled models, but has not been fully checked out in observations. More tests need to be made of this promising concept.

The motivation for attempting to predict El Niño is based on the premise that the state of the upper ocean contains information that would allow a forecast on timescales well beyond the limit of predictability of an atmospheric model by itself. There is a history of semi-empirical attempts to forecast El Niño on the basis of sea surface temperature anomalies alone. Recently more systematic programs in El Niño forecasting are being undertaken, making full use of subsurface ocean data as well (ENACT, 2005). Enough data exist for the tropical Pacific so that hindcasts can be made for a reasonable number of El Niño events. To combine the different types of data available from satellite measurements of SST, sea surface height, and the much less abundant *in situ* subsurface data into a coherent assimilation system is an exceedingly challenging task. The ENACT report summarizes the combined effort of several European institutions, led by the ECMWF. For hindcasts carried out in this program, the target is the SST averaged over the Niño3 area of the Pacific (90 W to 150 W and 5 N to 5 S). This is the region of the maximum El Niño SST anomaly. Predictions are judged by the improvement over a simple persistence forecast of the Niño3 temperature. Typically, normalized errors of purely persistence forecasts remain under 0.5 for about two months. The forecast errors of the

ENACT models remain under 0.5 for about four months. With this possible exception, coupled models are not able to produce significantly superior seasonal forecasts of the El Niño than simpler empirical methods (McPhaden *et al.*, 2006), but it is encouraging that the community is organizing itself (Guilyardi *et al.*, 2009) to address this challenge in a systematic way.

In the western equatorial Pacific and Indian Oceans another phenomenon has been discovered which has a very important effect on local SST and rainfall on seasonal and longer time-scales. It is called the Indian Ocean Dipole and its fluctuations appear to be nearly independent of El Niño. Progress is being made to develop a long-range, forecasting ability for this region (Luo *et al.*, 2008, 2009) using coupled models.

7.6 Multi-decadal climate variability

The understanding of natural variability on longer time-scales than El Niño is an exciting, relatively new application of coupled models, which is made possible by the more powerful computers now available. The new computers allow realistic climate models to be integrated for periods of centuries and millennia. Knowledge of natural variability on multi-decadal time-scales is extremely important in developing the ability to discriminate anthropogenic climate change from the background. Potentially, coupled models could make as important contributions in this area, as they already have to the study of El Niño. Research in multi-decadal climate variability is greatly hampered by the fact that the instrumental record of climate is only a century and a half long, and this record has many gaps in space and time. Since the best climate record is for the Atlantic region, it is not surprising that the most progress has been made in the study of the AMO (Atlantic Multi-decadal Oscillation). The AMO is defined as the variation of SST averaged over the whole North Atlantic. Seminal papers on AMO are contributions by Bjerknes (1964), Folland *et al.* (1986a,b), and Schlesinger and Ramankutty (1994). It is remarkable that Professor Bjerknes made a very important initial contribution to the AMO as well as to El Niño. Bjerknes (1964) had only a very small amount of data to work with, but he made a key distinction between decadal and multi-decadal, North Atlantic climate variability . He also conjectured that the multi-decadal climate variability was connected with variations of poleward heat transport by the oceans. Folland made a fundamental contribution in organizing and analyzing the global, historic SST records. His statistical analysis showed a global pattern of SST variability with multi-decadal time-scale which was dominated by the North Atlantic. Schlesinger and Ramankutty (1994) pointed out that this pattern of SST variability exhibited a nearly oscillatory behavior with a period of 50–70 years and linked this pattern to changes in the meridional overturning circulation of the North Atlantic.

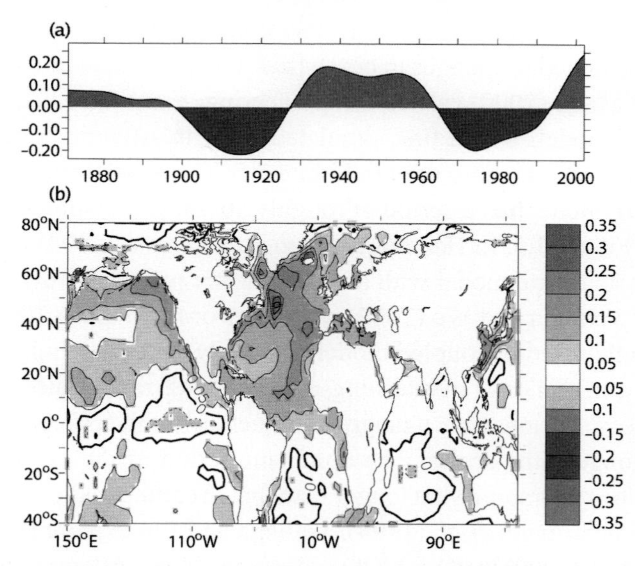

Figure 7.5 (From Sutton and Hodson, 2005. Reprinted with permission from AAAS) (a) Observed AMO temperature index, which is averaged SST over the North Atlantic from the equator to 60 N, and 75 W to 7.5 W. (b) Regression of SST to the AMO time series normalized by one unit of variance. The regression is calculated over the same time period shown in (a). See also color plate.

A paper by Sutton and Hodson (2005) provides a summary of some of the most important observational evidence for the AMO. Figure 7.5 (a) shows a version of the AMO index, the SST averaged between the equator and 60 N, and 75 W to 7.5 W. The index time series in Figure 7.5 (a) is low-pass filtered and detrended. The detrending removes a general warming during the entire instrumental record, which is believed to be caused by external anthropogenic warming and decreased volcanic cooling in recent decades. What remains in the record are two cold periods, separated by three warm episodes at the beginning, middle, and end of the record. The rise in AMO index between 1915 and 1935 is the feature that attracted Bjerknes' (1964) attention. Figure 7.5 (b) shows the global SST coherence pattern to the AMO, which was first shown in early versions of the Met Office Data by Folland *et al.* (1991). Most of the response is in the North Atlantic and North Pacific. A possible link of the AMO to long-term El Niño modulations is suggested by a weak negative response in the equatorial eastern Pacific, see Figure 7.5(b).

How have coupled models contributed to an understanding of the AMO? The circulation of the ocean is very difficult to measure directly, and we have little data on past changes in the ocean circulation. Models have provided some means to make a quantitative check on the inferences drawn from indirect evidence, such as changes in SST. Most important, coupled models (Knight *et al.*, 2005) have lent strong support to Bjerknes' (1964) original conjecture that multi-decadal climate variations were linked to changes in ocean heat transport. Delworth *et al.* (1993) and Delworth and Mann (2000) carried out pioneering, extended calculations with the GFDL climate model, which first simulated a North Atlantic climate oscillation, similar to the observed record seen in Figure 7.5. The quasi-periodic oscillations in the early GFDL model had a period of roughly 50–60 years, somewhat shorter than suggested by Figure 7.5. In Hamburg, Timmermann *et al.* (1998) found AMO-like oscillations in a climate model with an even shorter period of only 35 years.

A study by Vellinga and Wu (2004) was noteworthy in that a 1600-year integration of the Hadley Centre coupled model was analyzed to determine the variability of the Atlantic meridional overturning on very long time-scales. The long time series showed oscillations over a range of frequencies with appreciable amplitude in century and longer time-scales. The dominant period of the Hadley model overturning variations was somewhat longer than that in earlier studies of Delworth *et al.* (1993) and Timmermann *et al.* (1998). Vellinga and Wu explore an earlier suggestion by Latif *et al.* (2000) that the AMO cycle in the North Atlantic was initiated by a salinity anomaly, which is implanted by anomalous surface forcing just north of the equator in the North Atlantic. In the HadCM3 model of the Hadley Centre, salinity anomalies were then carried northward by the Atlantic "conveyor". At low latitudes, the salinity anomalies were associated with a negligible density anomaly, but in the colder waters of high latitudes the associated density anomaly became large enough to modulate the overturning. Vellinga and Wu suggested that the timing of the variations of the Atlantic meridional circulation in their model was determined by the speed of the conveyor belt. The faster the conveyor belt the shorter the transit time from the tropics to the subarctic region of water mass production. Recently the Vellinga and Wu mechanism was called into question by Hawkins and Sutton (2007) in a reanalysis of the same 1600-year Hadley Centre integration. Hawkins and Sutton noted that the poleward flow of the conveyor belt in the Hadley model in the Nordic Seas had a maximum below the surface. Relatively warm, salty water flowed northward below colder and fresher water. Hawkins and Sutton showed that as this northward flow intensified it triggered local convection. The convection in turn, increased the salinity at the surface. This created a salinity anomaly at the surface unrelated to lateral surface advection. It will be interesting to see if the Hawkins and Sutton mechanism is present in simulations of the AMO being carried out in other laboratories.

Most of the long-term simulations of climate models showed variations in North Atlantic meridional overturning and SST variations with a spectral peak, although the dominant frequency may be different. Can this behavior be verified from proxy data for longer periods than the relatively short instrumental record shown in Figure 7.5? Using data based on tree-rings from different locations around the

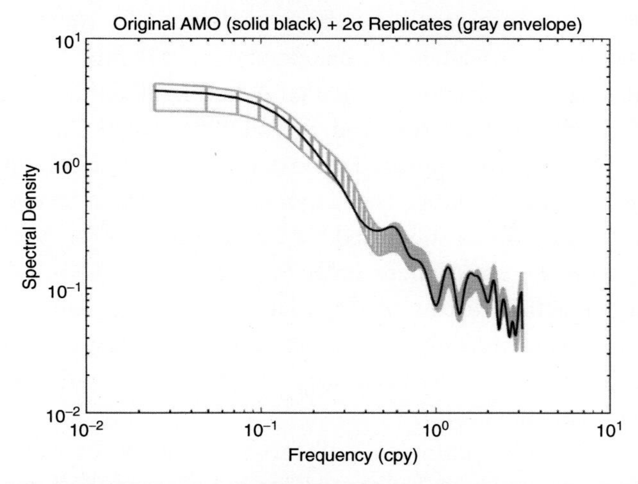

Figure 7.6 (From Enfield and Cid-Serrano, 2006). Black curve: Auto spectrum of the unsmoothed Gray *et al.* (2004) index. Light-shaded envelope: mean ±2 standard deviations of the spectral energy at each frequency for 50 synthetic versions of the Gray *et al.* index based on proxy data for the AMO.

North Atlantic basin, Gray *et al.* (2004) have extended the AMO time series over four centuries. Surprisingly, the proxy AMO time series did not show evidence for the 60–70 year cycles in the three centuries before the instrumental record. Enfield and Cid-Serrano (2006) have analyzed the Gray *et al.* (2004) time series and synthesized an analogue time series with similar statistics. The result is shown in Figure 7.6. There was no spectral peak at a particular frequency, but a continuum with a suggestion of a shoulder at a frequency of 1 cycle per century or less. The proxy data suggested that the designation "Atlantic Multi-Decadal Oscillation" could be a misleading result of the relatively short instrumental record.

Putting aside the question whether the AMO is really an oscillation, as suggested by coupled models, or a more amorphous, saturated red-noise phenomenon, as indicated in Figure 7.6, it is interesting to explore the modeling results in more detail. Griffies and Bryan (1997a, b) suggested that the AMO could be thought of as a resonant response to nearly random forcing by the atmosphere, without specifying what determined the resonant frequency. Delworth and Greatbatch (2000) carried out a series of carefully designed numerical experiments to investigate a similar concept using the GFDL R30 climate model. Their aim was to determine if the AMO-like variability in the model was the result of a truly coupled air–sea interaction, or a resonant response of the ocean to atmospheric forcing with very weak feedback. They concluded that for their model the latter seemed to be true. This result suggested that the AMO is a very different kind of air–sea interaction than El Niño. This view of Delworth and Greatbatch (2000) has been widely accepted in the community, but it has been challenged by more recent modeling results and data. The AMO coherence pattern shown in Figure 7.5 (b) shows a weak response in the eastern equatorial Pacific. A similar Pacific response has been demonstrated in models by Dong and Sutton (2007) and Zhang and Delworth (2007). The Pacific response appeared to be due to a climate teleconnection through the atmosphere from the Atlantic. The response is too rapid to be caused by a signal carried through the ocean as suggested by Cessi *et al.* (2004). Such an interbasin interaction through the atmosphere indicates that the AMO is a more complex, global air–sea interaction linking the high latitudes and the tropics, which is much more complicated than envisioned in the simpler, almost passive resonant response without large atmospheric feedback confined to the Atlantic as suggested by Delworth and Greatbatch (2000).

The AMO is much too complex a phenomenon to be explained by a simple theory, but some insight can be gained by a box model of the thermohaline circulation proposed by Stommel (1961), Rooth (1982), and simplified by Rahmstorf (1996). Overturning in the Rahmstorf box model is assumed to be a flow driven by the north–south density gradient between the Southern Ocean and the subarctic North Atlantic. The density gradient has two components, one which is due to the north–south temperature difference and another which is due to the north–south salinity difference. In the present climate these two components oppose each other. It is possible to combine the equations for the model in such a way as to derive a single quadratic equation for the north–south density gradient and the freshwater forcing that drives the salinity component. Figure 7.7 from Rahmstorf (1996) includes the results from the simplified box model and the Atlantic meridional overturning transport of a full three-dimensional ocean climate model. The abscissa is the fresh water flux added in both models to the upper branch of the conveyor belt. To balance this flux at the surface the circulation must have a net transport of fresh water to the south. The curves in Figure 7.7 correspond to hysteresis loops generated by the box model (dashed) and the full ocean model (solid), when the surface water flux to the surface branch of the conveyor belt is slowly increased and subsequently decreased. The solutions of the box model are indicated by a dotted line. The box model has a stable upper solution, and an unstable lower solution. The upper solution corresponds to the present climate in which the north–south density gradient is dominated by temperature. The lower unstable solution corresponds to the other case in which the north–south gradient is dominated by salinity. The box model has two constants which are tuned to fit the GCM, but nevertheless the correspondence in the transient solutions of the box model and the full numerical model supports the simple theoretical concept of an unstable Atlantic thermohaline circulation.

Bryan and Hansen (1993) and Cessi (1994) applied stochastic forcing to box models to simulate the white-noise forcing of the ocean by the atmosphere. The

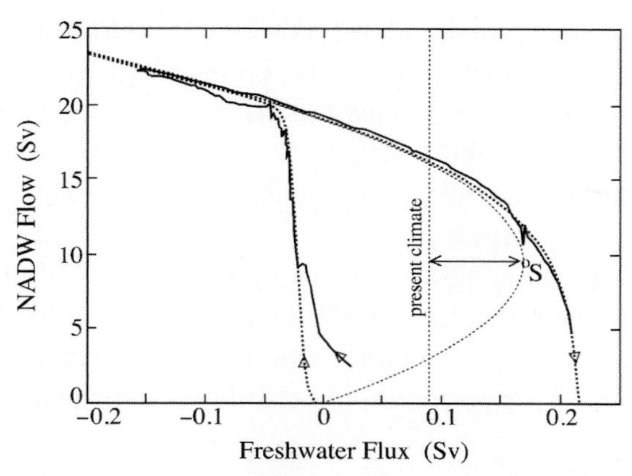

Figure 7.7 (from Rahmstorf, 1996. Reprinted with permission from Springer Science+Business Media) Hysteresis curves of a box model and a global ocean circulation model (GCM, solid lines) for increasing and subsequently decreasing freshwater flux, which modifies the north–south salinity gradient from the subpolar North Atlantic to the Southern Ocean. The ordinate is the strength of the Atlantic meridional overturning circulation in Sverdrups (10^6 tons/s). The abscissa is the amount of fresh water added to the upper branch of the meridional overturning circulation as it transits from the Southern Ocean to the subarctic North Atlantic. The dashed lines correspond to the box-model solution for the same rate of change of freshwater flux.

models are different in detail from the Rahmstorf (1996) model shown in Figure 7.7, but illustrate the same basic physics. The conveyor transport spectrum of the box models forced by white noise does not show a distinct peak like the coupled general circulation models of Delworth *et al.* (1993) or Knight *et al.* (2005), but a simpler damped red-noise spectrum, more like that of Enfield and Cid-Serrano (2006), shown in Figure 7.6. This different type of spectrum suggests that there is some dynamic element in the complete, coupled models, perhaps a more realistic ocean bathymetry (ocean topography), perhaps a more detailed treatment of advection, that allows a near resonance, which ranges between 50–200 years. The wide range of preferred frequencies in different climate models has not been explained.

In recent years the use of coupled ocean–atmosphere models to predict future climate change on both El Niño and multi-decadal time-scales has come into its own. A requirement for these predictions is knowledge of the initial state of the ocean as well as of the atmosphere. Although there is no way to determine in advance the exact data requirements for initializing a model of the World Ocean, it would appear that the limited data available in the past from satellites, drifting buoys, and ships of opportunity are far from sufficient to make useful forecasts on the decadal scale, and are only barely sufficient for El Niño time-scales. Data availability is improving rapidly, but the most interesting results to date from coupled models are studies of decadal "predictability" rather than actual predictions. The concept of model predictability is based on work by Lorenz (1969). The basic idea is to use models to determine how fast errors in the initial conditions will grow in time. Since it is an impossibility to have error-free initial conditions, the growth of errors even in a "perfect" model will be the ultimate limitation to practical prediction of either weather or the climate.

As pointed out by Boer (2001, 2004), classical predictability (Lorenz, 1969) measures the separation of solutions of a given system due to small differences in the initial conditions. There is another type of "potential predictability" which is simpler to determine. This potential predictability is defined as the ratio, r,

$$r = \frac{s_s^2}{s_s^2 + s_e^2}, \tag{7.5}$$

where s_s^2 is the variability of the frequency band to be predicted, and $s_s^2 + s_e^2$ is the total variability. If r is small, the signal-to-noise ratio is small and the potential predictability is negligible. Results based on the SST generated by an ensemble of coupled experiments of 21 models comprising more than 8000 years of simulation are shown in Figure 7.8. In the tropical oceans the scaled decadal signal is less than

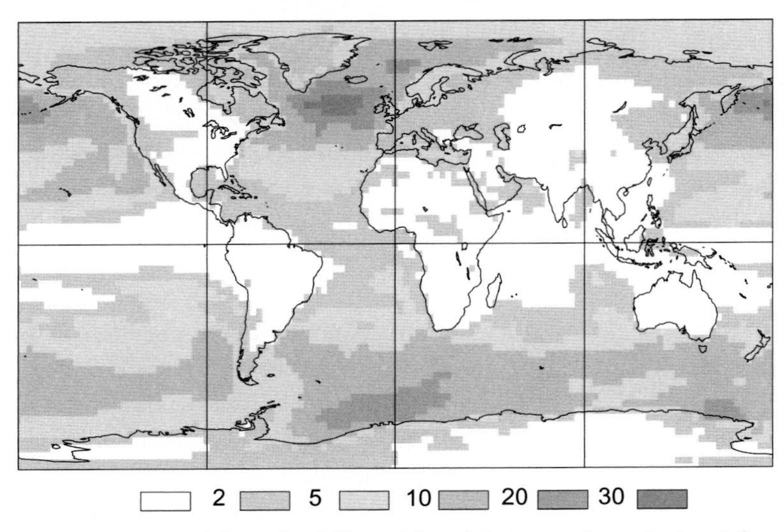

Figure 7.8 The potential predictability of decadal time-scales calculated from the output of several coupled models. Color shading indicates the ratio of variance of decadal means to the total variance, which is shown as a percentage (from Boer and Lambert, 2008). Note that the highest potential predictability is in the polar seas of both hemispheres. See also color plate.

10% of the large variance associated with higher-frequency signals such as El Niño. Potential predictability for periods of decades is large only in the North Atlantic and the Sea of Okhotsk in the Northern Hemisphere. In the Southern Ocean, there are large areas where r is greater than 25%, and in areas close to the Antarctic continent it is over 50%. It is clear from these model results that potential predictability is largest in high latitudes. Coupling at the ocean surface is the local ratio of an anomalous heat flux to an anomalous SST perturbation, which is strongly dependent on ambient temperature. A problem with the measure of predictability given by Eq. 7.5, however, is that it does not take into account the coupling between the atmosphere and the ocean. Very significant temperature perturbations at high latitudes may give rise to only small changes in heat flux at the ocean surface.

Modeling studies by Delworth *et al.* (1993) and Knight *et al.* (2005) have amply demonstrated the close connection between the Atlantic meridional overturning circulation and the surface SST signal of the AMO index. The Atlantic meridional overturning circulation (AMOC) is now being monitored at a single latitude (http://www.noc.soton.ac.uk/rapidmoc/), but our observational picture of it is still very incomplete. Knight *et al.* (2005) found that in the HadCM3 model of the Hadley Centre the AMOC has a 0.70 correlation to the model AMO SST index in a run of 1200 years. One of the very earliest coupled ocean–atmosphere global models to illustrate the mechanics of the AMO was the GFDL R15 climate model (Manabe *et al.*, 1991), which used flux adjustments at the ocean surface. Griffies and Bryan (1997a,b) used this global model to investigate "perfect model predictability" as defined by Lorenz (1969) in the North Atlantic sector of the World Ocean. Griffies and Bryan examined the growth of perturbations in the initial conditions. At the time this study was made, a 1000-year archive of the R15 model was available, which showed marked evidence of an AMO variability. Several ensemble runs were carried out. In all cases, the ocean initial conditions of the ensemble were identical, but the initial conditions for the atmosphere were taken from various parts of the record. From the first 200 years of the archive, empirical orthogonal functions were calculated for the North Atlantic region. The amplitudes of these empirical orthogonal functions were used to measure the ensemble predictions of the model. More recently Collins *et al.* (2006) have compared the predictability of several much higher-resolution coupled models. In Figure 7.9 the normalized mean-squared error of a measure of the AMOC is plotted as a function of the length of the forecast based on the older Griffies and Bryan results and more recent predictability estimates from Collins *et al.* (2006) for the HadCM3 and ECHAMP5/MPI-OM models. The variance of the ensemble predictions of meridional overturning is normalized by the variance of the model AMOC climatology. When the normalized variance is equal to unity, all forecast skill is lost. A dashed line is shown at 0.5. When the normalized spread of the ensemble is 0.5 of the climatological variance the practical

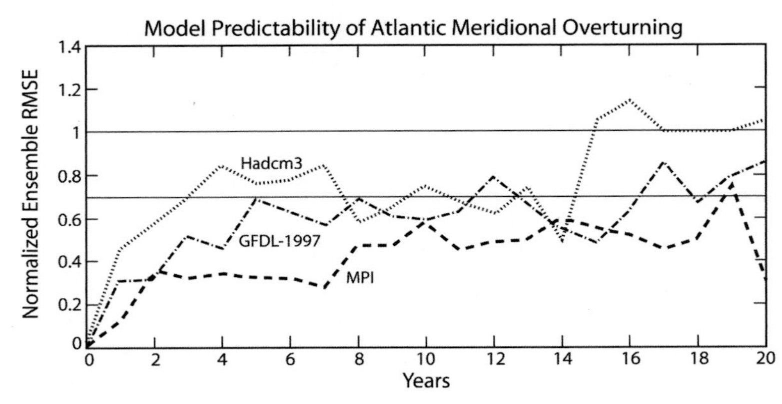

Figure 7.9 (modified from Collins, *et al.*, 2006) The ideal model predictability of Atlantic meridional overturning for three models: dashed-dot, the GFDL R15 model (Griffies and Bryan, 1997a); dotted, the HadCM3 model; and dashed, the ECHAMP5/MPI-OM coupled model. All the models used the same method for initialization of the ensemble.

limit of forecast usefulness has been reached (Griffies and Bryan, 1997a,b). The HadCM3 and ECHAMP5/MPI-OM are models with a nominal resolution of 1.25 degrees latitude and longitude in the ocean model component and approximately 4 by 2.5 degrees in longitude and latitude for the atmospheric component. The R15 model has much lower resolution and depends on flux adjustment at the air–sea boundary. All the models have pronounced AMOC variability with the largest amplitude in the ECHAMP5/MPI-OM model. The greatest predictability is shown for the ECHAMP5/MPI-OM model, with a useful predictability extending for nearly two decades, whereas the other models only indicate useful predictability of a decade or less.

What general conclusions can be drawn from the results in Figure 7.9? Two advanced climate models, the Hadley Centre model and the Hamburg ECHAMP5/MPI–OM model, give very different results for the growth of errors in initial conditions. The Hamburg model gives the most optimistic measure of the potential for prediction, while the Hadley Centre model gives a more pessimistic measure. The coarse-resolution version of the GFDL model lies somewhere in between. These differences seem to be a function of intrinsic variability in the model. The noise in the climate of the Hadley Centre model causes a more rapid spread of the ensemble forecasts so that any memory of initial conditions is lost more quickly than in the case of the ECHAMP5/MPI-OM model. One can only conclude that at this point different climate models have very different signal-to-noise ratios with respect to the meridional overturning circulation. The noise levels allow small differences in the initial conditions to grow at faster or slower rates. Until these model characteristics are better understood, "model predictability" can only be a limited guide to actual prediction on multi-decadal time-scales.

7.7 Future challenges and opportunities for coupled models

In this chapter we have only attempted to give a sampling of the very broad range of research and development using coupled ocean–atmosphere models. The use of coupled models to make projections of climate for various greenhouse gas scenarios is perhaps the most important application, but this topic is omitted, since it is adequately covered in IPCC reports. There are three factors that hold great promise for the future of coupled modeling. One is the nearly universal demand from society for more information on climate trends, and willingness to support coordinated international effort. Second is the growth of new technology to observe the ocean and atmosphere globally, which is filling in serious gaps in our previous data for the oceans. Third are the continual advances in large-scale computing that allow for high enough resolution to really resolve the important time and space scales of the coupled climate system. Thanks to the ARGO network, data below the ocean surface has increased by a factor of 10 (Roemmich *et al.*, 2009).

In the past, simulations of the ocean circulation were seriously inadequate in many coupled models, mainly due to inadequate numerical resolution. Modern supercomputers allow much higher resolution in ocean models and deficiencies in the physics become more obvious. It is exciting that some high-resolution ocean-only models are showing more, rather than less, energy than observations suggest (Smith *et al.*, 2000). This shows the need to revise the physics of ocean models to include effects that were often overlooked or ignored in coarser-resolution models. Frontier areas are a more detailed treatment of lateral mixing in the surface mixed layer of the ocean and the vertical transfer of momentum by internal gravity waves. The present generation of ocean modelers have a special opportunity to use the power of modern computers, and the greater availability of ocean data to bring ocean models to a level of maturity nearer to that of atmospheric models. This will mean coupling atmospheric models to ocean models which fully resolve mesoscale eddies in the ocean.

Up to this point the major effort in operational modeling has been focused on forecasting the El Niño. These forecasts on the time-scale of seasons have tremendous practical importance. Predictive skill is still only a little superior to that of empirical methods (McPhaden *et al.*, 2006), but has the potential to go far beyond that level. This field stands to benefit in a very direct way from additional ocean measurements in the tropical Indian and Atlantic Oceans and the application of more powerful computers which will allow the detailed treatment of all the delicate feedbacks important for the El Niño. The use of more detailed resolution should reduce the common biases in model simulation of the seasonal cycle in the tropical oceans. These biases must be reduced for the new additional ocean data to make a maximum positive impact on model prediction. Already the community is moving to models of higher resolution than those of IPCC-4 with positive results in eliminating persistent biases such as the "double ITCZ" (A. Rosati, personal communication).

Variations in the water masses of the deep ocean, which can now be monitored with modern technology, provide great potential for the new field of forecasting climate changes over time-scales longer than the El Niño. Recent studies by Smith *et al.* (2007), Keenlyside *et al.* (2008), and Pohlmann *et al.* (2009) are pioneering forecasts of AMO trends over the coming 10–20 years, taking into account both changes in radiative forcing and natural variability related to changes in ocean circulation. For real progress in forecasting global climate trends we must use a new generation of coupled models, in which the ocean component has a much higher resolution. The reason for this is that low-resolution ocean models have systematic circulation biases, which make it impossible to make use of the subsurface data provided by the new ocean datasets. Predictability studies show that anomalies in water mass structure provide the basis for decadal predictability, so it is essential to use models which can simulate ocean circulation climatology accurately. Such models are now under development. This new field of decadal forecasting has the possibility of greatly increasing the understanding of low-frequency climate variations and great practical benefits.

Acknowledgments

I thank Stephen Griffies and three anonymous reviewers for very helpful comments.

References

Adkins, J. F., McIntyre, K., and Schrag, D. P. (2002). The salinity, temperature, and oxygen-18 of the Glacial Deep Ocean. *Science*, 289, 1769–1773.
Anderson, W. G., Gnanadesikan, A., Hallberg, R., Dunne, J., and Samuels, B. (2007). Impact of ocean color on the maintenance of the Pacific Cold Tongue. *Geophysical Research Letters*, 34, L11609.
Arakawa, A. (1966). Computational design for long-term numerical integration of the equations of fluid motion: Two-dimensional incompressible flow. Part I. *Journal of Computational Physics*, 1, 119–143.
Battisti, D. S. (1988). Dynamics and thermodynamics of a warming event in a coupled tropical atmosphere–ocean model. *Journal of the Atmospheric Sciences*, 45, 2889–2919.
Bjerknes, J. (1964). Atlantic air-sea interaction. *Advances in Geophysics*. 10, 1–82.
Bjerknes, J. (1969). Atmospheric teleconnections from the equatorial Pacific. *Monthly Weather Review*, 97, 163–172.
Bleck, R. (2002). An oceanic general circulation model in hybrid isopycnic-cartesian coordinates. *Ocean Modeling*, 4, 55–88.
Boer, G. J. (2001). Decadal potential predictability in coupled models, *CLIVAR Exchanges*, 19, no. 3, International CLIVAR Project Office, Southampton, United Kingdom, 3 pp.

Boer, G. J. (2004). Long timescale potential predictability in an ensemble of coupled climate models. *Climate Dynamics*, 23, 29–44.

Boer, G. J. and Lambert, S. J. (2008). Multi-model decadal potential predictability of precipitation and temperature. *Geophysical Research Letters*, 35, L05706.

Bryan, K. (1969). A numerical method for the study of the circulation of the World Ocean. *Journal of Computational Physics*, 4, 347–376.

Bryan, K. (1984). Accelerating the convergence to equilibrium of ocean-climate models. *Journal of Physical Oceanography*, 14, 666–673.

Bryan, K. (1997). A numerical method for the study of the circulation of the world ocean. *Journal of Computational Physics*, (Special Issue, Commemoration of the 30th Anniversary), 135, 154–169.

Bryan, K. and Hansen, F. C. (1993). A toy model of North Atlantic climate variability on a decade to century time-scale. In *The Natural Variability of the Climate System on 10–100 Year Time Scales*. U.S. National Academy of Science, Washington D.C.

Cessi, P. (1994). A simple box model of stochastically forced thermohaline flow, *Journal of Physical Oceanography*, 24, 1911–1920.

Cessi, P., Bryan, K., and Zhang, R. (2004). Global seiching of thermocline waters between the Atlantic and the Indian-Pacific Ocean Basins. *Geophysical Research Letters*, 31, L04302.

Chang, P., Yamagata, T., Schopf, P. *et al.* (2006). Climate fluctuations of tropical coupled systems – The role of ocean dynamics. *Journal of Climate*, 19, 5122–5172.

Chang, P., Zhang, L., Saravanan, R. *et al.* (2007). Pacific meridional mode and El Niño-southern Oscillation. *Geophysical Research Letters*, 34, L16608.

Collins, M., Botzet, M., Carril, A. F. *et al.* (2006). Interannual to decadal climate predictability in the North Atlantic, A multimodel-ensemble study. *Journal of Climate*, 19, 1195–1203.

Courant, R., Friedrichs, K., and Lewy, H. (1928). Über die partiellen differenzengleichungen der mathematischen physik. *Mathematische Annalen*, 100, 32–74.

Cox, M. D. (1970). A mathematical model of the Indian Ocean. *Deep-Sea Research*, 17, 47–75.

Cox, M. D. (1980). Generation and propagation of 30-day waves in a numerical model of the Pacific. *Journal of Physical Oceanography*, 10, 1168–1186.

de Szoeke, S. P. and Xie, S. -P. (2008). The tropical eastern Pacific seasonal cycle, Assessment of errors and mechanisms in IPCC AR4 coupled ocean-atmosphere general circulation models. *Journal of Climate*, 21, 2573–2590.

Delworth, T. L. and Greatbatch, R. J. (2000). Multidecadal thermohaline circulation variability driven by atmospheric surface flux forcing. *Journal of Climate*, 13, 1481–1495.

Delworth, T. L. and Mann, M. E. (2000). Observed and simulated multidecadal variability in the Northern Hemisphere. *Climate Dynamics*, 16, 661–676.

Delworth, T. L., Manabe, S., and Stouffer, R. J. (1993). Interdecadal variations of the thermohaline circulation in a coupled ocean-atmosphere model. *Journal of Climate*, 6, 1993–2011.

Delworth, T. L., Stouffer, R. J., Dixon, K. W. *et al.* (2002). Review of simulations of climate variability and change with the GFDL R30 coupled climate model. *Climate Dynamics*, 19, 555–574.

Dong, B. -W. and Sutton, R. T. (2007). Enhancement of El Niño-Southern Oscillation (ENSO) variability by a weakened Atlantic thermohaline circulation in a coupled GCM. *Journal of Climate*, 20, 4920–4939.

Dukowicz, J. K. and Smith, R. D. (1994). Implicit free surface method for the Bryan-Cox-Semtner model. *Journal of Geophysical Research*, 99, 7991–8014.

Duplessy, J. C., Shackleton, N. J. , Fairbanks, R. G., Labeyrie, L., Oppo, D., and Kallel, N. (1988). Deepwater source variations during the last climatic cycle and their impact on the global deepwater circulation. *Paleoceanography*, 3, 343–360.

ENACT (2005). Enhanced ocean data assimilation and climate prediction, Final Report, project coordinator, Davey, M. (www.lodyc.jussieu.fr/ENACT/inex.html)

Enfield, D, B. and Cid-Serrano, L. (2006). Projecting the risk of future climate shifts, *International Journal of Climatology*, 26, 885–895.

Folland, C. K., Palmer, T. N., and Parker, D. E. (1986a). Sahel rainfall and worldwide sea temperatures. *Nature*, 320, 602–606.

Folland, C. K., Parker, D. E., Ward, M. N., and Colman, A. W. (1986b). Sahel rainfall, northern hemisphere circulation anomalies and worldwide sea temperature changes. Memorandum # 7a. *Long Range Forecasting and Climate*. Meteorological Office. Bracknell. UK., 49 pp.

Folland, C. K., Owen, J., Ward, M. N., and Colman, A. W. (1991). Prediction of seasonal rainfall in the Sahel region using empirical and dynamical methods. *Journal of Forecasting*, 10, 21–56.

Gent, P. R. and McWilliams, J. C. (1990). Isopycnal mixing in ocean circulation models. *Journal of Physical Oceanography*, 20, 150–155.

Gray, S. T., Graumlich, J. L., Betancourt, J. L., and Pederson, G. T. (2004). A tree-ring based reconstruction of the Atlantic Multidecadal oscillation since 1567 A.D. *Geophysical Research Letters*, 31, L12205.

Griffies, S. M. (2004). *Fundamentals of Ocean Climate Models*. Princeton, NJ, Princeton University Press, 518 pp.

Griffies, S. M. and Bryan, K. (1997a). Predictability of North Atlantic multidecadal climate variability. *Science*, 275, 181–184.

Griffies, S. M. and Bryan, K. (1997b). A predictability study of simulated North Atlantic multidecadal variability. *Climate Dynamics*, 13, 459–487.

Guilyardi, E., Wittenberg, A., Fedorov, A. *et al.* (2009). Understanding El Niño in ocean-atmosphere general circulation models: Progress and challenges. *Bulletin of the American Meteorological Society*, 90, 325–340.

Hawkins, E. and Sutton, R. (2007). Variability of the Atlantic thermohaline circulation described by three-dimensional empirical orthogonal functions. *Climate Dynamics*, 29, 745–762.

Keenlyside, N. S., Latif, M., Jungclaus, J., Kornblueh, L., and Roeckner, E. (2008). Advancing decadal-scale climate prediction in the North Atlantic sector. *Nature*, 453, 84–88.

Knight, J. R., Allan, R. J., Folland, C. K., Vellinga, M., and Mann, M. E. (2005). A signature of persistent natural thermohaline circulation cycles in observed climate. *Geophysical Research Letters*, 32, L20708.

Latif, M. E., Roeckner, E., Mikolajewitz, U., and Voss, R. (2000). Tropical stabilization of the thermohaline circulation in a greenhouse warming simulation. *Journal of Climate*, 13, 1809–1813.

Levitus, S., (ed) (2005). *World Ocean Atlas 2005, Volume 1, Temperature*, NOAA Atlas NESDIS vol. 61, U.S. Government Printing Office, Washington, D.C., 182 pp.

Lorenz, E. (1969) Three approaches to atmospheric predictability. *Bulletin of the American Meteorological Society*, 50, 345–349.

Luo, J. J., Masson, S., Roeckner, E., Madec, G., and Yamagata, T. (2005). Reducing climatology bias in an ocean–atmosphere CGCM with improved coupling physics. *Journal of Climate*, 18, 2344–2360.

Luo, J. -J., Behera, S., Masumoto, Y., Sakuma, H., and Yamagata, T. (2008). Successful prediction of the consecutive IOD in 2006 and 2007. *Geophysical Research Letters*, 35, L14S02.

Manabe, S. and Bryan, K. (1969). Climate calculations with a combined ocean-atmosphere model. *Journal of the Atmospheric Sciences*, 26, 786–789.

Manabe, S., Stouffer, R. J., Spelman, M. J., and Bryan, K. (1991). Transient responses of a coupled ocean-atmosphere model to gradual changes of atmospheric CO2. Part I, Annual mean response. *Journal of Climate*, 4, 785–818

Marchuk, G. I. (1974). *Numerical Methods in Weather Prediction*, Academic Press, New York, 277 pp.

McPhaden, M. J., Zebiak, S. E., and Glantz, M. H. (2006). ENSO as an integrating concept in Earth Science. *Science*, 314, 1740–1745.

Moore, D. W. (2006). Reflections of an Equatorial Oceanographer. In *Physical Oceanography, Developments Since 1950*. ed. Jochum, M. and Murtugudde, R., Springer, 250 pp.

O'Brien, J. J. and Hurlburt, H. E. (1974). Equatorial jet in the Indian Ocean. *Science*, 124, 1075–1077.

Otto-Bliesner, B. L., Hewitt, C. D., Marchitto, T. M. *et al.* (2007). Last Glacial Maximum ocean thermohaline circulation, PMIP2 model inter comparisons and data constraints. *Geophysical Research Letters*, 34, L12706.

Pacanowski, R. C. (1987). Effect of equatorial currents on surface stress. *Journal of Physical Oceanography*, 17, 833–838.

Philander, S. G. H., Lau, N. C., Pacanowski, R. C., and Nath, M. J. (1989). Two different simulations of the Southern Oscillation and El Niño with coupled ocean-atmosphere general circulation models. *Philosophical Transactions of the Royal Society of London, A*, 329, 167–178.

Pohlmann, H., Jungclaus, H. J., Koehl, A., Stammer, D., and Marotzke, J, (2009). Initializing decadal climate predictions with the GECCO oceanic synthesis: Effects on the North Atlantic. *Journal of Climate*, 22, 3926–3938.

Rahmstorf, S. (1996). On the freshwater forcing and transport of the Atlantic Thermohaline Circulation. *Climate Dynamics*, 12, 799–811.

Randall, D. A., Wood, R. A., Bony, S. *et al.* (2007). Climate Models and Their Evaluation. In: *Climate Change 2007: The Physical Science Basis. Contribution of Working Group I to the Fourth Assessment Report of the Intergovernmental Panel on Climate Change* [Solomon, S., Qin, D., Manning, M., Chen, Z., Marquis, M., Averyt, K. B., Tignor, M., and Miller H. L. (eds.)]. Cambridge University Press, Cambridge, United Kingdom and New York, NY, USA.

Reichler, T. and Kim, J. (2008). How well do coupled models simulate today's climate? *Bulletin of the American Meteorological Society*, 89, 303–311.

Richardson, L. F. (1922). *Weather Prediction by Numerical Process*. Cambridge University Press. London, Reprinted 1965 (with an introduction by Sydney Chapman) Dover, New York.

Richter, I. and Xie, S. -P. (2008). On the origin of equatorial Atlantic biases in coupled general circulation models. *Climate Dynamics*, 31, 587–598.

Roemmich, D., Johnson, G. C., Riser, S. *et al.* (2009). The Argo Program, Observing the global oceans with profiling floats. *Oceanography*, 22, 24–33.

Rooth, C. (1982). Hydrology and ocean circulation. *Progress in Oceanography*, 11, 131–149.

Ruttiman, J. (2006). Milestones in scientific computing. *Nature*, 440, 399–405.

Saji, N. H., Xie, S. -P., and Yamagata, T. (2006). Tropical Indian Ocean variability in the IPCC twentieth-century climate simulations. *Journal of Climate*, 19, 4397–4417.

Sausen, R., Barthel, K., and Hasselmann, K. (1988). Coupled ocean-atmosphere models with flux correction. *Climate Dynamics*, 2, 145–163.
Schlesinger, M. and Ramankutty, R. (1994). An oscillation in the global climate system of period 65–70 years. *Nature*, 367, 723–726.
Schneider, B., Latif, M., and Schmittner, A. (2007). Evaluation of different methods to assess model projections of the future evolution of the Atlantic Meridional Overturning Circulation. *Journal of Climate*, 20, 2121–2132.
Schopf, P. S. and Suarez, M. J. (1988). Vacillations in a coupled ocean–atmosphere model. *Journal of the Atmospheric Sciences*, 45, 549–566.
Schopf, P. S. and Suarez, M. J. (1990). Ocean wave dynamics and the timescale of ENSO. *Journal of Physical Oceanography*, 20, 629–645.
Shin, S. -I., Liu, Z., Otto-Bliesner, B. L., Kutzbach, J. E., and Vavrus, S. J. (2003). A simulation of the Last Glacial Maximum climate using the NCAR CSM. *Climate Dynamics*, 20, 127–151.
Smith, D. M., Cusack, S., Colman, A. W., Folland, C. K., Harris, G. R., and Murphy, J. M. (2007). Improved surface temperature prediction for the coming decade for a global climate model. *Science*, 317, 796–799.
Smith, R. D., Maltrud, M. E., Bryan, F. O., and Hecht, M. W. (2000). Numerical simulation of the North Atlantic Ocean at 1/10°. *Journal of Physical Oceanography*, 30, 1532–1561.
Stommel, H. (1961) Thermohaline convection with two stable regimes of flow. *Tellus*, 13, 224–228.
Sutton, R. T. and Hodson, D. L. R. (2005), Atlantic Ocean forcing of North American and European summer climate. *Science*, 309, 115–118.
Timmermann, A., Latif, M., Voss, R., and Grötzner, A. (1998). Northern Hemispheric interdecadal variability, A coupled air–sea mode. *Journal of Climate*, 11, 1906–1931.
Trenberth, K. E. and Carron, J. M. (2001). Estimates of meridional atmospheric and Oceanic Heat Transports. *Journal of Climate*, 14, 3433–3443.
Vallis, G. K. (2006). *Atmospheric and Oceanic Fluid Dynamics, Fundamentals and Large-Scale Circulation*. Cambridge University Press, Cambridge, 745 p.
van Oldenborgh, G. J., Philip, S. Y., and Collins, M. (2005). El Niño in a changing climate, a multi-model study. *Ocean Science*, 1, 81–95.
Vellinga, M. and Wu, P. (2004), Low-latitude fresh water influence on centennial variability of the thermohaline circulation. *Journal of Climate*, 17, 4498–4511.
Veronis, G. (1975). The role of models in tracer studies. In *Numerical Models of the Ocean Circulation*, Natl. Acad. Sci., pp. 133–146.
Williams, G. P. and Bryan, K. (2006). Ice age winds: an aquaplanet model. *Journal of Climate*, **19**, 1706–1715.
Xie, S. -P. and Philander, S. G. H. (1994). A coupled ocean-atmosphere model of relevance to the ITCZ in the eastern Pacific. *Tellus*, 46a, 340–350.
Zebiak, S. E. and Cane, M. (1987). A model El Niño–Southern Oscillation. *Monthly Weather Review*, 115, 2262–2278.
Zhang R. and Delworth, T. (2007). Impact of the Atlantic Multidecadal Oscillation on North Pacific climate variability. *Geophysical Research Letters*, 34, L23708.

8

Coupling atmospheric circulation models to biophysical, biochemical, and biological processes at the land surface

ROBERT E. DICKINSON

8.1 Beginnings of modeling the land surface as a component of GCMs

The land surface is a major component of the climate system. It provides directly the environment that sustains much of life on Earth and so human welfare. It influences weather and climate patterns, cloudiness, and aerosols, and in turn is influenced by these and other aspects of the climate system. Over the last four decades understanding of the processes involving the land surface that must be included in climate models has advanced considerably. This chapter reviews the development of this understanding and where it is currently.

Land was initially viewed as a prescribed boundary condition for the integration of atmospheric general circulation models (GCMs). However, modelers soon realized that for GCMs to be used as climate models, the surface had to be modeled as interactive, but maintaining the key constraints, i.e. conservation of water and energy. Precipitation delivers the water and incident solar radiation and downward longwave energy; land must be modeled to respond to these inputs. The first generation of land models were very simple, e.g. a slab that stored water, distributing it into evapotranspiration (ET) and runoff; and a surface temperature that balanced the energy input with black body, sensible, and ET fluxes (Manabe, 1969a,b). Surface parameters for albedo, soil moisture release, and aerodynamic roughness were uniform over land. The black body and sensible fluxes depend on surface temperature, so more ET acts to lower temperature locally. Namias (1955) called attention to the role of land boundary conditions and, in particular, soil moisture for the occurrence of drought. Shukla and Mintz (1982) showed through a GCM simulation that a very large difference in temperature would occur for wet versus completely dry continental land surfaces, and that very large differences in precipitation would also occur.

Soil moisture and surface temperatures work together in response to precipitation and radiative inputs. Vegetation influences these terms through its controls on energy and water fluxes, and through these fluxes, precipitation. It also affects radiative heating. Clouds and precipitation are affected through modifications of the temperature and water vapour content of near-surface air.

The energy from ET is released elsewhere in the atmosphere as latent heat of precipitation. Consequently, shifts in cloud cover and precipitation patterns provide further feedback, but these feedbacks may be muted when, as in Shukla and Mintz and many later studies, the ocean temperatures are assumed fixed. Another control is the fraction of incident solar radiation that is reflected, i.e. the albedo. The largest contrasts in albedo occur between vegetation and brighter underlying surfaces, e.g. snow or light soils. Charney (1975) first discussed the response of climate over North Africa to changing from the albedo of vegetation to that of the Sahara Desert. His study emphasized that a reduction of solar input from higher albedo would impose a cooling in the atmospheric column that would be largely balanced by subsidence (i.e. adiabatic warming), hence reducing precipitation and cloudiness. It illustrates a common theme that shifts in surface energy balance can have implications for the energy balance of the overlying atmosphere with resulting dynamic adjustments modifying cloud properties and precipitation. Berkofsky (1976) first attempted a GCM simulation of Charney's hypothesis.

The review of Sellers *et al.* (1997) characterized land models as comprising three generations. The previously mentioned first generation of models consistently conserved energy and water. These models demonstrated the important climatic role of ET and its dependence on soil moisture. The flux of water from the surface can be formulated as depending on a network of resistances. The first generation included only a soil resistance r_g, and aerodynamic resistance r_a. Their ET was proportional to the difference between v_g, the saturated water vapour concentration at the soil temperature and v_a, the atmospheric near-surface water vapour concentration. Formulated mathematically and taking F_q, to denote the water flux, their ET with typical units of kg m^{-2}s^{-1} was given by:

$$F_q = (v_g - v_a)/(r_a + r_g) \tag{8.1}$$

Ecophysiological studies in the 1970s (e.g. Jarvis, 1976) had already demonstrated that leaf stomata provided a major resistance to the passage of water through vegetation. Following Deardorff (1978), Dickinson (1984), Dickinson *et al.* (1986), and Sellers *et al.* (1986) introduced a second generation of land models whose most essential innovation was to include a canopy resistance r_c added into the denominator of Eq. 8.1 and inferred by scaling from parameterization of leaf stomatal and boundary-layer resistance, e.g. Gates, (1980). In addition, they introduced the geographic variation of other surface parameters, in part by use of land cover datasets, with tables converting land cover (e.g. tropical evergreen forest) to some of the parameters needed by the model (e.g. roughness-length, fractional vegetation cover). Soil properties (e.g. porosity and albedo) were independently specified. Datasets targeted at climate models (Matthews, 1983; Wilson and Henderson-Sellers, 1985; as also the study of Olson *et al.*, 1983) were key contributors to this generation of land models. Further substantial improvement became possible with the advent of satellite-derived information, especially that from MODIS (Friedl *et al.*, 2002).

The resistances in (8.1) are parameterized in terms of a) r_a depending on temperature differences between ground and air as well as the ecosystem inferred roughness-length, b) r_g depending on layered models of soil moisture, and c) the r_c introduced in the second generation of models depending on absorption of visible radiation, leaf temperature, and vapor pressure deficit from saturation. Another important advance was the recognition that the heterogeneity of vegetation within a model grid square needed to be explicitly modeled. Thus, grid squares were subdivided into "tiles" giving the fractional coverage of the major vegetation and non-vegetation land cover components (Avissar and Pielke, 1989; Koster and Suarez, 1992).

Since the 1980s, interest developed in the mechanisms of terrestrial carbon cycling. In particular, leaves assimilate carbon through intake of CO_2 and its photosynthesis. This carbon flux also passes through the stomata so leaf transpiration is proportional to the carbon flux. On this basis, Sellers *et al.* (1996a), introduced a third generation of land surface models that jointly modeled the leaf and carbon fluxes. The previous empirical model dependencies for water flux were subsumed into treatment of the carbon assimilation terms, based on the formulation of Farquhar, von Caemmerer, and Berry (1980), and implemented in Collatz *et al.* (1991).

Another important component of land models is the treatment of surface radiation. Sellers (1985) developed the currently most detailed such treatment. Leaves are assumed to be randomly distributed scatterers that have a distribution of orientations and prescribed optical properties (i.e. reflection and transmission coefficients). Dickinson (1983) in describing an earlier version of this model pointed out that it might have serious deficiencies because of its neglect of organization of leaf locations within a canopy. Approaches to account for this organization are being developed (e.g. Pinty *et al.*, 2006).

The next section (Section 8.2) summarizes the evolution of understanding the role of land in climate variability and change; Section 8.3 summarizes the inclusion of dynamic vegetation; Section 8.4 describes how carbon cycling has been included in GCMs; and Section 8.5 summarizes what has been discussed and what has been left out.

8.2 Land's role in climate variability and change

Climate has always changed on inter-annual and decadal time-scales as a consequence of long-term variability of the oceans as manifested in shifting patterns of ocean surface temperatures, discussed in detail in Chapter 7. These patterns connect to the atmosphere and the latter carries these changes to land. However, human activities have grown over the last century to impose additional changes of comparable magnitude but more persistent. In particular, increasing consumption of fossil fuels has led to increased concentrations of carbon dioxide in the atmosphere. Other greenhouse gases of shorter atmospheric lifetimes have also contributed to the ongoing global radiative heating and the consequent long-term trend in global temperatures. In addition, terrestrial ecosystems have been heavily managed in ways that both affect atmospheric greenhouse gases (Section 8.4) and have direct climate consequences.

Land cover changes from a combination of human land use and climate variability. Such changes, as occurred in the dust bowl era, have long been speculated to have impacts on regional climates and local microclimates, but only recently has climate model quantification become possible (e.g. Cook *et al.*, 2009). Much early science examined the various surface energy exchange processes, e.g. Monteith and Szeicz (1961) and Fritschen (1967) related incident solar radiation to net radiation. With the development of climate models it has become possible to quantify these effects.

The simplifications of early climate models engendered skepticism from the micrometeorological community. Idso and Deardorff (1978) pointed out that the growth of vegetation had other equally important consequences besides lowering albedo, the only effect considered by Charney and Berkofsky. Land surfaces involve large radiative exchanges (in the order of 100 W m^{-2} or more) but small temperature changes (less than 10 K), that have provided skeptics with contrarian estimates of sensitivity of global temperature to radiative forcing, commonly by their assuming local land surface processes were representative of global changes.

Land cover use/change has many consequences (compare the review of Foley *et al.*, 2005) including modification of climate through land cover impacts on surface temperatures and hydrology. These impacts result from some combination of changing net radiation and evapotranspiration. Regional modeling studies have for a long time demonstrated that changes in evapotranspiration could change regional temperatures (e.g. by removal of the Amazon forest, Dickinson and Henderson-Sellers, 1988). Such studies demonstrated important coupling to the atmosphere, i.e. modification of precipitation and changes of radiative fluxes at the surface from cloud changes comparable to those from the surface albedo change (e.g. Dickinson and Kennedy, 1992). However, interactive oceans were not included, hence excluding the possibility of compensating changes from ocean temperature changes.

The connections between climate, vegetation, and humans were given widespread visibility through the Gaia hypothesis of James Lovelock (compare Watson and

Lovelock, 1983; Schneider, 2004). These connections and the role of climate modeling in the context of Amazonia were highlighted in an international workshop in San Jose dos Campos, Brazil, in 1985, and by the resulting book (Dickinson, 1987). Recent reviews of the climate effects of land cover change include: Jackson *et al.* (2005), Salmun and Molod (2006), Bonan (2008).

Effects of anthropogenic land cover change on global temperature have until recently been addressed in GCMs primarily in terms of changes of global albedo. The first such analysis (Sagan *et al.*, 1979) suggested substantial contributions from this mechanism. Henderson-Sellers and Gornitz (1984), recalculated anthropogenic albedo change allowing for the masking effects of clouds (hence, halving the albedo change at the top of the atmosphere) and found a maximum reduction over the previous 30 years corresponding to a global radiative forcing of -0.1 W m^{-2}.

More recently, attention has been brought to the distinction between direct and diffuse solar radiation. Radiation is made diffuse by scattering by clouds, aerosols, and molecules. Direct radiation can only strike a limited fraction of the leaves, but diffuse radiation is incident on all the leaves in a canopy, hence can promote more ET and carbon assimilation for the same amount of radiation. Roderick *et al.* (2001) reported the first observational tower evidence for this effect in the tropics. Diffuse radiation resulting from the Mt. Pinatubo eruption, through this mechanism, may have created an enhanced terrestrial carbon sink.

For estimating climate impacts of land cover change over time, it is necessary to describe not only current land conditions, but also how they may have changed from the past. Land devoted to agriculture has greatly increased since preindustrial times. The understanding of what land cover and albedo changes have occurred over time is being advanced. In particular, the historical dataset by Ramankutty and Foley (1999) for agricultural land use has now been widely used for modeling the impacts of land cover change. Historically, land cover change has been more complicated than a simple uniform increase in agriculture. In particular, in some parts of the world such as New England in the USA, cleared forests have been allowed to grow back. Elsewhere, large areas of grazing have been developed, initially from natural grasslands but now mostly either from modification of shrublands or forests. Hurtt *et al.* (2006) have provided a global estimate of land cover change over the last three centuries (1700–2000). These changes in recent decades have largely been concentrated in the tropics. Tropical shrub-lands and forests have been cleared for ranches, for agro-forestry, e.g., oil palms, and for export crops, e.g., soybeans and sugar cane. These changes are now of equal interest for analyses of the global carbon budget, as addressed further in Section 8.4.

Betts (2001) estimated a global radiative forcing of -0.2 W m^{-2} for replacing "natural vegetation" with the agricultural land use classification of Wilson and Henderson-Sellers (1985). This estimate applies to change over a period of at least several hundred years, not the 30 years that earlier estimates addressed. This estimate was taken as a central value by the Intergovernmental Panel on Climate Change AR4 (FPCC AR4) assessment report. Use of historical data (1750–1950) for changing agriculture gave a similar result (-0.18 W m^{-2}, Betts *et al.*, 2007). Any such estimate depends on the tabulated values of land cover change, the albedo for each land cover class assumed in the model, as well as the masking and feedback by clouds. Myhre *et al.* (2005) found that the albedo measured by the MODIS satellite showed less contrast between agriculture and natural vegetation than had been assumed by climate models and that changing land albedo since prehistoric times may have contributed only -0.1 W m^{-2} to global radiative forcing.

In moist warm regions, the fraction of energy going into water fluxes can undergo large changes, for example, from changes in vegetation cover or in precipitation, and hence in soil moisture. Climate impacts from land cover change, at least local ones, commonly depend as much or more on the changes in these water fluxes as on the changes in net radiation. The changes of water flux depend not only on the net radiation at the surface but also on changes of the partitioning of this net radiation between sensible and latent heat fluxes, and usually to a lesser degree, on changes in surface roughness and the diurnal storage of heat. Such changes may result either from changes in vegetation structure or from changes in the extraction of soil moisture. Vegetation structure parameters e.g. cover, height, leaf area, largely control the surface aerodynamic roughness used in Eq. 8.1; they also determine various resistances that connect the molecular and turbulent diffusion of moisture from leaves to the overlying boundary layer. Shading of soil by vegetation reduces its daytime storage of thermal energy as does its lower heat capacity when dry.

The climate change from past land cover change is arguably as large or larger on local scales where land cover change has occurred as that from increasing greenhouse gases (e.g. Pielke *et al.*, 1998). Thus, over the last decade, substantial efforts have been made to better quantify the former through GCM modeling studies. As a generality, forest removal has been found by many studies to promote warming in the tropics through reduction of ET and, in high latitudes, cooling from increased albedo. Quantitative results depend on what details are assumed about the forest, what it is replaced by, what is the albedo of underlying soil, whether or not an interactive ocean is used, and what models are used. Mid-latitude climate can be nearly equally affected by ET changes and albedo changes and consequently, if these are of opposite sign, even the sign of change can depend on time of year and modeling details such as mentioned above.

Forests in early spring act as especially dry surfaces with consequent large sensible fluxes that mix the atmosphere to a great depth (e.g. Betts *et al.*, 2001), but increase their water fluxes with spring green-up. Water fluxes are less in cooler climates. The boreal forest exerts a strong resistance to water movement and has been characterized as a "green desert" because of its small release of water to the atmosphere (Gamon *et al.*, 2004).

Change of mid-latitude forest to agriculture causes cooling if both albedo and ET are modeled to increase (e.g. Bonan, 1997; Oleson *et al.*, 2004a). The consequent reduction of surface roughness promotes some warming, but this is too small to compensate. Cooling in spring is most pronounced because snow is no longer being shaded by the forest with its lower albedo.

Two recent studies have constrained their albedo change with MODIS satellite data and included interactive oceans (Kvalevåg *et al.*, 2009; Lawrence and Chase, 2010). Both used new versions of the NCAR climate model and changed vegetation from natural to current day. They found similar results for this change. Substantial warming occurred in the tropics from deforestation driven by reductions in ET but cooling occurred in high latitudes driven by increased albedos. The global temperature changes were small, 0.04 °C and 0.02 °C, respectively, but somewhat larger warming was found with prescribed ocean temperatures, primarily from less high-latitude cooling. The computed warming demonstrates the possibility of global temperature changing in the opposite direction to that of direct radiative forcing because of the impact of other adjustments. Lawrence and Chase reported that global average land albedo was increased by 0.1%. Clouds were reduced by a similar amount, consistent with the reduction in ET. Consequently, the net solar flux at the top of the atmosphere slightly increased, i.e. was of opposite sign to that implied by the land albedo change.

Feedbacks of surface moisture on precipitation, as also on clouds, may act differently on different time-scales (Dickinson *et al.*, 2003). Leaves initially intercept much of the precipitation over vegetation, and a significant fraction of this leaf water re-evaporates in an hour or less. This loss reduces the amount of water stored in the soil for use by plants and depends inversely on the intensity of the precipitation, which is larger at smaller temporal and spatial scales. The evaporation of intercepted precipitation can be more than 40% in a cold drizzle (e.g. Gash *et al.*, 1980), or as low as 10% in a tropical storm (Shuttleworth, 1988). Modeling results such as the timing of evaporative cooling can be wrong either through neglect of or through exaggeration of the magnitude of the exchange with the fast time-scale moisture stores of the leaves.

Evaporative transfer from leaves can also change with atmospheric composition, in particular CO_2. Current biophysical modeling of ET from leaves makes leaf stomatal resistance (used to derive canopy resistance) proportional to the ratio of the concentration of CO_2 to its assimilation. This ratio generally increases with increasing CO_2. A number of modeling studies demonstrated by directly increasing stomatal resistance or with physiological models (e.g. Henderson-Sellers *et al.*, 1995; Pollard and Thompson, 1995; Sellers *et al.*, 1997), that increasing stomatal resistance implies a substantial reduction of ET, hence a surface warming that adds to the radiative effects of increasing greenhouse gases.

8.3 The introduction of dynamic vegetation

Ecologists have long correlated the occurrence of different ecosystems with climate statistics, most commonly annual temperatures and precipitation, e.g. Holdridge (1967), Box (1981). In addition, early models were developed for the stochastic dynamics of forest growth over a single patch, or "gap," that changed with local climate conditions (e.g. Botkin *et al.*, 1972). Woodward (1987) described how growth of global vegetation depended on climate parameters through physiological principles corresponding to those parameterized in GCM land models. He related the distribution of ecosystems to availability of solar radiation, soil water, and temperature minima. Prentice *et al.* (1992) developed a steady-state model for global vegetation following a logic similar to that described by Woodward. It limited boreal and tropical forests by the mean temperature of their coldest months and minimum moisture availability, and grasses and shrubs by moisture availability, growing degree days, and minimum temperatures. The current emphasis on dynamic vegetation in climate models was initiated by a conference in 1988 at the Internationaal Institute of Applied Systems Analysis (IIASA) and summer workshops in 1988 and 1989 (Solomon and Shugart, 1992).

Scientists at the time of the IIASA workshops recognized that the direct effect of increasing CO_2 might be important for determining carbon storage, but judged it less important than climate change for changing the distribution of global vegetation. Solomon (1986) and Overpeck *et al.* (1990) considered how the forests of the eastern United States would change with step function changes of climate, the former from a climate model simulation, and the latter from simple scenarios, respectively. Both studies used a gap model and found large changes on a time-scale of hundreds of years. However, extending such modeling to the rest of the world was seen to be impractical because of the dependence of the gap model on data for individual species. Thus, a scaling to the globe was seen to require a simpler approach corresponding to the equilibrium models of Holdridge and Box. Emanuel *et al.* (1985) had already pioneered such an approach by projecting the changes of global ecosystems in steady state responding to a climate-model simulation for changing from present climate to a 2x CO_2 world. Henderson-Sellers (1993) developed a simplified description of the global distribution of vegetation from the output of two versions of the NCAR Community Climate Model (CCM) and analyzed the change expected from a 2x CO_2 climate. Her study was a first attempt to calculate the change of vegetation interactively with the climate change (for 67 months) and to determine how this coupling modified the model's net carbon uptake (see also Claussen, 1994).

Foley *et al.* (1996, 1998) developed the first land model (IBIS) suitable for coupling to a GCM that combined transient dynamics and carbon assimilation of vegetation with a biophysical model. Because IBIS included an understorey of herbaceous grass vegetation, it was able to represent the competition between grass and overlying trees interacting within a climate model. Trees adjusted to changing climate on a time-scale of centuries. The Foley *et al.* approach highlights the role of vegetation dynamics on annual and longer time-scales. Their vegetation assimilated and allocated carbon annually to determine plant growth, with leaf presence determined by simple phenology rules. Leaves in a climate model function on an hour by hour basis. Dickinson *et al.* (1998) demonstrated a model for leaf dynamics operating on this scale interactively with a climate model and extended it to include nitrogen cycling (Dickinson *et al.*, 2002). Another model for leaf dynamics was described by Arora and Boer (2005). Such models must describe what fraction of assimilated carbon goes into leaf growth, the dependence of assimilation on canopy light levels, and effects of water stress. They also describe leaf turnover and its dependence on cold and water stress. Realism of such modeling may be limited in part by lack of understanding of details of fast canopy level processes.

Cox *et al.* (2000) developed another dynamical model, TRIFFID, with vegetation fractions that compete based on their relative heights through negative feedback on each other (a "Lotka–Volterra" equation). The formulation of dynamic vegetation in climate models has been influenced by models initially designed to be forced by monthly climate statistics, e.g. the Lund–Potsdam–Jenna (LPJ) model (e.g. Sitch *et al.*, 2003). On a long time-scale, it grows forests by adjustment to a climate–ecosystem correlation (e.g. Prentice *et al.*, 1992). Other such models are mentioned in the next section.

Bonan *et al.* (2002) argued that global models should use "plant functional types" or pfts to represent generalized vegetation evolving dynamically and used this basis (Bonan *et al.*, 2003) for a new global dynamic vegetation model based on concepts and algorithms from the IBIS and LPJ models. Arora and Boer (2006) have developed an advanced scheme for including competition between different pfts that allows multiple pfts to coexist. Currently, dynamic vegetation models developed for GCMs use these models as part of their carbon cycling, the issue addressed in the next section.

8.4 Coupling to the carbon cycle

Many of the current approaches to the carbon cycle for use in climate models have their origins in the 1960s and 1970s, and were initially concerned largely with agricultural applications, e.g. Monteith (1972) related the productivity of tropical ecosystems to incident solar radiation. Extensive laboratory studies of photosynthetic pathways and leaf gas exchanges were used as a basis for various models for carbon assimilation by leaves that were not too dissimilar to those used today, e.g. Tenhunen *et al.* (1976).

The terrestrial contribution to the global carbon cycle was first discussed in terms of global net primary production (NPP), i.e. the net uptake of carbon by plants allowing for respiratory losses. It was characterized by Whittaker and Likens (1975), Lieth (1973, 1984), based on surveys of observed growth rates in various ecosystems. Agriculturists have long suggested that the NPP of crops should increase with the increase in atmospheric CO_2 concentrations (e.g. Rosenberg, 1981; Leuning *et al.*, 1993). More recently, impacts of increased CO_2 on overall growth of forests have also been established (e.g. Eamus and Jarvis, 1989; Norby *et al.*, 2005).

Ecosystems are modeled as divided into a number of carbon components that exchange carbon with each other or with the atmosphere. The live carbon is approximated as leaves, roots, and above-ground woody material. Leaves assimilate carbon and use it to construct all the plant components. They use some of this carbon for respiration. The gross primary productivity (GPP) is the initial carbon assimilation through leaves that provides both the NPP and plant respiration. Respiration is the CO_2 returned from the plants to the atmosphere resulting from oxidation of photosynthate to construct the molecular compounds that the plants consist of, and to maintain these compounds against thermodynamic degradation. Lieth and Box (1977) first modeled the GPP. More detailed ecosystem biogeochemical models were initially designed to be driven by monthly climate data but have had an influence on the design of the carbon-cycle components of contemporary climate models, e.g. CENTURY (Parton *et al.*, 1993); TEM (Melillo *et al.*, 1993) CASA (Potter *et al.*, 1993), and Biome-BGC (Running and Coughlan, 1988).

Terrestrial vegetation currently stores approximately the same amount of carbon as contained in the atmosphere. Thus, by changing this storage, it has a substantial capability to have a large impact on how much carbon is contained in the atmosphere. Consequently, land cover change has been a major component of terrestrial carbon exchange with the atmosphere. Bolin (1977) suggested that global deforestation was contributing to the rapid rate of increase of atmospheric CO_2 being seen at that time. More recent studies emphasize net ecosystem exchange (NEE), i.e. the carbon losses from dead plant materials accumulated at the surface or incorporated into soils minus NPP, giving the net exchange with the atmosphere as needed in a climate model. Its observational characterization has been local and global. Locally, it has been measured micro-meteorologically at a large number of flux towers (e.g. Baldocchi *et al.*, 2001), and globally, by inversion from the flask network using atmospheric transport models (pioneered by Fung *et al.*, 1983).

Early monitoring of atmospheric CO_2 revealed large seasonal variability that was attributed to the seasonal growth of vegetation in the Northern Hemisphere (e.g. Keeling *et al.*, 1976). Fung *et al.* (1983) developed a gridded dataset for NPP that they used to estimate the seasonably varying NEE. They used their data and a transport model derived from the GISS GCM to reproduce this observed variability. Tans *et al.* (1990) combined estimates of oceanic net carbon uptake with the Fung *et al.* (1983) terrestrial carbon and transport model to estimate the global spatial distribution of atmospheric CO_2. They accounted for the fossil fuel addition of 5.3 Gt C yr^{-1}, added primarily in the Northern Hemisphere. Atmospheric CO_2 increases by less than that being added by burning of fossil fuel (i.e. it was increasing by 3.0 Gt C yr^{-1} at the time of their paper). The difference is taken up by oceans and land. Fung *et al.* found that to reduce the interhemispheric gradient of atmospheric CO_2 to that observed, the terrestrial system had to take up more carbon than the oceans. Denning *et al.* (1995) incorporated the seasonably varying terrestrial uptake of carbon inferred by Tans *et al.* (1990) into a GCM, hence modeling transport more realistically. They discussed the role of seasonal and diurnal variations of moist convection in determining the hemispheric gradients of atmospheric CO_2. They showed that the seasonality of land carbon sources combined with seasonality of vertical transport should further enhance the interhemispheric gradient and so exacerbate the requirement for a Northern Hemisphere sink for carbon.

Bonan (1995) developed a detailed model of terrestrial carbon cycling that he integrated as part of a GCM, and showed that its NPP by ecosystems was comparable to that observed. Denning *et al.* (1996) similarly determined NPP and, by including an atmospheric transport model, were able to discuss the simulation of the atmospheric distribution of CO_2. Spatial details of this distribution have little effect on atmospheric radiation, but are, nevertheless, of great interest as an observational means to diagnose spatial details of CO_2 sources and sinks. Over the last decade, many additional climate models have attempted to determine atmospheric carbon exchange with land and ocean, some using dynamic vegetation models, following the striking results obtained by Cox *et al.* (2000), who found in the Hadley Centre GCM that the Amazon rainforest would be destroyed with future climate change, and that its huge store of carbon would be released to the atmosphere (corresponding to several decades of fossil fuel use). The primary driver of this catastrophic event was the establishment of a strong permanent ENSO that suppressed Amazon precipitation (Cox *et al.*, 2004), but feedbacks from vegetation and CO_2 change also contributed (Betts *et al.*, 2004). (See Chapter 7 for further discussion of ENSO.)

The release of CO_2 by combustion of fossil fuels and deforestation has been causing global climate change only because some fraction of this CO_2 has remained in the atmosphere. After averaging out inter-annual variability, this fraction has been between 0.4 and 0.5, but it appears to have increased by about 0.05 since 1959

(Canadell *et al.*, 2007) apparently primarily from declining ocean uptake (Le Quere *et al.*, 2007). Climate models can now simulate the changes of terrestrial carbon storage with climate change and variability. Variability of this storage is more susceptible to observational comparisons. Jones *et al.* (2001) examined the carbon cycle response to ENSO in the Hadley coupled climate model. They found a release from the terrestrial biosphere during an El Niño of 1.8 Gt C yr^{-1} from plant and soil respiration but an ocean uptake of 0.5 Gt C yr^{-1}, similar to that observed. Large decreases of NPP in water-stressed tropical regions were compensated by large increases in southern USA and southern South America. Their model did not include fires or many other possibly important factors.

Details of the terrestrial carbon cycle are still highly model-dependent. Consequently, global constraints from observations are desirable. Flask network inversions are able to distinguish between the contributions of continental-scale regions on decadal time-scales, providing land-use contributions are separately prescribed. Flask observations show a strong increase in CO_2 with ENSO as first noted by Bacastow (1976). Since fluxes from the ocean decrease because of reduced upwelling (e.g. Winguth *et al.*, 1994), this signal must be explained by an increase in terrestrial carbon sources or by a reduction in sinks, i.e. more respiration and/or less NPP. Flask network observations cannot distinguish between continents on this time-scale but have shown that the variability of carbon sources with ENSO is nearly entirely tropical (Baker *et al.*, 2006).

Hashimoto *et al.* (2004) have used satellite observations (1982–1999) to estimate the ENSO-correlated variability of NPP and so to deduce the contributions of respiration, including fires as a residual. They find this residual term to contribute about half the variability. Their estimated NPP variability was uncorrelated with global temperature, but the respiration component was strongly correlated. Randerson *et al.* (2005) examined the contribution of fires to carbon emission during the 1997/1998 El Niño event, using a biogeochemical model constrained by observations of C^{13}, CO, and CO_2. They found that burning of tropical forests contributed more than half the increase in atmospheric CO_2 during this period. Randerson *et al.* (2009) recommend a suite of observational tests that are appropriate for testing the carbon cycling component of a GCM and illustrate their application with two carbon models forced with observed climate variations.

The extent to which CO_2 from anthropogenic generation can be stored in the land or oceans is of considerable practical economic importance. Storage of 1 Pg C would equate to one hundred billion dollars if the cost were 10¢ per kg of carbon so stored. Climate models have generated a wide range of answers for the change in terrestrial carbon with increasing CO_2 and climate change because of different climate simulations and different assumptions about carbon-cycle processes. Some of these differences are a consequence of different model temperature sensitivities

Table 8.1 *Changes in airborne and land-borne fractions of carbon by climate change in several GCM simulations of climate from 1850 to 2100 (based on Friedlingstein* et al.*, 2006). The change is derived from the difference between coupled and uncoupled simulations.*

Model	Airborne fraction change	Land-borne fraction change
Had CM3LC[1]	0.24	−0.26
IPSL-CM4-LOOP[2]	0.02	−0.01
LLNL[3]	0.04	−0.05
CSM1[4]	0.01	−0.01

The Had CM3LC[1] model is that reported by Cox *et al.* (2000), IPSL-CM4-LOOP[2] uses the ORCHIDEE dynamic vegetation model (Krinner *et al.*, 2005), the LLNL[3] model is described by Thompson *et al.*, (2004) and uses the IBIS model. The CSM1[4] model (Fung *et al.*, 2005) uses a version of CASA coupled to the CCSM3 climate model.

to radiative forcing. However, more are from the different approaches to carbon modeling, especially the terrestrial component. Friedlingstein *et al.* (2006) intercompare a number of GCM simulations of global carbon cycling. All their models show a decline in the fraction taken up by the terrestrial system with future climate change. Table 8.1 summarizes for several of these GCMs how much their airborne and land-borne fractions of CO_2 were found to change with climate change. It shows a range of terrestrial uptake from −0.01 to −0.26. Drying of forest regions in the tropics generally reduces the carbon being stored by land, but elsewhere storage may increase. Table 8.1 predicts that reduced storage by land is largely compensated by increased storage in the atmosphere.

The uptake and storage of terrestrial carbon depends on leaf-level carbon assimilation, respiration of soil organic matter, and land-use change. The now standard modeling of leaf-level carbon assimilation (e.g. Oleson *et al.*, 2004b, in this reference, Eqs. 8.2, 8.3, 8.4, 8.8) involves three limiting rates. The first two of these increase with increased CO_2 concentration internal to the leaf, and the first and third depend on V_{max}, i.e., the rate of carbon assimilation in the limit of very large CO_2 concentrations; V_{max} depends on the activity of enzymes, especially RUBISCO, that speed up with temperature. These enzymes require nitrogen for their construction and because of their thermodynamic instability impose a large energy cost on leaves for their maintenance respiration. The increase of carbon uptake with increased CO_2 has been referred to as "CO_2 fertilization".

Leaf assimilation may have a weaker dependence on CO_2 than that of the enzyme-limited rate because of control by the third rate referred to as "the export limited rate of carboxylation", which is commonly and simply parameterized as $0.5\ V_{max}$, i.e. no dependence on CO_2. The control of this rate becomes stronger for colder temperatures and more CO_2. How it varies between plants is not known and its omission in some models (e.g. Kucharik *et al.*, 2000; Maayar *et al.*, 2006; Arora *et al.*, 2009) increases a land model's sensitivity to atmospheric CO_2 increase (as seen in Friedlingstein *et al.*, 2006).

To further clarify how biogeochemical models determine additional terrestrial storage of carbon with increasing atmospheric CO_2 we assume a simple "one-box" model. Let C_t be the additional terrestrial carbon, satisfying

$$dC_t/dt + C_t/\tau = \Delta NPP. \tag{8.2}$$

The effective decay rate $1/\tau$ would be obtained as a spatial and seasonal average of local effective decay rates. Multi-box models can be similarly discussed by decomposition into modes. Then $1/\tau$ would be diagnosed as an average over individual modal rates weighted by the projection of ΔNPP onto those modes. The term ΔNPP is incremental NPP resulting from increases in atmospheric CO_2 and assumed given by

$$\Delta NPP = (NPP)_o \gamma \log_e(C_a/C_{ao}), \tag{8.3}$$

where C_a is atmospheric carbon, C_{ao} its preindustrial (1850) value, and γ is the correlation of Bacastow and Keeling (1973), Friedlingstein *et al.* (1995) their Eq. 15, Arora *et al.* (2009).

Assume $C_a = C_{ao} \exp(rt)$ where t is time measured from 1850. Then for $t >> \tau$, Eq. 8.2 integrates to

$$C_t = (NPP)_o \gamma r \tau t. \tag{8.4}$$

if we assume as approximate values, $(NPP)_o = 69$ Pg yr^{-1}, $\gamma = 0.7$ (non-dimensional), $r = 0.005$ (prior to 1960), and $\tau \approx 20$ yr,

$$\frac{\Delta C_t}{\Delta t} = 0.4 \text{ Pg yr}^{-1}. \tag{8.5}$$

The role of γ has been extensively discussed in previous analyses but Eq. 8.4 has the advantage over complex numerical modeling in also showing that the land sink has a near-linear dependence on growth rate of atmospheric CO_2 and on the time-scale of terrestrial carbon decay. In the last half-century, r has nearly tripled from its earlier values, suggesting that the terrestrial sink from "fertilization" has grown from the estimate of Eq. 8.5 to over 1 Pg yr^{-1}. The growth-rate dependence of the sink implies a negative feedback on attempts to control the growth of atmospheric CO_2. Modeled values of τ may reduce in a warmer world by as much as 5–10% per degree of warming from a speed-up of decomposition, contributing further to reduction of terrestrial uptake.

In summary, the CO_2 fertilization land sink scales with the sensitivity of NPP to CO_2 increase, the rate that atmospheric CO_2 is increasing, and an average residence time for carbon pools. The value assumed for the sensitivity parameter γ in Eq. 8.4 is several times smaller than the value that would be derived from leaf-level assimilation and a constant V_{max}. We return later to discussion of mechanisms for "downregulation" of V_{max}.

The change in land carbon with climate change is somewhat more complicated and geographically variable. Berthelot *et al.* (2005) demonstrated a very wide range of such changes in driving a single terrestrial model from the output of 14 GCMs. Such results can only in part be diagnosed with simple analysis. The major dependencies on climate change are:

(a) the dependence of NPP on temperature,
(b) the dependence of the carbon residence time on temperature,
(c) the dependence of NPP time on soil moisture, and
(d) the dependence of carbon residence time on soil moisture.

Both NPP and plant respiration increase with temperature. The NPP increase should dominate in a cooler climate but respiration, microbial decomposition, and forms of heat damage may dominate in tropical climates.

The rate at which carbon pools of dead plants decay is modeled to increase with temperature by approximately a factor of two every 10 °C. This assumption is plausible when the carbon decay is limited by metabolic rates of micro-organisms, but is questionable when transport limited (e.g. Davidson *et al.*, 2006; McCarthy *et al.*, 2008). If τ in Eq. 8.2 changes, $\tau = \tau_o - \Delta\tau$, then the changed decay rate will act on the total carbon pool, approximated by a constant reference value, C_0. We further approximate by $\tau^{-1} \equiv \tau_0^{-1} + \Delta\tau/\tau^2$. The net result is that τ adds to the right-hand side of Eq. 8.2 an additional term $C_0\,\Delta\tau/\tau^2$. If we further approximate by $\Delta\tau = (-d\tau/dT)w$ where w is the long-term warming trend, then corresponding to Eq. 8.4, we have a perturbation in land storage C_t

$$\frac{\Delta C_t}{\Delta t} = C_0 a w, \tag{8.6}$$

where for approximate values $a = d\log(1/\tau)/dT \approx 0.09\,°C^{-1}$, $C_0 = 700$ Pg, and $r = 0.02\,°C\ yr^{-1}$, Eq. 8.2 implies a 1 Pg yr^{-1} loss of terrestrial carbon, i.e. the current rate of warming could be causing that much additional loss of soil carbon to the atmosphere, if all the assumptions made are correct.

The change of NPP and soil carbon decay from changing soil moisture stress is expected to have considerable geographic variability. It should not contribute in regions where the ratio of precipitation to potential evaporation is larger than 1. At a local site, soil moisture stress on plants sets in only near wilting point with almost a step function dependence on soil moisture. However, at the resolution of a climate model, soil moisture stress, but not complete wilting, may occur over a much broader range of soil moisture. Possibly because of this scaling issue, there is no widely accepted treatment of moisture stress and a wide variety of choices have been made in different models. This variety of different modeling options that have been used has been a major contributor to differences between models in their release of terrestrial carbon from increased moisture stress as well as the dependence of their ET on soil moisture conditions.

Modelers understand even less how to relate the decay of soil carbon to soil moisture. Some reduction is expected with extremely dry or wet conditions but how much reduction occurs may depend on details of the microbiota that can be active under those conditions. Most of the soil carbon, perhaps more than 80% (e.g. Potter and Klooster, 1997), is locked into the soil as old carbon, i.e. with lifetimes of centuries. Much of this old carbon consists of charcoal and carbon locked within soil microaggregates (e.g. McCarthy *et al.*, 2008). Eqs. 8.4 and 8.5 are equivalent to neglecting the time derivative in Eq. 8.1, which would be included for this old pool. If as little as 10% of annual NPP were put into this pool, it would have a large contribution to the modeled terrestrial carbon sink but with a very long adjustment time. Unfortunately, the dynamics of this old carbon and overall what might destabilize it are still very poorly known (e.g. Trumbore, 2006).

The impact of a warming and drying climate on ET, hence carbon uptake is substantially reduced if the leaf stomatal resistance increases with increasing CO_2, as occurs for downregulation of V_{max}. Many but not all plants are observed to reduce their V_{max} and corresponding leaf N with increased concentration of CO_2. This downregulation is important in determining both the biophysical estimation of increased temperature from reduced ET and the terrestrial carbon storage. Its mechanisms are not well understood but some combination of two explanations may apply. The first explanation is that with increasing CO_2, the enzyme-limited carbon assimilation increases more rapidly than the light-limited assimilation. Hence, V_{max} should reduce to insure optimal energy efficiency (e.g. Franklin, 2007). The second, more widely employed explanation, is that of nitrogen (N) limitation. Nitrogen cycling is included in many biogeochemical models that commonly fix the ratio of carbon C to N. Thornton *et al.* (2007, 2009) has explored how such inclusion modifies the dynamics of terrestrial carbon storage. Not all observations support this approach. For example, Crous *et al.* (2009) studied a mixture of C-3 grass and forbs and found that contrary to concepts being modeled, the grass did not downregulate, and the addition of N did not reduce the downregulation of the forbs.

Thornton *et al.* (2007) find in their model that nitrogen cycling strongly weakens the dependence of carbon-flux changes not only on growth of CO_2 concentrations but also on temperature and moisture variability, e.g. as modeled for El Niño. Introduction of nitrogen into a GCM adds multiple feedbacks with climate to its carbon cycle, mostly negative. For example, increased soil carbon decomposition will release more N that in turn, incorporated into leaves, will increase a canopy's carbon assimilation. Soil respiration depends on a microbial pool that has high N requirements. Increases of soil-carbon decomposition can be limited by shrinkages of these pools. Nutrients taken up by plants are dissolved in soil water and pass into the roots with the water supplying ET (e.g. Dickinson *et al.*, 2002). Thus water stress should reduce the uptake of nutrients by adding an additional positive feedback with further reduction of leaf carbon assimilation. For additional carbon energy costs, plants can promote soil sources of nitrogen e.g. N-fixing bacteria, for more efficient nutrient uptake. Thornton *et al.* (2009) find with inclusion of N that the variability of carbon fluxes from soil respiration is largely compensated by variability of plant growth.

Much of the terrestrial storage of carbon occurs in the Arctic zone. The terrestrial processes determining carbon storage in this region such as permafrost and wetlands (compare McGuire *et al.*, 2009) are still poorly characterized in climate models, but some improvements are being made (e.g. Lawrence and Slater, 2008).

8.5 Concluding discussion

In discussing the history of modeling the terrestrial system component of climate models, I have tried to also provide an up-to-date assessment of the inclusion in GCMs of land-use change, dynamic vegetation, and carbon cycling. In doing so, I have omitted or given short shrift to other important topics.

In particular, the sources of other major greenhouse gases besides carbon, in particular methane and nitrous oxide are modeled with components of terrestrial processes, and land is a source or sink for many other important chemical constituents of the atmosphere.

Land has historically been integrated at the same resolution as the GCM. However, much of the description and modeling, in principle, should be done on a much finer scale. Some aspects of such heterogeneities have been demonstrated to simply average to climate-model grid squares. Others involving ET do not, as has been demonstrated many times (e.g. Jarvis and McNaughton, 1986; Seth *et al.*, 1994).

The advent of satellite datasets with vegetation-cover information has provided an observational basis for including in climate models information at a scale of 1 km or finer (e.g. Zeng *et al.*, 2000). However, many aspects of land in a GCM occur below ground and are poorly characterized on a global basis, e.g. soil characteristics and rooting depths (e.g. Kleidon and Heimann, 2000), microbial decomposition, water table levels, and depth to bedrock.

Other aspects of land models that require scaling of atmospheric inputs, e.g. occurrence and intensity of precipitation, are very scale dependent (e.g. Eltahir and Bras,

1993; Gao and Sorooshian, 1994; Chen *et al.*, 1996), but GCMs have not yet advanced their convective precipitation parameterizations sufficiently to provide the needed information.

The global community that is now contributing to the formulation of land in GCMs is large, and the issues, complex. Thus, all activities (beginning with PILPS, of Henderson-Sellers, 1993; Yang *et al.*, 1995; Desborough *et al.*, 1996, for early such activities) allowing different model versions to be compared have proved very valuable.

A major and currently active research topic, in part connected to scaling, is the influence that land processes have on atmospheric precipitation. Estimation of this influence varies widely between models. Perhaps simpler, but equally important, is the connection between land processes and boundary-layer clouds (e.g. Freedman *et al.*, 2001; Ek and Holtslag, 2004).

The issue of mitigating climate change from increasing CO_2 has now raised many issues relating tradeoffs between carbon storage and other biophysical climate impacts (e.g. Betts, 2000) that require GCM simulations to quantify their impacts.

In conclusion, the inclusion of land in GCMs has evolved over the last half-century from a very simple to a very complex area of science that now requires climate modelers to exchange ideas with many other disciplines.

References

Arora, V. K. and Boer, G. J. (2005). A parameterization of leaf phenology for the terrestrial ecosystem component of climate models. *Global Change Biology*, 11, 39–59.

Arora, V. K. and Boer, G. J. (2006). Simulating competition and coexistence between plant functional types in a dynamic vegetation model. *Earth Interactions*, 10, 1–30.

Arora, V. K., Boer, G. J., Christian, J. R. *et al.* (2009). The effect of terrestrial photosynthesis down regulation on the twentieth-century carbon budget simulated with the CCCma Earth System Model. *Journal of Climate*, 22, 6066–6088.

Avissar, R. and Pielke, R. A. (1989). A parameterization of heterogeneous land surfaces for atmospheric numerical models and impact on regional meteorology. *Monthly Weather Review*, 117, 2113.

Bacastow, R. B. (1976). Modulation of atmospheric carbon dioxide by the Southern Oscillation. *Nature*, 261, 116–118.

Bacastow, R. B. and Keeling, C. D. (1973). Atmospheric carbon dioxide and radiocarbon in the natural carbon cycle, II, Changes from A.D. 1700 to 2070 as deduced from a geochemical reservoir, In *Carbon and Biosphere*, eds. Woodwell, G. M. and Pecan, E. V. U.S. Dep. Of Comm., Springfield, Va, pp. 86–135.

Baker, D. F., Law, R. M., Gurney, K. R. *et al.* (2006). TransCom 3 inversion intercomparison: Impact of transport model errors on the interannual variability of regional CO_2 fluxes, 1988–2003. *Global Biogeochemical Cycles*, 20, GB1002.

Baldocchi, D., Falge, E., Gu, L. H. *et al.* (2001). FLUXNET: A new tool to study the temporal and spatial variability of ecosystem-scale carbon dioxide, water vapor, and energy flux densities. *Bulletin of the American Meteorological Society*, 82, 2415–2434.

Berkofsky, L. (1976). The effect of variable surface albedo on the atmospheric circulation in desert regions. *Journal of Applied Meteorology*, 15, 1139–1144.

Berthelot, M., Friedlingstein, P., Ciais, P., Dufresne, J.-L., and Monfray, P. (2005). How uncertainties in future climate change predictions translate into future terrestrial carbon fluxes. *Global Change Biology*, 11, 959–970.

Betts, R. A. (2000). Offset of the potential carbon sink from boreal forestation by decreases in surface albedo. *Nature*, 408, (6809), 187–190.

Betts, R. A (2001). Biogeophysical impacts of land use on present-day climate: near-surface temperature change and radiative forcing. *Atmospheric Science Letters*, 2, Issue 1, 39–51.

Betts, A. K., Ball, J., and McCaughey, J. (2001). Near-surface climate in the boreal forest. *Journal of Geophysical Research*, 106, 33 529–33 541.

Betts, R. A., Cox, P. M., Collins, M., Harris, P. P., Huntingford, C., and Jones, C. D. (2004). The role of ecosystem–atmosphere interactions in simulated Amazonian precipitation decrease and forest dieback under global climate warming. *Theoretical and Applied Climatology*, 78, 157–175.

Betts, R. A., Falloon, P. D., Goldewijk, K. K., and Ramankutty, N. (2007). Biogeophysical effects of land use on climate: Model simulations of radiative forcing and large-scale temperature change. *Agricultural and Forest Meteorology*, 142, 216–233.

Bolin, B. (1977). Changes of land biota and their importance for the carbon cycle. *Science*, 196, No. 4290, 613–615.

Bonan, G. B. (1995). Land-atmosphere CO_2, exchange simulated by a land surface process model coupled to an atmospheric general circulation model. *Journal of Geophysical Research*, 100, 2817–2831.

Bonan, G. B. (1997). Effect of land use on the climate of the United States. *Climactic Change*, 37, 449–486.

Bonan, G. B. (2008). Forests and climate change: Forcings, feedbacks, and the climate benefits of forests. *Science*, 320(5882), 1444–1449.

Bonan, G. B., Levis, S., Kergoat, L., and Oleson, K. W. (2002). Landscapes as patches of plant functional types: an integrating concept for climate and ecosystem models. *Global Biogeochemical Cycles*, 16, 1021.

Bonan, G. B., Levis, S., Sitch, S., Vertenstein, M., and Oleson, K. W. (2003). A dynamic global vegetation model for use with climate models: concepts and description of simulated vegetation dynamics. *Global Change Biology*, 9, 1543–1566.

Botkin, D. B., Janak, J. F., and Wallis, J. R. (1972). Some ecological consequences of a computer model of forest growth. *Journal of Ecology*, 60, 849–872.

Box, E. O. (1981). *Macroclimate and Plant Forms: An Introduction to Predictive Modeling in Phytogeography.* Springer, The Hague, Netherlands, 258pp.

Canadell, J. G., Le Quere, C., Raupach, M. R. *et al.* (2007). Contributions to accelerating atmospheric CO_2 growth from economic activity, carbon intensity, and efficiency of natural sinks, *Proceedings of the National Academy of Sciences*, 104, No. 47, 18 866–18 870.

Charney, J. G. (1975). Dynamics of deserts and drought in the Sahel. *Quarterly Journal of the Royal Meteorological Society*, 101, no. 428, 193–202.

Chen, M., Dickinson, R. E., Zeng, X., and Hahmann, A. N. (1996). Comparison of precipitation observed over the continental United States to that simulated by a climate model. *Journal of Climate*, 9, 2233–2249.

Claussen, M. (1994). Coupling global biome models with climate models. *Climate Research*, 4, 203–221.

Collatz, G. J., Ball, J. T., Grivet, C., and Berry, J. A. (1991). Physiological and environmental regulation of stomatal conductance, photosynthesis and transpiration:

A model that includes a laminar boundary layer. *Agricultural and Forest Meteorology*, 54, 107–136.

Cook, B. I., Miller, R. L., and Seager, R. (2009). Amplification of the North American "Dust Bowl" drought through human-induced land degradation. *Proceedings of the National Academy of Sciences*, 106, Issue 13, 4997–5001.

Cox, P. M., Betts, R. A., Jones, C. D., Spall, S. A., and Totterdell, I. J. (2000). Acceleration of global warming due to carbon-cycle feedbacks in a coupled climate model. *Nature*, 408, 184–187.

Cox, P. M., Betts, R. A., Collins, M., Harris, P. P., Huntingford, C., and Jones, C. D. (2004). Amazonian forest dieback under climate-carbon cycle projections for the 21st century. *Theoretical and Applied Climatology*, 78, 137–156.

Crous, K. Y., Reich, P. B., Hunter, M. D., and Ellsworth, D. S., (2009). Maintenance of leaf N controls the photosynthetic CO_2 response of grassland species exposed to nine years of free-air CO_2 enrichment. *Global Change Biology*, (in press), doi: 10.1111/j.1365–2486.2009.02058.x.

Davidson, E. A., Janssens, I. A., and Luo, Y. (2006). On the variability of respiration in terrestrial ecosystems: moving beyond *Q*10. *Global Change Biology*, 12, 154–164.

Deardorff, J. (1978). Efficient prediction of ground surface temperature and moisture with inclusion of a layer of vegetation. *Journal of Geophysical Research*, 83, 1889–1903.

Denning, A. S., Fung, I. Y., and Randall, D. A. (1995). Gradient of atmospheric CO_2 due to seasonal exchange with land biota. *Nature*, 376, 240–243.

Denning, A. S., Randall, D. A., Sellers, P. J., and Collatz, G. J. (1996). Simulations of terrestrial carbon metabolism and atmospheric CO_2 in a general circulation model: Part 2: Simulated CO_2 concentrations. *Tellus B*, 48, No. 4, 543–567(22).

Desborough, C. E., Pitman, A. J., and Irannejad, P. (1996). Analysis of the relationship between bare soil evaporation and soil moisture simulated by 13 land surface schemes for a simple non-vegetated site. *Global and Planetary Change*, 13, 47–56.

Dickinson, R. E. (1983). Land surface processes and climate – Surface albedos and energy balance. In *Theory of Climate, Advances in Geophysics, B.*, ed. Saltzman, Academic Press, New York, **25**, pp. 305–353.

Dickinson, R. E. (1984). Modelling evapotranspiration for three dimensional global climate models. In *Climate Processes and Climate Sensitivity*, ed. Hansen, J. E. and Takahashi, T., Geophysical Monograph 29, Maurice Ewing, *American Geophysical Union*, **5**:58–72.

Dickinson, R. E. (1987). *The Geophysiology of Amazonia*: *Vegetation and Climate Interactions*. John Wiley & Sons, Ltd.

Dickinson, R. E. and Henderson-Sellers, A. (1988). Modelling tropical deforestation: a study of GCM land-surface parameterizations. *Quarterly Journal of the Royal Meteorological Society*, 114(B), 439–462.

Dickinson, R. E. and Kennedy, P. (1992). Impacts on regional climate of Amazon deforestation. *Geophysical Research Letters*, 19, 1947–1950.

Dickinson, R. E., Henderson-Sellers, A., Kennedy, P. J., and Wilson, M. F. (1986). Biosphere/atmosphere transfer scheme (BATS) for the NCAR community climate model. *NCAR Technical Note, TN-275 + STR*, 69.

Dickinson, R. E., Shaikh, M., Bryant, R., and Graumlich, L. (1998). Interactive canopies for a climate model. *Journal of Climate*, 11, 2823–2836.

Dickinson, R. E., Berry, J. A., Bonan, G. B. *et al.* (2002). Nitrogen controls on climate model evapotranspiration. *Journal of Climate*, 15, No. 3, 278–295.

Dickinson, R. E., Wang, G., Zeng, X., and Zeng, Q.-C. (2003). How does the partitioning of evapotranspiration and runoff between different processes affect the variability and predictability of soil moisture and precipitation? *Advances in Atmospheric Science*, 20(3), 475–478.

Eamus, D. and Jarvis, P. G. (1989). The direct effects of increase in the global atmospheric CO_2 concentration on natural and commercial temperate trees and forests. *Advances in Ecological Research*, 19, 1–55.

Ek, M. B and Holtslag, A. A. M. (2004). Influence of soil moisture on boundary layer cloud development. *Journal of Hydrometeorology*, 5, 86–99.

Eltahir, E. A. B. and Bras, R. L. (1993). A description of rainfall interception over large areas. *Journal of Climate*, 6, 1002.

Emanuel, W. R., Shugart, H. H., and Stevenson, M. (1985). Climate change and the broad-scale distribution of terrestrial ecosystem complexes-response. *Climatic Change*, 7(4), 457–460.

Farquhar, G. D., von-Caemmerer, S., and Berry, J. A. (1980). A biochemical model of photosynthetic CO_2 species. *Planta*, 149, No. 1, 78–90.

Foley, J. A., Prentice, I. C., Ramankutty, N. *et al.* (1996). An integrated biosphere model of land surface processes, terrestrial carbon balance, and vegetation dynamics. *Global Biogeochemical Cycles*. 10, 603–628.

Foley, J. A., Levis, S., Prentice, I. C., Pollard, D., and Thompson, S. L. (1998). Coupling dynamic models of climate and vegetation. *Global Change Biology*, 4, Issue 5, 561–579.

Foley, J. A., DeFries, R., Asner, G. P. *et al.* (2005). Global consequences of land use. *Science*, 309, 570–574.

Franklin, O. (2007). Optimal nitrogen allocation controls tree responses to elevated CO_2. *New Phytologist*, 174, 811–822.

Freedman, J. M., Fitzjarrald, D. R., Moore, K. E., and Sakai, R. K. (2001). Boundary layer clouds and vegetation-atmosphere feedbacks. *Journal of Climate*, 14, 180–197.

Friedl, M. A., McIver, D. K., Hodges, J. C. F. *et al.* (2002). Global land cover mapping from MODIS: algorithms and early results. *Remote Sensing of Environment*, 83, 287–302.

Friedlingstein, P., Fung, I., Holland, E. *et al.* (1995). On the contribution of CO_2 fertilization to the missing biospheric sink. *Global Biogeochemical Cycles*, 9, No. 4, 541–556.

Friedlingstein, P., Cox, P., Betts, R. *et al.* (2006). Climate–carbon cycle feedback analysis: Results from the C4MIP Model Intercomparison, *Journal of Climate*, 19, 3337–3353.

Fritschen, L. J. (1967). Net and solar radiation relations over irrigated field crops. *Journal of Applied Meteorology*, 2, 55–62.

Fung, I., Prentice, K., Matthews, E., Lerner, J., and Russell, G. (1983). Three-dimensional tracer model study of atmospheric CO_2: Response to seasonal exchanges with the terrestrial biosphere. *Journal of Geophysical Research*, 88, No. C2, 1281–1294.

Fung, I. Y., Doney, S. C., Lindsay, K., and John, J. (2005). Evolution of carbon sinks in a changing climate. *Proceedings of the National Academy of Sciences*, 102, No. 32, 11 201–11 206.

Gamon, J. A., Huemmrich, K. F., Peddle, D. R. *et al.* (2004). Remote sensing in BOREAS: Lessons learned. *Remote Sensing of Environment* 89, 139–162.

Gao, X. and Sorooshian, S. (1994). A stochastic precipitation disaggregation scheme for GCM applications. *Journal of Climate*, 7, 238–247.

Gash, J. H. C., Wright, I. R., and Lloyd, C. R. (1980). Comparative estimates of interception loss from the three coniferous forests in Great Britain. *Journal of Hydrology*, 48, 89–105.

Gates, D. M. (1980). *Biophysical Ecology*. USA: Springer-Verlag, pp. 1–603.

Hashimoto, H., Nemani, R. R., White, M. A. *et al.* (2004). El Niño–Southern Oscillation–induced variability in terrestrial carbon cycling. *Journal of Geophysical Research*, 109, D23110.

Henderson-Sellers, A. (1993). Continental vegetation as a dynamic component of a global climate model – A preliminary assessment. *Climate Change*, 23, 337–377.

Henderson-Sellers, A. and Gornitz, V. (1984). Possible climatic impacts of land cover transformations, with particular emphasis on tropical deforestation. *Climate Change*, 6, No. 3, 231–257.

Henderson-Sellers, A., McGuffie, K., Zhang, L., and Gross, C. (1995). Sensitivity of global climate model simulations to increased stomatal resistance and CO_2 increases. *Journal of Climate*, 8, 1738–1756.

Holdridge, L. R. (1967). *Life Zone Ecology*. Tropical Science Center, San Jose, Costa Rica, 206pp.

Hurtt, G. C., Frolking, S., Fearon, M. G. *et al.* (2006). The underpinnings of land-use history: three centuries of global gridded land-use transitions, wood-harvest activity, and resulting secondary lands. *Global Change Biology*, 12, 1208–1229.

Idso, S. B. and Deardorff, J. W. (1978). Comments on the effect of variable surface albedo on the atmospheric circulation in desert regions. *Journal of Applied Meteorology*, 17, 560.

Jackson, R. B., Jobbágy, E. G., Avissar, R. *et al.* (2005). Trading water for carbon with biological carbon sequestration. *Science*, 310,1944–1947.

Jarvis, P. G. (1976). The interpretation of the variations in leaf water potential and stomatal conductance found in canopies in the field. *Philosophical Transactions of the Royal Society of London*, Series B, 273, 593–610.

Jarvis, P. G. and McNaughton, K. G. (1986). Stomatal control of transpiration scaling up from leaf to region. *Advances in Ecological Research*, 15, 1–49.

Jones, C. D., Collins, M., Cox, P. M., and Spall, S. A. (2001). The carbon cycle response to ENSO: A coupled climate–carbon cycle model study. *Journal of Climate*, 14, Issue 21, 4113–4129.

Keeling, C. D., Bacastow, R. B., Bainbridge, A. E., Ekdahl, C. A., Guenther P. B., and Waterman, L. S. (1976). Atmospheric carbon dioxide variations at Mauna Loa Observatory, Hawaii. *Tellus*, 28, 538–551.

Kleidon, A. and Heimann, M. (2000). Assessing the role of deep rooted vegetation in the climate system with model simulations: mechanism, comparison to observations and implications for Amazonian deforestation. *Climate Dynamics*, 16, 183–199.

Koster, R. D. and Suarez, M. J. (1992). Modeling the land surface boundary in climate models as a composite of independent vegetation stands. *Journal of Geophysical Research*, 97**(D3)**, 2697.

Krinner, G., Viovy, N., Noblet-Ducoudré, N. *et al.* (2005). A dynamic global vegetation model for studies of the coupled atmosphere-biosphere system. *Global Biogeochemical Cycles*, 19, GB1015.

Kucharik, C. J., Foley, J. A., Delire, C. *et al.* (2000). Testing the performance of a dynamic global ecosystem model: Water balance, carbon balance, and vegetation structure. *Global Biogeochemical Cycles*, 14, No. 3, 795–825.

Kvalevåg, M. M., Myhre, G., Bonan, G., and Lewis, S. (2009). Anthropogenic land cover changes in a GCM with surface albedo changes based on MODIS data. *International Journal of Climatology*. 10.1002/joc.2012.

Lawrence, D. M. and Slater, A. G. (2008). Incorporating organic soil into a global climate model. *Climate Dynamics*, 30, 145–160.

Lawrence, P. J. and Chase, T. N. (2010). Investigating the climate impacts of global land cover change in the community climate system model (CCSM). *International Journal of Climatology*, 10.1002/joc.2061.

Le Quere, C., Rodenbeck, C., Buitenhuis, E. T. *et al.* (2007). Saturation of the southern ocean CO_2 sink due to recent climate change. *Science*, 316, 1735–1738.

Leuning, R., Wang, Y. P., Pury, D. de *et al.* (1993). Growth and water use of wheat under present and future levels of CO_2. *Journal of Agricultural Meteorology*, 48, 807–810.

Lieth, H. (1973). Primary production: terrestrial ecosystems. *Human Ecology*, 1, 303–331.

Lieth, H. (1984). Net primary production deduced with the Hamburg model from climate change prediction with GCMs for elevated CO_2 scenarios. *Interaction between Climate and Biosphere*, eds Lieth, H., Fantechi, R., and Schnitzler, H., Progress in Biometeorology, Vol. 3, Swets and Zeitlinger, Lisse, the Netherlands, pp. 335–343.

Lieth, H. and Box, E. (1977). The gross primary productivity pattern of the land vegetation: A first attempt. *Tropical Ecology*, 18, 109–115.

Maayar, M. E., Ramankutty, N., and Kucharik, C. J. (2006). Modeling global and regional net primary production under elevated atmospheric CO_2: On a potential source of uncertainty. *Earth Interactions*, 10, 1–20.

Manabe, S. (1969a). Climate and the ocean circulation. I. The atmospheric circulation and the hydrology of the Earth's surface. *Monthly Weather Review*, 97, Issue 11, 739–774.

Manabe, S. (1969b). Climate and the ocean circulation. II. The atmospheric circulation and the effective of heat transfer by ocean currents. *Monthly Weather Review*, 97, Issue 11, 775–805.

Matthews, E. (1983). Global vegetation and land use: New high-resolution data bases for climate studies. *Journal of Climate and Applied Meteorology*, 22, 474–487.

McCarthy, J. F., Ilavsky, J., Jastrow, J. D., Mayer, L. M., Perfect, E., and Zhuang, J. (2008). Protection of organic carbon in soil microaggregates via restructuring of aggregate porosity and filling of pores with accumulating organic matter, *Geochimica et Cosmochimica Acta*, 72, 4725–4744.

McGuire, A. D., Anderson, L. G., Christensen, T. R. *et al.* (2009). Sensitivity of carbon cycle in the Arctic to climate change. *Ecological Monographs*, 79(4), 523–555.

Melillo, J. M., McGuire, A. D., Kicklighter, D. W., Moore, B., Vorosmarty, C. J., and Schloss, A. L. (1993). Global climate change and terrestrial net primary production. *Nature*, 363, 234–240.

Monteith, J. L. (1972). Solar radiation and productivity in tropical ecosystems. *Journal of Applied Ecology*, 9, 747–66.

Monteith, J. L. and Szeicz, G. (1961). The radiation balance of bare soil and vegetation. *Quarterly Journal of the Royal Meteorological Society*, 92, 128–140.

Myhre, G., Kvalevag, M. M., and Schaaf, C. B. (2005). Radiative forcing due to anthropogenic vegetation change based on MODIS surface albedo data. *Geophysical Research Letters*, 32,1–4.

Namias, J. (1955). Some empirical aspects of drought with special reference to the summers of 1952–54 over the United States. *Monthly Weather Review*, 83, 199–205.

Norby, R. J., DeLucia, E. H., Gielen, B. *et al.* (2005). Forest response to elevated CO_2 is conserved across a broad range of productivity. *Proceedings of the National Academy of Sciences*, 102, 18 052–18 056.

Oleson, K. W., Bonan, G. B., Levis, S., and Vertenstein, M. (2004a). Effects of land use change on North American climate: Impact of surface datasets and model biogeophysics. *Climate Dynamics*, 23, 117–132.

Oleson, K. W., Dai, Y., Bonan, G. *et al.* (2004b). Technical description of the community land model (CLM). NCAR technical note, NCAR/TN-461+STR.

Olson, J. S., Watts, J. A., and Allison, L. J. (1983). *Carbon in Live Vegetation of Major World Ecosystems. Oak Ridge National Lab*, Oak Ridge, TN.

Overpeck, J. T., Rind, D., and Goldberg, R. (1990). Climate-induced changes in forest disturbance and vegetation. *Nature*, 343, 51–53.

Parton, W. J., Scurlock, J. M. O., Ojima, D. S. *et al.* (1993). Observations and modeling of biomass and soil organic matter dynamics for the grassland biome worldwide. *Global Biogeochemical Cycles*, 7, No. 4, 785–809.

Pielke, R. A., Avissar, R., Raupach, M., Dolman, H., Zeng, X., and Denning, S. (1998). Interactions between the atmosphere and terrestrial ecosystems: Influence on weather and climate. *Global Change Biology*, 4, 461–475.

Pinty, B., Lavergne T., Dickinson, R. E., Widlowski, J.-L., Gobron, N., and Verstraete M. M (2006). Simplifying the interaction of land surfaces with radiation for relating remote sensing products to climate models. *Journal of Geophysical Research*, 111, D02116.

Pollard, D. and Thompson, S. L. (1995). Use of land-surface-transfer scheme (LSX) in a global climate model: the response to doubling stomatal resistance. *Global and Planetary Change*, 10, Issues 1–4, 129–161.

Potter, C. S. and Klooster, S. A. (1997). Global model estimates of carbon and nitrogen storage in litter and soil pools: response to changes in vegetation quality and biomass allocation. *Tellus*, 49B, 1–17.

Potter, C. S., Randerson, J. T., Field, P. A. *et al.* (1993). Terrestrial ecosystem production: A process model based on global satellite and surface data. *Global Biogeochemical Cycles*, 7, No. 4, 811–841.

Prentice, I. C., Cramer, W., Harrison, S., Leemans, R., Monserud, R. A., and Solomon, A. M. (1992). A global biome model based on plant physiology and dominance, soil properties and climate. *Journal of Biogeography*, 19, 117–134.

Ramankutty, N. and Foley, J. A. (1999). Estimating historical changes in global land cover: Croplands from 1700 to 1992. *Global Biogeochemical Cycles*, 13, No. 4, 997–1027.

Randerson, J. T., van der Werf, G. R., Collatz, G. J. *et al.* (2005). Fire emissions from C_3 and C_4 vegetation and their influence on interannual variability of atmospheric CO_2 and $\delta^{13}CO_2$. *Global Biogeochemical Cycles*, 19, GB2019.

Randerson, J. T., Hoffman, F. M., Thornton, P. E. *et al.* (2009). Systematic assessment of terrestrial biogeochemistry in coupled climate-carbon models. *Global Change Biology*, 15, Issue 10, 2462–2484.

Roderick, M. L., Farquhar, G. D., Berry, S. L., and Noble, I. R. (2001). On the direct effect of clouds and atmospheric particles on the productivity and structure of vegetation. *Oecologia*, 129, 21–30.

Rosenberg, N. J. (1981). The increasing CO_2, concentration on the atmosphere and its implications on agricultural productivity. *Climatic Change*, 3, 265–279.

Running, S. W. and Coughlan, J. C. (1988). A general model of forest ecosystem processes for regional applications, I. Hydrologic balance, canopy gas exchange and primary production processes. *Ecological Modeling*, 42, 125–154.

Sagan, C., Toon, O. B., and Pollack, J. (1979). Anthropogenic albedo changes and the Earth's climate. *Science*, 206, No. 4425, 1363–1368.

Salmun, H. and Molod, A. (2006). Progress in modeling the impact of land cover change on the global climate. *Progress in Physical Geography*, 30, 737–749.

Schneider, S. H. (2004). A goddess of the earth?: The debate on the Gaia Hypothesis – An editorial. *Climatic Change*, 8, No. 1, 1–4.

Sellers, P. J. (1985). Canopy reflectance, photosynthesis and transpiration. *International Journal of Remote Sensing*, 6, 1335–1372.

Sellers, P. J., Mintz, Y., Sud, Y. C., and Dalcher, A. (1986). A simple biosphere model (SiB) for use within general circulation models. *Journal of the Atmospheric Sciences*, 43, 505–531.

Sellers, P. J., Randall, D. A., Collatz, G. J. *et al.* (1996a). A revised land surface parameterization (SiB2) for atmospheric GCMs. Part I: Model formulation. *Journal of Climate*, 9, 676–705.

Sellers, P. J., Bounoua, L., Collatz, G. J. *et al.* (1996b). Comparison of radiative and physiological effects of doubled atmospheric CO_2 on climate. *Science*, 8, 1402–1406.

Sellers, P. J., Dickinson, R. E., Randall, D. A. *et al.* (1997). Modeling the exchanges of energy, water, and carbon between continents and the atmosphere. *Science*, 275, 502–509.

Seth, A., Giorgi, F. and Dickinson, R. E. (1994). Simulating fluxes from heterogeneous land surfaces: Introducing a vectorized version of Biosphere-Atmosphere Transfer Scheme (VBATS). *Journal of Geophysical Research*, 99, 18,651.

Shukla, J. and Mintz, Y. (1982). Influence of land-surface evapotranspiration on the earth's climate. *Science*, 215, 1498–1501.

Shuttleworth, W. J. (1988). Evaporation from Amazonian rainforest. *Proceedings of the Royal Society of London, Series B*, 233, 321–346.

Sitch, S., Smith, B., Prentice, I. C. *et al.* (2003). Evaluation of ecosystem dynamics, plant geography and terrestrial carbon cycling in the LPJ dynamic global vegetation model. *Global Change Biology*, 9, No. 2, 161–185.

Solomon, A. M. (1986). Transient response of forests to CO_2-induced climate change: simulation modeling experiments in eastern North America. *Oecologia*, 68, 567–579.

Solomon, A. M. and Shugart, H. H. (1992). *Vegetation Dynamics & Global Change*. Chapman & Hall, New York, 338pp.

Tans, P. P., Fung, I. Y., and Takahashi, T. (1990). Observational constraints on the global atmospheric CO_2 budget. *Science*, 247, 1431–1438.

Tenhunen, J. D., Yocum, C. S., and Gates, D. M. (1976). Development of a photosynthesis model with an emphasis on ecological applications. I. Theory. *Oecologia*, 26, 89–100.

Thompson S. L., Govindasamy, B., Mirin, A. *et al.* (2004). Quantifying the effects of CO_2-fertilized vegetation on future global climate and carbon dynamics. *Geophysical Research Letters*, 31, L23211.

Thornton, P. E., Lamarque, J.-F., Rosenbloom, N. A., and Mahowald, N. M. (2007). Influence of carbon-nitrogen cycle coupling on land model response to CO_2 fertilization and climate variability. *Global Biogeochemical Cycles*, 21, GB4018.

Thornton, P. E., Doney, S. C., Lindsay, K. *et al.* (2009). Carbon-nitrogen interactions regulate climate-carbon cycle feedbacks: results from an atmosphere-ocean general circulation model. *Biogeosciences*, 6, 2099–2120.

Trumbore, S. (2006). Carbon respired by terrestrial ecosystems – recent progress and challenges. *Global Change Biology*, 12, 141–153.

Watson, A. J. and Lovelock, J. E. (1983). Biological homeostasis of the global environment: the parable of Daisyword. *Tellus B*, 35, 284–289.

Whittaker, R. H. and Likens, G. E. (1975). The biosphere and man. In *The Primary Production of the Biosphere*, Springer-Verlag, NY, pp. 302–328.

Wilson, M. F. and Henderson-Sellers, A (1985). Land cover and soils data sets for use in general circulation climate models. *Journal of Climatology*, 5, 119–143.

Winguth, A. M. E., Heimann, M., Kurz, K. D., Maier-Reimer, E., Mikolajewicz, U., and Segschneider, J. (1994). El-Niño – Southern Oscillation related fluctuations of the marine carbon cycle. *Global Bigeochemical Cycles*, 8(1), 39–63.

Woodward, F. I. (1987). *Climate and Plant Distribution*. Cambridge University Press, Cambridge, MA, 174pp.

Yang, Z.-L., Dickinson, R. E., Henderson-Sellers, A., and Pitman, A. J. (1995). *Journal of Geophysical Research*, 100, 16 553–16 578.

Zeng, X., Dickinson, R. E., Walker, A., Shaikh, M., DeFries, R. S., and Qi, J. G. (2000). Derivation and evaluation of global 1-km fractional vegetation cover data for land modeling. *Journal of Applied Meteorology*, 39, 826–839.

9

The evolution of complexity in general circulation models

DAVID RANDALL

9.1 Introduction

We all value simplicity. Einstein famously remarked that a theory should be as simple as possible, but not simpler. On the other hand, the myriad phenomena of the global atmosphere are undeniably complex. The majestic Hadley and Walker cells, monsoons and planetary waves, mid-latitude baroclinic waves and tropical typhoons, squall lines and thunderstorms, and the turbulent eddies of the boundary layer perpetually interact across a huge range of space and time-scales, and across the full range of the Earth's weather regimes, without spectral gaps, and without the slightest regard for our human preference for simplicity.

Atmospheric general circulation models (AGCMs) are intended to simulate the many emergent phenomena of the global circulation by starting from fundamental physical principles that apply on small scales. AGCMs are among humanity's most elaborate creations. The trend has been towards ever more complex AGCMs, because the amazing intricacy of the real atmosphere motivates continuing refinement, because the relentless growth of computing power makes possible increasingly comprehensive simulations, and because society's appetite for more detailed and quantitatively accurate predictions can never be satisfied (WCRP, 2009).

In the context of AGCMs, it is useful to distinguish three types of complexity:

- The *conceptual complexity* of AGCMs is a measure of the intellectual effort needed to understand their formulations. It has been increasing rapidly (Claussen *et al.*, 2002), driven by the deepening subtlety of the underlying ideas, the growing level of physical detail, the increasing sophistication of the mathematical methods, and the sheer size of the computer codes that embody the models.
- *Coupling complexity* arises because AGCMs are including ever more coupled processes, and are linked to an increasingly wide variety of similarly elaborate models representing other components of the Earth system.

- Finally, the spatial and temporal resolutions of the models are being dramatically refined, giving rise to *numerical complexity*, which can be measured by the sheer number of numbers needed to represent the state of a simulation.

Complexity creates challenges. Conceptual complexity raises the ante in terms of the talent and training needed by modelers, while slowing the publication process that is key to the reward system for scientific professionals. Coupling complexity makes the results of a model vulnerable to the deficiencies of its weakest component, regardless of the merits of its stronger components. This also slows the publication process, because the strengths of a model component may be hidden by the weaknesses of other components. Numerical complexity makes simulations expensive and time consuming, even on the fastest available computers, and it renders the model output difficult to store, transport, analyze, visualize, and interpret.

Increasing the number of lines of code in a model does not necessarily increase our confidence in the model's predictions,[1] and the results produced by longer codes are often harder to interpret. Complex models sometimes have impressive predictive power, but only simple models have true explanatory power. For example, suppose that a complex AGCM is used to forecast a severe winter storm on the east coast of North America, and that the prediction turns out to be accurate. Does the numerical forecast explain why the storm developed? Most scientists would say no. We look for explanations in the form of simple but physically based interpretations, often in the form of idealized models, which distill the essence of complex observations or simulations while omitting or glossing over the details (e.g. Held, 2005). *Useful explanations are simple, more or less by definition.*

This chapter discusses the complexity of AGCMs as it has evolved up to now, analyzes the tradeoffs among the different types of complexity, and projects future developments, especially in light of the continuing rapid increase in computer power and society's dramatically increasing reliance on model-generated predictions of weather and climate.

9.2 In the beginning

It is useful to compare today's models with their esteemed ancestors. The early predecessors of today's models are discussed in Chapters 2 and 3. The first true AGCMs, created during the early 1960s, are discussed in Chapter 3 by Washington and Kasahara. They were actually similar in many ways to today's models. It is perhaps slightly embarrassing that some of today's AGCMs still contain a few

[1] Here I am roughly quoting Bjorn Stevens of the Max Planck Institute for Meteorology.

snippets of code that, like highly conserved genes, have survived unchanged from that primordial era. The ancestral models were:

- *The GFDL[2] model, created by Joseph Smagorinsky, Syukuro Manabe, and colleagues (Smagorinsky et al.,1965; Manabe et al., 1965).* The early GFDL AGCM was distinguished by its relatively high vertical resolution: nine glorious layers, in an age of two-layer models. Today, GFDL is still very active in the AGCM arena.
- *The UCLA[3] model, also called the Mintz–Arakawa model, created by Akio Arakawa in collaboration with Yale Mintz (Mintz, 1965).* This two-layer model was the first to overcome the nonlinear numerical instability discovered by Norman Phillips (1956, 1959), through the use of energy-conserving numerical methods (Arakawa, 1966). It was also the first AGCM to predict the distribution of radiatively active clouds. Drastically updated versions of the UCLA model are still in use today.
- *The Livermore model (Leith, 1965), created by Cecil Leith of the Livermore Radiation Laboratory[4].* This five-layer model had a short lifetime, because its creator moved on to the National Center for Atmospheric Research (NCAR) which had its own AGCM development effort, as discussed below.
- The NCAR[5] model, created by Akira Kasahara, Warren Washington, and colleagues (Kasahara and Washington, 1967; Washington, 2007). This began as another two-layer model. The first version did not predict the distribution of water vapor, but assumed saturation everywhere, so that latent heat was released wherever and whenever the air moved upward. Chapter 3 discusses this model in some detail.

The four ancestral models listed above shared several common elements. All were developed in the United States, although the developers of three of them (GFDL, UCLA, and NCAR) included scientists who had immigrated to the USA from Japan (Figure 9.1). All were developed primarily for the sake of their scientific or academic utility, rather than for any immediate practical applications, although of course the potential for such applications was apparent to the model-builders, and to scientific leadership more broadly, including Harry Wexler, as discussed in Chapter 4.

All four models used finite differences to represent the three-dimensional structure of the atmosphere. All used horizontal grids based on spherical coordinates. Leith's model included a particularly strong horizontal smoothing that suppressed realistic variability (Charney *et al.*, 1966). The sigma coordinate proposed by Norman Phillips (1957) was used in the GFDL and UCLA models. Leith used a pressure coordinate, with a predicted surface pressure, despite the complexities

[2] Geophysical Fluid Dynamics Laboratory. [3] University of California at Los Angeles.
[4] Now called the Lawrence Livermore National Laboratory.
[5] U.S. National Center for Atmospheric Research.

Figure 9.1 The creators of the world's first four AGCMs. Top left: Syukuro Manabe (from the AIP Emilio Segrè Visual Archives, with permission from

associated with pressure surfaces whose intersections with the Earth's surface move as the model runs. A unique feature of the early NCAR model is that, to this day, it is the only AGCM that has ever used height as its vertical coordinate. The impending era of nonhydrostatic AGCMs, discussed later, may see the return of the height coordinate.

All four models included suites of physical parameterizations to represent solar and terrestrial radiation, the effects of latent heat release, and the surface fluxes of sensible heat, moisture, and momentum. The specifics of the parameterizations differed drastically among the models, but in general they were much simpler than the parameterizations in today's AGCMs. The early GFDL model included moist convective adjustment, which is still occasionally used today, and also the simple "bucket model" representation of the land surface. The bucket model was quickly adopted by other modeling groups; e.g. it was used in the UCLA model. The UCLA AGCM used the earliest "mass flux" representation of cumulus convection (Arakawa, 1969).

Further discussion of the early days of AGCMs can be found in Chapter 3 and in the papers by Smagorinsky (1983), Lewis (1998), Arakawa (2000), Edwards (2000), and the books by Washington (2007) and Edwards (2010). Additional information is available on the following websites:

http://www.aip.org/history/sloan/gcm/
http://www.aip.org/history/climate/GCM.htm#L000

9.3 Numerics

The components of an AGCM that solve the equations of fluid motion, including the thermodynamic energy equation and various constituent-advection equations, are conventionally referred to as the "dynamical core" of the model. Modeling enthusiasts can be overheard comparing the designs of their "dycores."

As mentioned above, the dynamical cores of all of the early AGCMs used finite difference methods in both the horizontal and vertical. The early UCLA model used what is now called the "finite-volume" approach, in which conservation principles

Caption for Figure 9.1 (cont.)
Syukuro Manabe). Top right: Joseph Smagorinsky of GFDL (Provided by Geophysical Fluid Dynamics Laboratory/NOAA, Princeton, New Jersey, USA). Middle: Yale Mintz and Akio Arakawa of UCLA (Courtesy of Michio Yanai). Bottom left: Cecil Leith of the Lawrence Radiation Laboratory (Copyright, University Corporation for Atmospheric Research). Bottom right: Akira Kasahara and Warren Washington of NCAR. (Copyright, University Corporation for Atmospheric Research).

for mass, momentum, and energy are emphasized (see the retrospective by Arakawa, 2000). Some of the early finite-difference models worked pretty well; others did not. A problem with finite-difference methods based on spherical coordinates is that the meridians converge at both poles. This "pole problem" necessitates the use of very short timesteps unless a filter of some kind is used to remove short zonal wavelengths at high latitudes (e.g. Arakawa and Lamb, 1977).

Spectral methods eliminate the pole problem for wave propagation, but not for advection. The basic idea is to represent the horizontal structure of the global atmosphere using truncated spherical harmonic expansions (Silberman, 1954; Platzman, 1960). By including a sufficient number of spherical harmonics, the resolution can be made as high as desired. A problem with this approach is that the quadratically nonlinear processes of a model, notably advection, are represented by sums in which the numbers of terms increase quadratically with the number of spherical harmonics kept, making high-resolution models impractical. A further difficulty is that it would be virtually impossible to formulate the physical parameterizations of a model in wavenumber space.

These obstacles were overcome by Orszag, and independently by Eliasen *et al.*, both in 1970. They proposed the "transform method," in which a grid is used to evaluate the products appearing in the nonlinear terms, while the spectral method continues to be used to evaluate the horizontal derivatives appearing in these terms. The physical parameterizations, many of which are also highly nonlinear, are also evaluated on the grid. Transforms are used to go back and forth between wavenumber space and the grid, as needed.

The spectral method was further developed and advocated by Bourke (1974), Baer (1972), and others. One of its strengths is that it can easily be adapted to semi-implicit time differencing for the linear terms of the dynamical core. The spectral method became popular after it was adopted by GFDL (Manabe and Hahn, 1981) and the European Centre for Medium Range Weather Forecasts (ECMWF; Jarraud and Simmons, 1983).

The implementation of the spectral transform method has become fairly standard, with just a few variations, and in many respects it works well. In contrast, there are infinitely many ways to construct a finite-difference scheme, almost all of them bad. From this point of view, the spectral approach is simpler than the finite-difference approach. The widespread adoption of spectral methods during the 1980s can therefore be viewed as a step towards simplification of the dynamical cores of AGCMs – a retreat from complexity. Note, however, that the finite-difference dynamical cores that worked well in 1980 were in many cases retained (and further refined) by their modeling groups (e.g. the United Kingdom Meteorological Office, and UCLA).

A serious drawback of the spectral transform method is that, when applied to advection, it has a strong tendency to produce spurious negative concentrations of atmospheric constituents such as water vapor (e.g. Williamson and Rasch, 1994). Spectral advection also requires the use of relatively short timesteps, because of the pole problem mentioned earlier. Both of these issues have been addressed through the introduction of the "semi-Lagrangian" advection method (Ritchie *et al.*, 1995), as a replacement for spectral-transform advection. Semi-Lagrangian advection is intrinsically stable and sign-preserving, and allows long timesteps. It is a gridpoint method, and its widespread adoption starting in the 1990s can be viewed as a partial retreat from the spectral method.

In fact, much of the recent work on AGCM dynamical cores has been based on finite-difference methods. This is motivated in part by the recent development of very high-resolution AGCMs, with horizontal grid spacings of just a few kilometers (e.g. Miura *et al.*, 2007a, b). Especially for these very high-resolution models, there has been a move towards finite-difference methods based on spherical grids that are not derived from spherical coordinates. These include "cubed sphere" grids (e.g. Adcroft *et al.*, 2004; Nair *et al.*, 2005) and icosahedral or "geodesic" grids (Sadourny *et al.*, 1968; Williamson, 1968; Heikes and Randall, 1995; Ringler *et al.*, 2000). Cubed sphere grids are being used at GFDL, MIT (Adcroft *et al.*, 2004),[6] and GISS.[7] Geodesic grids are being used at Colorado State University, Deutsche Wetterdienst (Majewski *et al.*, 2002), the Frontier Research Center for Global Change (e.g. Tomita *et al.*, 2001, 2005; Tomita and Satoh, 2004; Miura *et al.*, 2005; Satoh *et al.*, 2005), NOAA's ESRL (http://fim.noaa.gov/),[8] and the Max Planck Institute for Meteorology (Bonaventura, 2004). A geodesic grid is also being evaluated for possible use in a global nonhydrostatic model under development at NCAR (Skamarock *et al.*, 2008). Figure 9.2 shows an example of a geodesic grid.

Further discussion is given by Warren Washington and Akira Kasahara in Chapter 3 of this volume.

9.4 Parameterizations

The fundamental principles of fluid dynamics, radiative transfer, etc., are simple to state, but not so simple to use. They apply locally, at a point. Limited computer resources make it necessary to formulate AGCMs in terms of averages over finite volumes. Because the governing equations are nonlinear, this averaging introduces new unknowns, which are essentially statistics characterizing relevant aspects of the unresolved processes. The fundamental principles cannot be directly applied to determine such statistics, except by revising the model to use much higher (and

[6] The Massachusetts Institute of Technology.
[7] The Goddard Institute for Space Studies of the National Aeronautics and Space Administration and Columbia University.
[8] The Earth System Research Laboratory of the National Oceanic and Atmospheric Administration.

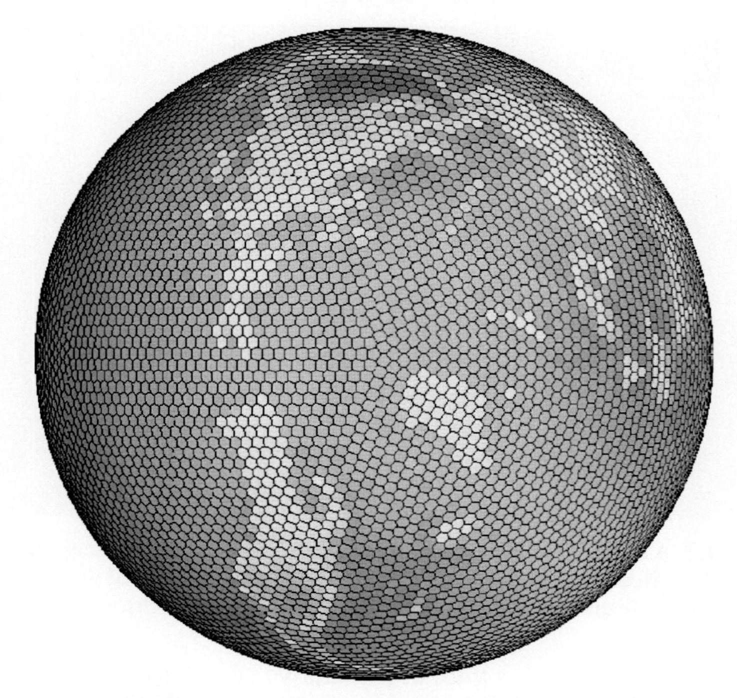

Figure 9.2 The distribution of surface elevation, plotted on a geodesic grid with about 10,000 cells. The figure has been made in such a way that the individual grid cells are visible. See also color plate.

unaffordable) spatial resolution. *The need to predict statistics over (large) finite volumes is a fundamental source of conceptual complexity.*

The earliest AGCMs included parameterizations of important processes that were not at all well understood. They had to be parameterized anyway, simply because they were too fundamental to neglect. Forty years later, we still struggle with many of the same parameterization challenges (Randall *et al.*, 2003).

Perhaps the most difficult of these problems is that of cumulus parameterization, which is designed to represent the collective effects of an "ensemble" of cumulus clouds on the larger-scale circulation of the atmosphere (Figure 9.3). Arakawa (2004) recently published an in-depth review of the status of cumulus parameterization. As mentioned earlier, the simple moist convective adjustment scheme of Manabe *et al.* (1965) captures some of the essential physics of cumulus convection. It eliminates conditional instability, releases latent heat, and produces precipitation. It also transports water and energy upward, although only through a simple layer-by-layer mixing, rather than through realistic penetrative cumulus towers.

Less than a decade later, Arakawa and Schubert (1974) published what is probably still the most conceptually complex cumulus parameterization ever

Figure 9.3 A field of cumulus congestus clouds. Models use parameterizations to include the effects of such clouds on the large-scale circulation of the atmosphere. (Copyright, University Corporation for Atmospheric Research).

devised. They included a spectrum of penetrative cumulus cloud "types" or sizes, each represented by a simple "entraining plume" cloud model. Their closure required solution of a system of equations for which the matrix of coefficients is computed through a very complicated procedure. Nevertheless, Arakawa and Schubert (1974) neglected many important processes, such as convective downdrafts and mesoscale organization (Randall *et al.*, 2003). Some of these missing ingredients have been added by later researchers. Simplified versions of the Arakawa–Schubert parameterization have also been proposed (e.g. Moorthi and Suarez, 1992; Pan and Randall, 1998).

The area-averaged non-radiative "apparent heat source" and "apparent moisture sink" defined by Yanai *et al.* (1973) are given by

$$Q_1 - \overline{Q_R} = L\overline{C} - \frac{1}{\rho}\frac{\partial}{\partial z}(\rho\overline{w's'}) - \frac{1}{\rho}\nabla_H \cdot (\rho\overline{\mathbf{v}'_H s'}),$$

$$Q_2 = -L\overline{C} - \frac{1}{\rho}\frac{\partial}{\partial z}(L\rho\overline{w'q_{v'}}) - \frac{1}{\rho}\nabla_H \cdot (L\rho\overline{\mathbf{v}'_H q'_v}).$$

The leading terms on the right-hand sides represent the effects of condensation, the next terms represent vertical divergences of the convective "eddy fluxes" of dry static energy and water vapor, respectively; and the last terms represent the horizontal divergences of the convective eddy fluxes, which are normally (and justifiably) neglected in large-scale models. *The overbars are very important*; an overbar represents an area average, which we interpret as a horizontal average over a grid cell. The expressions given above for Q_1 and Q_2 are valid regardless of how large or small the grid cells are; the grid spacing can be 100 km, or 100 m.

- As pointed out by Jung and Arakawa (2004), the roles and relative magnitudes of the various terms of Q_1 and Q_2 systematically change as the grid spacing becomes finer: the vertical transport terms become less important. Later horizontal averaging does not alter this.
- The horizontal transport terms become more important locally. Horizontal averaging over a sufficiently large area renders them negligible, however.
- The phase-change terms become more important, and ultimately become dominant at high resolution.

Models intended for use with a coarse grid spacing, which includes all existing climate models, must use a parameterization that can represent the area-averaged effects of vertical eddy flux divergences due to cumulus clouds. "Mass-flux" parameterizations (e.g., Arakawa and Schubert, 1974) are designed with this in mind. In contrast, models with much higher resolution appropriately focus on phase changes (as represented by microphysical processes), and place much less emphasis on the parameterization of eddy fluxes, which on fine grids are due only to turbulence.

A model that includes parameterizations of both deep convection and grid-scale saturation contains the elements of both low-resolution and high-resolution physics, as discussed above. If the grid spacing of such a model is incrementally refined from, say, 200 km to 2 km, the results may show that the deep convection parameterization is active with coarse resolution but "goes to sleep" and is supplanted by the grid-scale saturation parameterization with high resolution (Molinari and Dudek, 1992). This is *qualitatively* the right answer, but we lack any theory of how this transition should occur, especially for intermediate grid spacings of the order of 20 km. The intermediate case is, not surprisingly, the hardest.

The parameterizations used in high-resolution models can be simpler than those of low-resolution models because, as discussed above, the need to predict statistics over (large) finite volumes is one of the main sources of conceptual complexity. The equations that are actually used in a fine-mesh model can and should be closer to the fundamental principles of our science than those of a coarse-mesh model. High-resolution models are conceptually simpler than low-resolution models, but of course their numerical complexity is greater.

Model components are usually formulated separately and coupled at the end. "Modularity" has been advocated as a goal of model design. In this view, an AGCM is a software system; it should be possible to test alternative parameterizations and/or numerical schemes by *simply* plugging them into a suitably designed modeling chassis, using a standardized programming framework. Simplicity is in fact one of the main goals motivating this approach, and that in itself is certainly laudable.

The trouble with modular models is that nature is not modular. In the real atmosphere, the various physical processes interact on a wide range of space and time-scales, and also nonlinearly across a wide range of scales, so that their effects are thoroughly entangled. It is impossible to formulate the processes independently and snap them together at the end.

As an example, cumulus clouds grow out of the boundary layer, and modify the boundary layer through the effects of convective downdrafts. Stratiform clouds are formed by cumulus detrainment. The stratiform clouds contain convective turbulence, which is driven in part by radiative cooling near the cloud tops. The microphysical processes in the cloud are influenced, on small scales, by both turbulence and the radiation. Meanwhile, all of these processes interact with the large-scale circulation, and the land-surface or ocean at the lower boundary. The various processes and their interactions are represented through numerical schemes, and the equations must be solved simultaneously. These various connections imply that the model's dynamical core and the various parameterized processes – boundary layer, cumulus clouds, stratiform clouds, turbulence in clouds, radiation, and microphysics – must be formulated in a coordinated, coherent, and coupled way. Such an architectural unity can only be achieved by increasing the conceptual and coupling complexities of the model.

9.5 From academia to enterprise: a loss of innocence

The early AGCMs were used only for scientific research. There was relatively little funding, but on the other hand the practical demands on the time and resources of modeling groups were modest. As the predictive power of the models increased and became well established, practical applications were quickly identified and undertaken. Now that AGCMs are being extensively used in the service of society, the idyllic era of purely academic modeling has ended. A portion of today's AGCM work has become quasi-industrial in character. This change has brought a major increase in funding for modeling centers, but it has also burdened modelers with new responsibilities, and with deadlines and metrics that measure success in practical terms. Global modeling has become, in part, an enterprise. Our field has experienced a loss of innocence.

Operational global numerical weather prediction (NWP) began in the late 1970s, when increasing computer power finally made it practical (Woods, 2006). Today, humanity relies on weather forecasts based on simulations with AGCMs, although of course most of the people in our society have no idea how the forecasts are generated. Operational global NWP is a very expensive business, especially considering the hugely expensive global observing system, which includes many orbiting sensors. The operational centers that do global NWP have the benefit of substantial public resources that support infrastructure, personnel, and computing, although they could certainly use more, and in some cases much more.

The need for better weather forecasts has been steadily driving the AGCMs that are used for NWP towards higher resolution; see Chapter 5. For example, over the past twenty years, the horizontal resolution of the ECMWF AGCM has increased from T106 (~190 km grid spacing) to T799 (~25 km grid spacing). In the near future, it will increase further to T1279 (~16 km grid spacing). The number of layers has increased from O(20) to O(100). The physical parameterizations of the models have also improved, and this has led to major improvements in forecast skill (Woods, 2006).

Climate change assessments in service to society are a newer development, and operational national climate services, analogous to weather services, are now being discussed. As discussed in Chapter 10 , the Intergovernmental Panel on Climate Change (IPCC) began delivering its assessments during the 1990s. As the most mature components of coupled global climate models, AGCMs have played a crucial role in all of this.

The first coupled ocean–atmosphere GCM was created by Manabe and Bryan (1969), but coupled ocean–atmosphere modeling reached a degree of maturity only during the 1990s; see Chapter 7. Land-surface modeling underwent a qualitative improvement during the 1980s, under the influence of Robert Dickinson and Piers Sellers; see Chapter 8. Today there is continuing pressure to couple with additional submodels; for example, the climate models used in the IPCC Fourth Assessment did not include submodels of the continental ice sheets, which are obviously very important for sea-level changes, among other things. As a second example, many of the AGCMs used for climate research have only recently incorporated parameterizations of the effects of aerosols on radiation and/or cloud microphysics. Work is underway to include more processes, and more interactions among processes, than ever before.

At operational NWP centers, models are used in complicated suites of runs on fixed schedules, with data assimilation, optimized ensembles, and obsessively tracked skill scores. Similarly, in the climate modeling arena, years of work are needed to prepare and thoroughly test improved model designs in support of the quasi-periodic IPCC assessments. To perform and analyze the required simulations by their due dates requires enormous amounts of computer time, a lot of people time, and attentive management. This is noble and important work, but it is demanding, and the climate modeling community is small.

We obviously need bleeding-edge models that incorporate the latest ideas from the physical and computational sciences that form the foundation of global modeling. On the other hand, the models used for NWP and IPCC assessments must adhere to somewhat conservative designs; it would not be appropriate to undertake such applications with bleeding-edge models that are still being tested. There is a danger that modeling innovation will be inhibited by a combination of staggering external demands, a small modeling workforce (Jakob, 2010), and inadequate resources.

One way to avoid stifling innovation, while still meeting operational requirements, is to create two-track modeling centers. Track 1 supports operational weather and/or climate services using well-tested models. Track 2 innovates aggressively. Over time, successful new ideas migrate from Track 2 to Track 1.

9.6 Peta-flops and giga-grids

Increases in resolution and, to a lesser degree, increases in conceptual complexity explain why the most elaborate of today's global NWP models are about 10^4 times as costly, in computational terms, as the AGCMs of the 1960s. Climate models are different. The AGCMs used for climate simulation today have horizontal and vertical resolutions only marginally finer than those of the 1960s. The parameterizations used in climate models have become more complex, but not drastically so. The bottom line is that today's AGCMs for climate simulation are, at most, only a few hundred times more expensive than their counterparts of the 1960s.

Weather prediction models maximize numerical complexity and conceptual complexity, with (so far) only moderate coupling complexity. Climate prediction makes use of models with a lot of coupling complexity and conceptual complexity, but less numerical complexity. It seems likely that climate models will continue to have more coupling complexity than NWP models.

While these modeling changes have been evolving, *computer power has increased by about a factor of a million* or more – from megaflops to teraflops and beyond, following Moore's Law, with roughly a 100-fold increase in FLOP-rate per decade.

It has always been a challenge to adapt AGCMs to evolving computer architectures. The earliest models executed on serial machines that had, at most, just a few megabytes of main memory. During the 1970s and 80s, vector machines such as the Cray-1 (see Figure 9.4) imposed a new style of programming, while available memory increased rapidly. The 1990s saw a transition to increasingly parallel, cache-based architectures. At the same time, the speed of individual processors continued to increase, although this rate of increase has slowed recently. An inexpensive laptop computer today is faster, has more memory and disk space, consumes much less power, is more reliable, and is considerably easier to use than the venerable, multi-million-dollar Cray-1 of the 1970s.

The annual cycle is a fundamental unit of climate. The first numerical simulations of full annual cycles with AGCMs were performed during the 1960s (e.g. Mintz,

Figure 9.4 NCAR takes delivery of the first production model of the CRAY-1 supercomputer, in 1977. This machine provided a dramatic increase in computer power at NCAR. (Copyright, University Corporation for Atmospheric Research).

1965), but such long runs became commonplace only after about 1980. Prior to that, limited computer power dictated that what was then called a "climate simulation" was typically nothing more than a perpetual January run of an AGCM coupled to a simple land-surface model and a prescribed ocean.

Simulations of decades, centuries, and millennia are recent achievements. The goal of understanding Milankovitch cycles leads us to contemplate future simulations of hundreds of thousands or even millions of years. To perform such long simulations with fixed (i.e. similar to current) resolution on million-processor machines, we have to decrease the number of grid cells per processor, which probably will not be feasible due to communication bottlenecks, or else we have to increase the processor speed, which appears to be ruled out by power consumption issues. Therefore, *the ongoing evolution of supercomputers towards greater and greater parallelism, with fixed processor speeds, favors simulations with a fixed number of timesteps and higher-resolution models, rather than longer simulations with models of fixed resolution.* With this strategy, as we add grid cells, we also add processors, so that (ideally) the amount of computation per processor per timestep remains constant. In this way, massive parallelism strongly favors increased numerical complexity. Unfortunately, however, the size of the timestep must decrease linearly as the resolution increases, so that the actual simulation length decreases for a fixed number of timesteps.

It used to be true that as computers got faster, the additional speed could be used to refine the grid or to make longer runs on the same grid. No more. Technology trends now encourage us to dramatically refine our grids, but are not compatible with dramatically longer runs on our existing grids. This situation will persist until a major, currently unforeseeable technology change occurs.

How did we use the million-fold increase in computer power that has been achieved since the 1960s? The answers are different for NWP and climate applications. For NWP, we have argued above that increases in resolution and other changes have increased computing requirements by about a factor of 10^4. The additional factor of 100 needed to go from 10^4 to a million can be accounted for in terms of shorter execution times (enabling operational use on a fixed schedule) and the introduction of forecast ensembles. For the AGCMs used for climate, on the other hand, we have argued that current models are only a few hundred times slower than their predecessors. The additional factor of several thousand needed to go from a few hundred to a million is accounted for by much longer simulations – centuries instead of months – and by an increase in the number of simulations performed.

Over the next few decades, computers are expected to speed up by another factor of a million. Here is a question for the young scientists reading this chapter: what are you going to do with that next million?

Within the past ten years, computer power has crossed a threshold, such that brief simulations can now be performed with global atmospheric models that have resolution sufficiently fine to resolve important processes that have previously been parameterized, including deep moist convection and vertically propagating waves produced by flow over mountains. To take this qualitative leap, we must, at first, drastically decrease simulation lengths, from centuries to days or weeks, as we increase the number of floating point operations per simulated day by a factor of 10^5 to 10^6. Running the new very-high-resolution global models requires $O(10^5)$ processors, whereas up to now most models have used only $O(100)$ processors. This huge jump in parallelism is challenging beyond any of the previous challenges mentioned above. It will undoubtedly take a few years to learn how to do this well.

Because they have grid spacings comparable to the native scales of large clouds, global cloud-resolving models (GCRMs) do not need and in fact cannot use the parameterizations of deep cumulus convection needed by conventional AGCMs. The global models also have no need for parameterizations of cloud overlap, or gravity-wave drag, because these can be explicitly simulated. They do still rely on parameterizations of cloud microphysics, turbulence, and radiation, but all three of these parameterization problems become much more tractable with the high resolution of a GCRM. In fact, a strength of GCRMs is that they are ideal vehicles for the implementation of more advanced parameterizations of microphysics, turbulence, and radiation.

A GCRM can simulate interactions across a very wide range of spatial scales, including large individual clouds, tropical cyclones, mid-latitude baroclinic waves, larger planetary waves and monsoons, and elements of the zonally averaged circulation such as the Hadley cells. As is well known, the atmospheric circulation does not contain any convenient spectral gaps in which the energy density is low. All scales contain energy, and interact through a wide variety of processes, many of them involving cloud systems in important ways.

The high-resolution statistics simulated by GCRMs can be compared directly with high-resolution observations. This greatly simplifies the diagnosis of model deficiencies.

From a computational point of view, GCRMs are intrinsically more amenable to modularization than today's AGCMs, simply because the various physical processes represented by the parameterizations are less interdependent on the cloud scale. This means that the conceptual complexity of GCRMs is less than that of lower-resolution AGCMs, although of course the numerical complexity of GCRMs is much greater.

A decrease in the horizontal grid spacing by another factor of ten, relative to today's models, together with the introduction of nonhydrostatic dynamics and major changes in parameterizations, will yield global NWP models with grid cells just a few km across, at the coarse end of the "cloud-resolving" range. Taking into account the decrease in timestep needed to permit simulation of the birth and death of individual large clouds, we can estimate that these cloud-resolving NWP models of the not-too-distant future will be about 1000 times as expensive as today's high-resolution global NWP models. Such GCRMs will become marginally practical for operational global NWP within the coming decade, which means that now is the time to begin their development.

A GCRM has already been developed at the Frontier Research Center for Global Change (FRCGC), in Japan. The model has been tested on the Earth Simulator, with very provocative results (Miura *et al.*, 2005; Tomita *et al.*, 2005). The FRCGC's GCRM is called NICAM, which stands for "Nonhydrostatic Icosahedral Atmospheric Model." It is constructed on a "geodesic" grid based on the icosahedron. At the highest tested resolution, NICAM has cells 3.5 km across.

Recent tests of NICAM represent a milestone in global modeling. They are analogous to the heroic early experiments in numerical weather prediction that were performed by Charney, Fjörtoft, von Neumann, and colleagues over 50 years ago, using the ENIAC computer (Charney *et al.*, 1950; Platzman, 1979). The model of Charney *et al.* was highly simplified compared to today's AGCMs, and the ENIAC was painfully slow compared to the computers we have now. What Charney *et al.* really wanted to do is what we can do now, but it was far beyond their reach. That didn't stop them. They did what they could at the time that they lived. All of the subsequent achievements of atmospheric modelers have been built on the foundations that they created.

Similarly, the FRCGC scientists would like to have a computer much faster than the Earth Simulator, and a model that is more advanced than the current version of NICAM. The GCRMs of the future will no doubt have many improvements, including even higher resolution, better numerics, and more advanced parameterizations of cloud microphysics, turbulence, and radiation. Using future exaflop computers, such improved GCRMs will eventually be run for simulated centuries, far beyond what is possible today. Although these future models and future computers are out of reach for now, the FRCGC scientists have done what they can with the modeling and computer know-how of the early 21st century. They have created a new foundation on which a large body of work will be built.

With a 4 km grid spacing and 128 layers, a GCRM contains approximately 3 billion grid cells. It uses a timestep of about 10 simulated seconds. Simulation of one day requires about 10^{17} floating point operations. A snapshot of the model's prognostic (i.e. time-stepped) fields occupies about 1 TB. The data volume written for archival per simulated day can easily reach 100 TB, depending of course on the frequency of output and the number of fields saved. In these ways, emerging petascale computational resources are enabling and will enable global atmospheric models that are *qualitatively* more advanced than the models of the past. *This is a transformative rather than incremental change,* because GCRMs directly simulate important processes that were previously parameterized.

Computational challenges of GCRMs include:

- Efficient execution on a very large number of processors, in order to achieve acceptably fast run times;
- Parallel I/O (especially O);
- Management and distribution of the voluminous archived model output; and
- Visualization of the results.

It is remarkable that three of these four challenges relate to working with the model output. For some applications of GCRMs, the fine-scale details of the flow will be of secondary interest, so that a spatially and temporally subsampled or filtered depiction of the simulation is all that is needed. This can drastically reduce the output volume. For other applications, e.g. analysis of a rapidly intensifying tropical cyclone, full spatial resolution and high time resolution are needed to depict the growth of individual convective cells. Typically such detailed output will only be needed in selected portions of the global domain, but the locations will not necessarily be known in advance. Intelligent algorithms are needed to provide the capability to dynamically adapt the high-resolution output to specific geographical regions of interest.

Conventional climate models generally write four types of output files:

- A small (a few kB per simulated day) "day file" that suffices to reveal pathological behavior, allowing the model user to monitor the vital signs of the simulation.
- Restart files, which can be used to restart the model at intervals during a long run. For the atmospheric component of a conventional climate model, individual restart files are typically O(100 MB) or smaller. They might be written once per simulated day, or once per simulated month, but in either case the output volumes are small.
- Diagnostic or "history" files, which provide detailed information on the progress of the simulation. Examples of fields included here would be three-dimensional distributions of atmospheric heating due to solar radiation, or the two-dimensional surface precipitation rate produced by the cumulus parameterization; a recommended list can be found at http://www-pcmdi.llnl.gov/projects/amip/OUTPUT/WGNEDIAGS/.
- The total number of diagnostic fields is of course under the control of the modeler, but the data volume of a single record will approach 1 Terabyte for a 3D field at 1-km resolution with 100 levels. In many cases, these files would be written once per simulated month, but depending on the nature of the study they might be written as often as once per simulated day. In the extreme, they will be written up to hourly or perhaps 3 hourly depending on the capacity of the IO and storage systems in use and the expected use of the data.
- "Point-by-point" files, which contain selected fields with high time resolution (e.g. every 30 simulated minutes) at O(100) selected grid points. This type of output has been in use since the 1970s, if not before, but it is occasionally reinvented by new generations of modelers.

Up to now, global atmospheric models have relied almost entirely on serial output. With a GCRM, highly parallel output is obviously essential, because using a single "pipe" to send all of the model output to disk creates an unacceptable bottleneck.

A low-resolution but global depiction of the atmospheric circulation can be created by subsampling or filtering "on the fly," inside the model. This dataset can be analyzed using conventional methods. At the same time, a high-resolution but local analysis is needed to identify important centers of action, such as tropical cyclones or major snowstorms. One possible approach is to automatically identify where and when such events occur, and to automatically create "chunks" of model output, of manageable size, that can be used to analyze the events at high resolution.

To do this, it will be necessary to create rule-based "agents" that live inside the model and monitor the simulated weather as the model runs. The agents will use prespecified criteria to flag events of interest for the postprocessing program mentioned above. The criteria will be user selectable and user programmable during the set-up of a simulation.

Depending on the user-selectable specifics of the GCRM output, following the strategy discussed above, simulation of one annual cycle will produce several petabytes of model output. Strategies are needed to catalog, browse, subsample, and transport the output rapidly and efficiently.

Data will be staged from disk to archival media as available on-line storage is currently limited to the 10s of TB range at most supercomputing centers. While it is obviously desirable to avoid transporting huge volumes of data around the country or the world, via network or otherwise, some long-distance data transportation will be necessary. Physically mailing disk drives is an option, but even this is currently impractical for petabytes of data. For these reasons, major portions of GCRM data analysis and visualization work will have to be carried out at the same center where the simulation is performed. Only subsamples of the GCRM output can be transported between geographically separated research centers.

With a 1 km global grid, the range of scales involved (in one dimension) is roughly 40,000 km to 1 km. Displays, printers, eyes, and brains cannot handle all of that at once. A zooming capability is therefore needed. Sampled or filtered data can be used to depict the larger scales over the entire globe. Full-resolution data is needed to study local weather systems, but it is needed only in selected limited regions. We need a software system that can, on demand, automatically supply visualizations of appropriately sampled model output, in appropriately sized regions, and with appropriate resolutions. "Google Earth" comes to mind as an application that works in much this way. Obviously Google Earth itself is just a visualization tool, but we need an analogous tool that can be used for both more elaborate visualization and physically meaningful analysis of model output.

9.7 Conclusions

I am writing this chapter in early 2009. It is a tumultuous time for global atmospheric modeling. Climate change has become a major societal issue. There are calls for large increases in funding for climate research, but the world economy is staggering into a deep recession. The conceptual complexity of our models has reached the point that no single human being can possibly comprehend one of them in its entirety. The number of coupled processes is rapidly growing; the carbon cycle and terrestrial ice sheets are particularly active areas of current research. New technology is enabling dramatic increases in model resolution, and thus driving major changes in model design, but with many serious practical challenges.

Looking ahead, we can dimly see an era in which exaflop computers will permit global cloud-resolving atmospheric models, coupled to ocean, land-surface, and ice-sheet models of comparable resolution. Such models will be used to produce highly detailed simulations of the Earth system on time-scales relevant to both weather and climate prediction. Their conceptual complexity may be less than that of today's models, but their numerical complexity will begin to rival that of the beautiful physical system that they represent.

Acknowledgments

This work has been supported by the National Science Foundation Science and Technology Center for Multi-Scale Modeling of Atmospheric Processes, managed by Colorado State University under cooperative agreement ATM-0425247. The author benefitted from discussions with Bjorn Stevens, John Drake, Leo Donner, Rodger Ames, and Wayne Schubert, as well as reviewer comments by Anthony DelGenio.

References

Adcroft, A., Campin, J. M., Hill, C., and Marshall, J. (2004). Implementation of an atmosphere-ocean general circulation model on the expanded spherical cube. *Monthly Weather Review*, 132, 2845–2863.

Arakawa, A. (1966). Computational design for long-term numerical integration of the equations of fluid motion: Two-dimensional incompressible flow. Part I. *Journal of Computational Physics*, 1, 119–143.

Arakawa, A. (1969). Parameterization of cumulus convection. *Proc. WMO/IUGG Symp. Numerical Weather Prediction*, Tokyo, 26 November – 4 December, 1968, Japan Meteor. Agency, IV, 8, 1–6.

Arakawa, A. (1970). Numerical Simulation of Large-Scale Atmospheric Motions. In *Numerical Solution of Field Problems in Continuum Physics* (SIAM-AMS Proceedings, American Mathematical Society) 2, 24–40.

Arakawa, A. (2000). Chapter 1: A Personal Perspective on the Early Years of General Circulation Modeling at UCLA. In *General Circulation Model Development – Past, Present, and Future*, ed. Randall, D. A., International Geophysics Series; Vol. 70, San Diego, CA, Academic Press, pp. 1–65.

Arakawa, A. (2004). The cumulus parameterization problem: past, present, and future. *Journal of Climate*, 17, 2493–2525.

Arakawa, A. and Lamb V. R. (1977). Computational design of the basic dynamical processes of the UCLA general circulation model. In *Methods in Computational Physics*, 17, Academic Press, New York, pp. 173–265.

Arakawa, A. and Schubert, W. H. (1974). Interaction of a cumulus cloud ensemble with large-scale environment, Part 1. *Journal of the Atmospheric Sciences*, 31, 674–701.

Baer, F. (1972). An alternate scale representation of atmospheric energy spectra. *Journal of the Atmospheric Sciences*, 29, 649–664.

Bonaventura, L. (2004). Development of the ICON dynamical core: modelling strategies and preliminary results. *Proceedings of the ECMWF-SPARC Workshop on Modelling and Assimilation for the Stratosphere and Tropopause*, ECMWF, Reading, UK.

Bourke, W. (1974). A multi-level spectral model. I. Formulation and hemispheric integrations. *Monthly Weather Review*, 102, 687–701.

Charney, J. G., and coauthors (1966). The feasibility of a global observation and analysis experiment. *Bulletin of the American Meteorological Society*, 47, 200–220.

Charney, J. G., Fjörtoft, R., and von Neumann, J. (1950). Numerical integration of the barotropic vorticity equation. *Tellus*, 2, 237–254.

Claussen, M., Mysak, L. A., Weaver, A. J. *et al.* (2002). Earth system models of intermediate complexity: closing the gap in the spectrum of climate system models. *Climate Dynamics*, 18, 579–586.

Edwards, P. (2000). Chapter 2: A Brief History of Atmospheric General Circulation Modelling. In *General Circulation Model Development – Past, Present, and Future*, ed. Randall, D. A., International Geophysics Series; Vol. 70, San Diego, CA, Academic Press, pp. 67–90.

Edwards, P. N. (2010): *A Vast Machine: Computer Models, Climate Data, and the Politics of Global Warming*. MIT Press, 528 pp.

Eliasen, E., Machenhauer, B., and Rasmussen, E. (1970). *On a Numerical Method for Integration of the Hydrodynamical Equations with a Spectral Representation of the Horizontal Fields* (Institut for Teoretisk Meteorologi, Köbenhavns Universitet, Denmark).

Heikes, R. and Randall, D. A. (1995). Numerical integration of the shallow-water equations on a twisted icosahedral grid. Part I: Basic design and results of tests. *Monthly Weather Review*, 123, 1862–1880.

Held, I. M. (2005). The gap between simulation and understanding in climate modeling. *Bulletin of the American Meteorological Society*, 86, 1609–1614.

Jablonowski, C. and Williamson, D. L. (2006). A baroclinic instability test case for atmospheric model dynamical cores. *Quarterly Journal of the Royal Meteorological Society*, 132, 2943–2975.

Jakob, C. (2010). Accelerating progress in global model development – Challenges, opportunities, and strategies. Submitted to *Bulletin of the American Meteorological Society*.

Jarraud, M. and Simmons, A. J. (1983). The spectral technique. *Seminar on Numerical Methods for Weather Prediction*, European Centre for Medium Range Weather Prediction, Reading, England, 1–59.

Jung, J. H. and Arakawa, A. (2004). The resolution dependence of model physics: Illustrations from nonhydrostatic model experiments. *Journal of the Atmospheric Sciences*, 61, 88–102.

Kasahara, A. and Washington, W. M. (1967). NCAR global general circulation model of atmosphere. *Monthly Weather Review*, 95, 389–482.

Leith, C. E. (1965). Numerical Simulation of the Earth's Atmosphere, in *Methods in Computational Physics*, eds. B. Alder, S. Fernbach, and M. Rotenberg, Academic Press, 1–28.

Lewis, J. M. (1998). Clarifying the dynamics of the general circulation: Phillips's 1956 experiment. *Bulletin of the American Meteorological Society*, 79, 39–60.

Majewski, D., Liermann, D., Prohl, P. *et al.* (2002). The operational global icosahedral-hexagonal gridpoint model GME: Description and high-resolution tests. *Monthly Weather Review*, 130, 319–338.

Manabe, S. and Bryan, K. (1969). Climate calculations with a combined ocean-atmosphere model. *Journal of the Atmospheric Sciences*, 26, 786–789.

Manabe, S. and Hahn, D. G. (1981). Simulation of atmospheric variability. *Monthly Weather Review*, 109, 2260–2286.

Manabe, S., Smagorinsky, J., and Strickler, R. F. (1965). Simulated climatology of a general circulation model with a hydrologic cycle. *Monthly Weather Review*, 93, 769–798.

Mintz, Y. (1958). Design of some numerical general circulation experiments. *Bulletin of the Reserch Council of Israel*, 76, 67–114.

Mintz, Y. (1965). Very long-term global integration of the primitive equations of atmospheric motion. *WMO Tech. Note No. 66; UCLA Dept. of Meteorology Contribution No. 111. WMO-IUGG Syrup. on Res. and Develop. Aspects of Long Range Forecasting*, Boulder, Colo., 1964. Geneva, 1965, pp. 141–167.

Miura, H., Tomita, H., Nasuno, T., Iga, S.-I., Satoh, M., and Matsuno, T. (2005). A climate sensitivity test using a global cloud resolving model under an aquaplanet condition. *Geophysical Research Letters*, 32, L19717.

Miura, H., Satoh, M., Nasuno, T., Noda, A. T., and Oouchi, K. (2007a). A Madden-Julian Oscillation event realistically simulated by a global cloud-resolving model. *Science*, 318, 1763–1765.

Miura, H., Satoh, M., Tomita, H., Noda, A. T., Nasuno, T., and Iga, S. (2007b). A short-duration global cloud-resolving simulation with a realistic land and sea distribution. *Geophysical Research Letters*, 34.

Molinari, J. and Dudek, M. (1992). Parameterization of convective precipitation in mesoscale numerical models: a critical review. *Monthly Wealther Review*, 120, 326–344.

Moorthi, S. and Suarez, M. J. (1992). Relaxed Arakawa-Schubert: A parameterization of moist convection for general circulation models. *Monthly Weather Review*, 120, 978–1002.

Nair, R. D., Thomas, S. J., and Loft, R. D. (2005). A discontinuous Galerkin transport scheme on the cubed sphere. *Monthly Weather Review*, 133, 814–828.

Orszag, S. A. (1970). Transform method for calculation of vector-coupled sums: Application to spectral form of vorticity equation. *Journal of the Atmospheric Sciences*, 27, 890–895.

Pan, D. M. and Randall, D. A. (1998). A cumulus parametrization with a prognostic closure. *Quarterly Journal of the Royal Meteorological Society*, 124, 949–981.

Phillips, N. A. (1956). The general circulation of the atmosphere : A numerical experiment. *Quarterly Journal of the Royal Meteorological Society*, 82, 123–164.

Phillips, N. A. (1957). A coordinate system having some special advantages for numerical forecasting. *Journal of Meteorology*, 14, 184–185.

Phillips, N. A. (1959). An example of non-linear computational instability. In *The Atmosphere and Sea in Motion*, (ed. Bolin, B.), Rockefeller Inst. Press, New York, pp. 501–504.

Platzman, G. W. (1960). The spectral form of the vorticity equation. *Journal of Meteorology*, 17, 635–644.

Platzman, G. W. (1979). The ENIAC computations of 1950 – Gateway to numerical weather prediction. *Bulletin of the American Meteorological Society*, 60, 302–312.

Randall, D., Khairoutdinov, M., Arakawa, A., and Grabowski, W. (2003). Breaking the cloud parameterization deadlock. *Bulletin of the American Meteorological Society*, 84, 1547–1564.

Ringler, T. D., Heikes, R. P., and Randall, D. A. (2000). Modeling the atmospheric general circulation using a spherical geodesic grid: A new class of dynamical cores. *Monthly Weather Review*, 128, 2471–2490.

Ritchie, H., Temperton, C., Simmons, A. *et al.* (1995). Implementation of the semi-lagrangian method in a high-resolution version of the ECMWF forecast model. *Monthly Weather Review*, 123, 489–514.

Sadourny, R., Arakawa, A., and Mintz, Y. (1968). Integration of nondivergent barotropic vorticity equation with an icosahedral-hexagonal grid for sphere. *Monthly Weather Review*, 96, 351–356.

Satoh M., Tomita, H., Miura, H., Iga, S., and Nasuno, T. (2005), Development of a global cloud-resolving model – A multi-scale structure of tropical convections. *J. Earth Simulator*, 3, 11–19.

Satoh M., Nasuno, T., Miura, H., Tomita, H., Iga, S., and Takayabu, Y. (2007). Precipitation statistics comparison between global cloud resolving simulation with NICAM and TRMM PR data. *High Resolution Numerical Modelling of the Atmosphere and Ocean*, eds. Hamilton, K. and Ohfuchi, W., Springer, pp. 99–109.

Satoh, M., Matsuno, T., Tomita, H., Miura, H., Nasuno, T., and Iga, S. (2008). Nonhydrostatic icosahedral atmospheric model (NICAM) for global cloud resolving simulations. *Journal of Computational Physics*, 227, 3486–3514.

Skamarock, W. C., Klemp, J. B., Ringler, T. D., and Thuburn J. (2008). A Hexagonal C-Grid Atmospheric Core Formulation for Multiscale Simulation on the Sphere. *American Geophysical Union, Fall Meeting 2008*, abstract #A31H-01

Silberman, I. S. (1954). Planetary waves in the atmosphere. *Journal of Meteorology*, 11, 27–34.

Smagorinsky, J. (1983). The beginnings of numerical weather prediction and general circulation modeling: Early recollections, *Advances in Geophysics*, 25, 3–37.

Smagorinsky, J., Manabe S., and Holloway J. L. (1965). Numerical results from a nine-level general circulation model of the atmosphere. *Monthly Weather Review*, 93, 727–768.

Tomita, H. and Satoh, M. (2004). A new dynamical framework of nonhydrostatic global model using the icosahedral grid. *Fluid Dynamics Research*, 34, 357–400.

Tomita, H., Tsugawa, M., Satoh, M., and Goto, K. (2001). Shallow water model on a modified icosahedral geodesic grid by using spring dynamics. *Journal of Computational Physics*, 174, 579–613.

Tomita, H., Miura, H., Iga, S., Nasuno, T., and Satoh, M. (2005). A global cloud-resolving simulation: Preliminary results from an aqua planet experiment. *Geophysical Research Letters*, 32, L08805.

Washington, W. M. (2007). *Odyssey in Climate Modeling, Global Warming, and Advising Five Presidents*. Lulu.com, 302 pp.

Williamson, D. L. (1968). Integration of barotropic vorticity equation on a spherical geodesic grid. *Tellus*, 20, 642–653.

Williamson, D. L. and Rasch, P. J. (1994). Water vapor transport in the NCAR CCM2. *Tellus A*, 46, 34–51.

Woods, A. (2006). *Medium-Range Weather Prediction: The European Approach*. Springer, 270 pp.

World Climate Research Programme (2009). *World Modeling Summit for Climate Prediction*. WCRP Report No. 131, WMO/TD No. 1468, 29 pp.

Yanai, M., Esbersen, S., and Chu, J.-H. (1973). Determination of bulk properties of tropical cloud clusters from large-scale heat and moisture budgets. *Journal of the Atmospheric Sciences*, 30, 611–627.

10

The co-evolution of climate models and the Intergovernmental Panel on Climate Change

RICHARD C. J. SOMERVILLE

Carbon dioxide absorbs long-wave radiation and so helps to maintain the temperature of the earth's surface above that at which it would otherwise be in equilibrium with solar radiation. The amount of CO_2 in the atmosphere must have varied greatly during geological time, being depleted by the formation of limestones (carbonates) and coal measures, and replenished by volcanic action. Ordinarily, the variation was slow, because a great reserve of CO_2 is dissolved in the oceans. Arrhenius and Chamberlin saw in this a cause of climatic changes, but the theory was never widely accepted and was abandoned when it was found that all the long-wave radiation absorbed by CO_2 is also absorbed by water vapour.

In the past hundred years the burning of coal has increased the amount of CO_2 by a measurable amount (from 0.028 to 0.030 per cent) and Callendar (1939) sees in this an explanation of the recent rise in world temperature. But during the past 7000 years there have been greater fluctuations of temperature without the intervention of man, and there seems to be no reason to regard the recent rise as more than a coincidence. This theory is not considered further. – Brooks (1951).

10.1 Historical development of global climate models in the IPCC context

As recently as the 1950s, global climate models, or GCMs, did not exist, nor did the supercomputers that they require, and the notion that human-made carbon dioxide might lead to significant climate change was not regarded as a serious possibility by most experts. The quotation that opens this chapter, by the distinguished British climatologist C. E. P. Brooks, accurately reflects the thinking of most climate scientists of that time. Today, of course, the prospect or threat of exactly this type of climate change dominates the science and ranks among the most pressing issues confronting all humankind.

Indeed, the prevailing scientific view throughout the first half of the twentieth century was that adding carbon dioxide to the atmosphere would have only a negligible effect on climate. Svante Arrhenius, who carried out the first detailed calculations in 1896, thought that doubling atmospheric carbon dioxide levels might take 3,000 years. He was wrong, how wrong remains to be seen, but perhaps wrong by a factor of about 20, because he did not foresee the explosive twentieth-century growth in population and fossil-fuel use. Like other scientists of that era, Arrhenius investigated the possible connection between carbon dioxide and climate in terms of an explanation for the ice ages. He apparently never regarded it as a candidate climate-change mechanism operating on shorter time-scales.

After Arrhenius, with very few exceptions, the few scientists who seriously considered the issue generally came to incorrect conclusions. Some erroneously thought that the carbon dioxide absorption bands for infrared radiation were already saturated, so that adding additional carbon dioxide would not increase absorption. Other scientists found other reasons to dismiss the effect of rising atmospheric carbon dioxide levels, believing, for example, that water vapor absorption of infrared radiation overwhelmed that of carbon dioxide, or that the ocean would speedily take up any additional carbon dioxide that puny humankind might add to the atmosphere. Today, with the wisdom of hindsight, we know that they were all mistaken.

Thus, in about half a century, the science of climate change caused by atmospheric carbon dioxide changes has undergone a genuine revolution. An extraordinarily rapid development of GCMs has also characterized this period, especially since about 1980. In only three decades, the number of GCMs has greatly increased, and their physical and computational aspects have both markedly improved. Modeling progress has been enabled by many scientific advances, of course, but especially by a massive increase in available computer power, with supercomputer speeds increasing by roughly a factor of a million from about 1980 to 2010. This technological advance has permitted a rapid increase in the physical comprehensiveness of GCMs (Figure 10.1) as well as in spatial computational resolution (Figure 10.2). In short, GCMs have dramatically evolved over time, in exactly the same recent period as popular interest and scientific concern about anthropogenic climate change have markedly increased.

In parallel, a unique international organization, the Intergovernmental Panel on Climate Change, or IPCC, has also recently come into being and also evolved rapidly. Today, the IPCC has become widely respected and globally influential. The IPCC was founded in 1988, and its history is thus even shorter than that of GCMs. Yet, its stature today is such that a series of IPCC reports assessing climate-change science has already been endorsed by many leading scientific professional societies and academies of science world-wide. These reports are considered as definitive summaries of the state of the science. In 2007, in recognition of its exceptional accomplishments, the IPCC shared the Nobel Peace Prize equally with Al Gore.

Figure 10.1 The complexity of climate models has increased over the last few decades. The additional physics incorporated in the models are shown pictorially by the different features of the modeled world. This is Figure 1.2 in Le Treut *et al.* (2007). See also color plate.

Future historians may come to regard the half-century from about 1960 to about 2010 as a period marked by an astounding coincidence in climate change and climate science. In this period, rapidly increasing atmospheric concentrations of greenhouse gases due to human activities led to important observed climate changes and the prospect of even more serious future climate changes. During the same period, the science of climate change underwent a profound transformation, driven and enabled by the invention of satellites, supercomputers, and climate models, among other advances. Thus, just at the moment in history when humankind inadvertently became capable of changing the climate, the science required to understand and predict this climate change also came of age.

There is no fundamental reason why humanity might not have evolved somewhat differently since, say, the year 1900. For example, it is perfectly conceivable that during the twentieth century, our species might have increasingly exploited the Earth's reserves of coal and oil and natural gas, while more than tripling global population since 1930, without, however, also inventing the technologies that led to

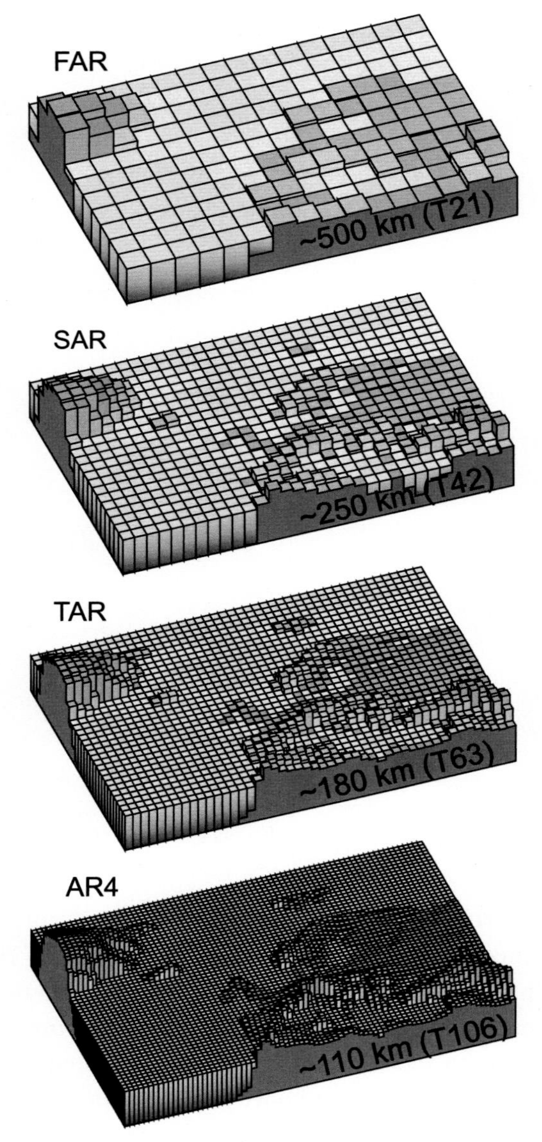

Figure 10.2 Geographic resolution characteristic of the generations of climate models used in the IPCC Assessment Reports: FAR (IPCC, 1990), SAR (IPCC, 1996), TAR (IPCC, 2001), and AR4 (IPCC, 2007). The figure shows how successive generations of these global models increasingly resolve northern Europe. These illustrations are representative of the most detailed horizontal resolution used for short-term climate simulations. The century-long simulations cited in IPCC Assessment Reports after the FAR were typically run with the previous generation's resolution. Vertical resolution in both atmosphere and ocean models is not shown, but it has increased comparably with the horizontal resolution, beginning typically with a single-layer slab ocean and ten atmospheric layers in the FAR and progressing to about thirty levels in both atmosphere and ocean. This is Figure 1.4 in Le Treut *et al.* (2007). See also color plate.

rapid progress in climate science, especially satellite remote sensing and digital electronic computers. In that hypothetical case, human-made global warming in the early years of the current century might have been underway to the same extent as it actually is today, but our ability to comprehend it scientifically and to describe its causes and project its future would be extremely limited.

In fact, however, the present era is characterized not only by the reality and seriousness of human-caused climate change, but also by a young yet powerful science that enables us to understand much about the climate change that has occurred already and that awaits in the future. The development of GCMs is a critical part of the scientific story, and the development of the IPCC is a key factor in connecting the science to the perceptions and priorities of the global public and policymakers. As we shall see, GCMs and the IPCC have co-evolved and strongly influenced one another, as both scientists and the world at large have worked to confront the challenge of climate change.

10.2 The beginnings of the Intergovernmental Panel on Climate Change (IPCC)

The authoritative history of the IPCC has been written by Bolin (2007). This book appeared shortly before the death of its author. Bert Bolin was a visionary scientist and pioneer in climate research who was the first chair of IPCC, serving from 1988 to 1997. The IPCC has produced four major climate assessment reports to date, appearing in 1990, 1996, 2001 and 2007. The first two of these reports were thus written under Bolin's chairmanship. Many of the principles and practices that have guided and characterized the IPCC, including its reliance on peer review and high scientific standards, its notable inclusiveness and transparency, and its steadfast insistence on absolute neutrality with respect to policy, took shape under Bolin's guidance. An appreciation of Bert Bolin as both a research scientist and a scientific statesman has been published by Somerville (2008).

Beginning in the 1970s, a growing scientific awareness of the potential importance of human-made climate change led to a series of important international meetings. The first World Climate Conference, held in Geneva in 1979, was followed by a second conference, in Villach, Austria in 1985. Both were landmark events that took stock of recent scientific developments that had increased our understanding of the risk of climate warming associated with an increase in the greenhouse effect. These conferences, together with growing awareness and concern on the part of governments and the scientific community, led to the creation in 1988 of the IPCC, under the joint auspices of the United Nations Environment Programme (UNEP) and the World Meteorological Organization (WMO). The mandated objective of the IPCC was to carry out, based on available scientific information, an assessment of the scientific, technical, and socio-economic aspects of the climate changes that might result from human activities. From the very beginning, the IPCC was organized into three Working Groups (WGs). In this chapter, only the first of these, WG I, which treats the physical science aspects of climate change, will be considered. The other two WGs are concerned with the consequences, adaptation, and vulnerability aspects of climate change (WG II), and with mitigation aspects of climate-change (WG III).

The IPCC thus occupies the interface between modern climate change science and public policy. Its function is to provide policymakers with reliable and intelligible scientific information and to assess climate-change science in an open and objective manner that is policy-relevant but not policy-prescriptive. The four IPCC Assessment Reports to date have all been powerful influences on international initiatives to respond to the prospect of human-made climate change.

The IPCC does not carry out research. It simply assesses the research performed by scientists throughout the world, research that is published in the peer-reviewed scientific literature. The IPCC has a small budget and staff. Its main function is to organize large numbers of scientists to perform the assessments and write the reports. Thousands of scientists throughout the world have contributed to the IPCC effort. The IPCC reports have expressed increasing certainty that human activity contributes significantly to global climate change, and that this contribution will increase in the foreseeable future.

10.3 The First Assessment Report (FAR) of the IPCC (1990)

In 1989, the United Nations General Assembly asked the fledgling IPCC to present its first report the following year. The IPCC complied, although the task of assessing all the relevant peer-reviewed scientific literature in such a short time was daunting. Published in 1990, this IPCC First Assessment Report (FAR) gave an account of what was known about the role of human activities in modifying the composition of the atmosphere through carbon dioxide, methane, and other gases contributing to the greenhouse effect. The FAR also described extremely worrisome predictions for the twenty-first century, with an average global warming that might reach 3 °C by 2100, accompanied by a sea-level rise of 65 cm. Although this report laid out the numerous uncertainties accompanying these climate projections based on computer models, and although it was careful to avoid attributing the warming of from 0.3 to 0.6 °C that had been observed over the preceding 100 years to a strengthened greenhouse effect, nevertheless the FAR would come to play an extremely significant role in the debate on climate change.

The IPCC FAR in 1990 was unquestionably a key factor in the establishment of the U.N. Framework Convention on Climate Change (UNFCCC) at the Earth Summit in Rio de Janeiro, Brazil, in 1992. This document, signed by virtually every nation in the world, has as its objective,

"to stabilize the concentrations of greenhouse gases in the atmosphere at a level that will prevent dangerous anthropogenic interference with the climate system."

The exact definition of "dangerous" in this context was not spelled out in the UNFCCC in 1992. In all the years since the establishment of the UNFCCC, no universally accepted definition of "dangerous" has ever been agreed by all the nations of the world. In fact, the concept of "dangerous" in this context is not a term that can be defined objectively using purely scientific standards, although science can inform such a definition. Risk tolerance and economic considerations, together with political concerns and other subjective judgments, must also play a role in deciding what degree of anthropogenic interference is to be considered "dangerous."

The FAR (IPCC, 1990) has aged well, a tribute to Bolin and the many other scientists who worked to produce it. Today, some twenty years after its publication, it is still well worth reading. The GCM experiments on which the FAR relied were generally devoted to exploring equilibrium climate change. That is, they were concerned with the idealized experiment of simulating the climate change arising from, say, a hypothetical instantaneous doubling of atmospheric carbon dioxide concentration, followed by waiting until the climate system had equilibrated. Thus, they did not address the "transient" aspects of how the climate would change before it reached equilibrium.

From the modeling point of view, therefore, these equilibrium GCM experiments assessed in the FAR were idealizations (Table 10.1). They provided insight into the sensitivity of the climate system to an externally imposed forcing, in this case the imposition of a sudden large addition of a greenhouse gas. However, they did not address many potentially important processes, such as, for example, feedbacks that might occur affecting the carbon cycle and thus the sources and sinks of carbon dioxide and other greenhouse gases. Furthermore, these early GCM experiments emphasized the atmospheric component of the climate system and simplified or omitted other components, such as the ocean and many biogeochemical aspects of climate.

In this sense, the FAR assessment of GCM results was largely process-oriented, devoted to exploring the behavior of an important but restricted part of the climate system. In many ways, this use of GCMs in climate research displays striking parallels with the early history of GCMs themselves. The celebrated first GCM integration, that of Phillips (1956), discussed in Chapters 2 and 3, was effectively a numerical experiment on baroclinic instability, designed to simulate an important physical process critical to the atmospheric general circulation. In retrospect, that experiment was a key step on the way to today's more comprehensive understanding of why the general circulation of the Earth's atmosphere has the properties that we observe.

The early state of development of GCMs at the time of the FAR was also a limiting factor of the climate science of the time. On the topic of time-dependent climate change simulated with coupled atmosphere–ocean GCMs, the FAR concluded:

Table 10.1 *Executive summary of equilibrium effects of doubling atmospheric levels of carbon dioxide, as simulated by climate models. Numbers of stars (*) indicate degree of confidence in each assertion, as determined subjectively from three factors: amount of agreement between models; 1990 IPCC report authors' understanding of model results; and authors' confidence in representation of relevant process in the models. Five stars indicates virtual certainty; one star indicates low confidence.*

Temperature	
*****	The lower atmosphere and Earth's surface warm
*****	The stratosphere cools
***	Near the Earth's surface, the global average warming lies between +1.5 °C and +4.5 °C, with a "best guess" of 2.5 °C
***	The surface warming at high latitudes is greater than the global average in winter but smaller than in summer (in time-dependent simulations with a deep ocean, there is little warming over the high-latitude southern ocean)
***	The surface warming and its seasonal variation are least in the tropics
Precipitation	
****	The global average increases (as does that of evaporation); the larger the warming, the larger the increase
***	Increases at high latitudes throughout the year
***	Increases globally by 3% to 15% (as does evaporation)
**	Increases at mid-latitudes in winter
**	The zonal mean value increases in the tropics, although there are areas of decrease. Shifts in the main tropical rain bands differ from model to model, so there is little consistency between models in simulated regional changes
**	Changes little in subtropical arid areas
Soil moisture	
***	Increases in high latitudes in winter
**	Decreases over northern mid-latitude continents in summer
Snow and sea ice	
****	The areas of sea ice and seasonal snow cover diminish

"Coupled ocean–atmosphere general circulation models, though still of coarse resolution and subject to technical problems such as the flux adjustment, are providing useful insights into the expected climate response due to a time-dependent radiative forcing. However, only a very few simulations have been completed at this time..."

Thus, the FAR assessed the use of GCMs to simulate equilibrium climate change due to doubling the atmospheric concentration of carbon dioxide and summarized the model results as shown in Table 10.1.

10.4 The Second Assessment Report (SAR) and Third Assessment Report (TAR)

The IPCC Second Assessment Report (SAR), dated 1995 but published in 1996, was a critical scientific element in the complex scientific and political process that ultimately resulted in the Kyoto Protocol in 1997. The drafting of both the FAR and the SAR attracted the participation of outstanding scientists from around the world, and it is certain that Bolin's own exceptional scientific reputation was a key reason for the willingness of other well-regarded scientific experts to contribute to the early work of the IPCC. Today, the IPCC enjoys a reputation for fairness, transparency, and inclusiveness in its work, and this too continues a tradition established under Bolin. The Third (TAR) and Fourth (AR4) Assessment Reports appeared in 2001 and 2007. As we shall see, they reached increasingly strong conclusions about the seriousness and anthropogenic origins of climate change.

Almost a decade before the establishment of the IPCC, one of the first uses of GCMs in climate research was to estimate the equilibrium sensitivity of the climate system to greenhouse-gas concentrations. A now-famous brief report on global warming (Charney *et al.*, 1979), on the basis of integrations of only two GCMs simulating the impact of doubled atmospheric CO_2 concentrations, concluded that a doubling of carbon dioxide might produce an equilibrium global-mean surface temperature increase of about 3 °C and a probable range for this important quantity of between 1.5 °C and 4.5 °C. It must be said that this range was initially estimated largely subjectively. Jule Charney, the senior author of this report, was a towering figure in twentieth-century theoretical meteorology. It is a tribute to the profound physical insight of Charney and his collaborators that this widely quoted range of equilibrium surface temperature sensitivity saw little change until at least as recently as the TAR.

These initial climate simulations, as well as many of those developed subsequently, including most of those presented in the FAR, were the results of atmospheric models coupled with simple "slab" ocean models. This jargon term signifies that the models used were coupled atmosphere–ocean models but were severely simplified, so that the model ocean could exchange heat and water with the model atmosphere, but ocean dynamics were not simulated at all. Effectively, the ocean being modeled was a shallow body of water that did not move.

The 1995 IPCC Second Assessment Report (SAR) was actually published a year later (IPCC, 1996). The SAR, referring once again to the Working Group I component alone, reinforced the main conclusions of the FAR with respect to the role of human activities in strengthening the greenhouse effect, as well as the observed climatic warming, the projected future changes, and the existence of uncertainties. It called attention to the role of aerosols (small particles in the atmosphere) of anthropogenic origin, which can sometimes partially compensate the greenhouse effect.

Space does not permit a detailed summary of the SAR. However, the most outstanding conclusion of the SAR was that

"*the balance of evidence suggests a discernible human effect on climate.*"

This sentence occurs in bold type in a 7-page Summary for Policymakers that appears in the front of the nearly 600-page report of Working Group I. This guarded and carefully worded conclusion of scientists had a substantial impact at that time. It is one of the factors which led in 1997 to the signing of the Kyoto Protocol.

This 11-word statement appears to have been composed so carefully that a reader seeing it for the first time might suspect it had been assembled by a committee. In fact, it was carefully negotiated in a formal plenary session, with each word requiring the unanimous consent of all participating governments. This fact illustrates part of the complex nature of the IPCC, which is both scientific and governmental. The SAR was written by scientists. The scientist authors also drafted the Summary for Policymakers. However, government representatives approved the Summary for Policymakers line by line, and effectively word by word. They could not choose a wording inconsistent with the science, but within that limit, they could express nuances of meaning. One result is that after such a Summary for Policymakers was finally approved, every participating government had effectively agreed to it, because any government had the option of vetoing any word to which it objected.

10.5 IPCC and the policy debate regarding human effects on climate change

In the view of many decision makers, the early questioning and doubt regarding the human effect on climate had, by the time the SAR appeared in 1996, given way to a greater level of scientific confidence, and this change of attitude played a key role in the creation and ultimately the ratification of the Kyoto Protocol. This famous protocol, negotiated in 1997, required the developed countries, during the period ending in the 2008–2012 time frame at the latest, to have reduced their greenhouse gas emissions by a weighted average of 5.2% with respect to 1990 levels. The developing countries, on the other hand, were not required by the Kyoto Protocol to take any action whatever. The emissions reduction goals also differed from one country to another. At Kyoto, for example, France, which already had a relatively low level of greenhouse gas emissions compared to many other developed countries, because of its heavy reliance on nuclear electrical power, was required simply to keep its emissions constant, while the United States accepted at Kyoto a reduction target of 7% in its emissions, again relative to a baseline of 1990 emissions levels.

For the Protocol to take effect, however, it was stipulated that it must be ratified by 55% of the member states of the United Nations, and that these ratifying states together must be responsible for at least 55% of global total emissions. The U.S. Senate, which is the governmental body in the United States responsible for ratifying such an agreement, announced its overwhelming opposition to the Protocol, citing the fact that developing countries were exempt from emissions reduction requirements. Despite the refusal of the United States to ratify the Protocol, the somewhat complex and arbitrary ratification criterion was eventually satisfied in November 2004 with the signature of the Russian Federation. Thus, three months after the Russian ratification, on 16 February 2005, the Kyoto Protocol officially entered into force.

The IPCC Third Assessment Report (TAR), published in 2001, concluded,

"*There is new and stronger evidence that most of the observed warming observed over the last 50 years is attributable to human activities*".

Once again, this statement appeared in a Summary for Policymakers and had been unanimously agreed to by all participating governments.

A very large change is obvious in comparing the summary statement of the SAR ("*the balance of evidence suggests a discernible human effect on climate*") with the above much more emphatic conclusion of the TAR. Clearly, much of the scientific uncertainty and hesitation at the time of the SAR in 1995 had become greatly reduced by the time the TAR appeared in 2001. As it turned out, one consequence of the TAR's much stronger conclusion was to largely relegate the scientific debate to the background of policy discussions. Thus, the TAR paved the way scientifically for the development of policy proposals to deal with the perceived threat of human-caused climate change.

The IPCC diagnosis set forth clearly in the TAR, that human activities were already modifying our climate, was not the only concern. The future outlook was also alarming. In particular, predictions of a large increase in global average surface temperature by 2100, arising from the inertia of the climate system, implied that the effects of warming would be felt long after the time when humanity succeeded in stabilizing the chemical composition of the atmosphere, and thus the human-caused augmentation of the natural greenhouse effect.

The TAR's major findings were examined in great detail by the scientific community, by scholarly organizations, by major corporations, and by the media. Greenhouse "skeptics" or "contrarians" (terms that came to be applied to those who dissented from the mainstream scientific view as expressed by the IPCC reports) questioned, among other things, the validity of paleoclimate reconstructions for the last millennium. Such reconstructions, based on proxy data such as tree rings in the pre-instrumental era, showed that temperatures in the most recent decades were significantly higher than those prevailing over the last thousand years or so. Skeptics also made much of an apparent disagreement between temperatures observed at the Earth's surface and those measured in the atmosphere far above the surface. These criticisms received attention, both in the scientific community and especially in the media, but it is fair to say today that the major conclusions of the IPCC in the TAR have been confirmed and repeatedly endorsed.

Among major endorsements, these IPCC TAR findings were supported in 2001 by the United States National Academy of Sciences in a report requested by the White House. In 2005, 11 national science academies of major countries, including the United States, Russia, China, and India, co-signed a declaration calling on world leaders and in particular the major developed ("G8") countries to "acknowledge that the threat of climate change is clear and increasing." Box 10.1 displays this declaration. Several prominent multinational corporations, including some that had previously been opposed, soon joined in acknowledging the scientific consensus. The ranks of the skeptics were gradually reduced as the awareness and acceptance of scientific findings as assessed by the IPCC slowly won over both the broader scientific community and the general public.

10.6 The IPCC process

Taken as a group, the four major IPCC assessment reports have come to be widely regarded as authoritative, both because of the quality of the scientists involved in writing them and because of the rigorous writing and reviewing process which the reports undergo. Each report is divided into chapters for which the initial drafting is entrusted to a team of researchers from several countries, the number of authors depending on the size assigned to the chapter by IPCC. These authors then also request input from scientists involved in the relevant disciplines. Currently, in addition to these extremely voluminous reports (nearly a thousand pages for the most recent Fourth Assessment Report of Working Group One), technical summaries of about fifty pages are also prepared, as well as a "Summary for Policymakers," which is shorter still and for which the goal is that it be written in a more accessible style than that of the full report. The entire assessment reports of all three Working Groups are brought together in a "Synthesis Report." These various documents are all extensively reviewed by both the scientific community (peer reviewers) and by representatives of government authorities. An iterative cycle of reviews and revisions of successive drafts takes place under strict guidelines stressing transparency and broad participation throughout the process.

A substantial specialized terminology has grown up to describe this process. The authors are called Lead Authors. Those Lead Authors who head the chapter writing

Box 10.1 **Joint statement issued in June, 2005 by the national academies of sciences of eleven countries, stating, "We recognize the international scientific consensus of the Intergovernmental Panel on Climate Change (IPCC)." Joint Science Academies' Statement: Global Response to Climate Change. Available at http://www.nationalacademies.org/onpi/06072005.pdf. Reprinted with permission from the National Academy of Sciences, courtesy of the National Academies Press, Washington, DC, USA.**

Joint science academies' statement: Global response to climate change

Climate change is real

There will always be uncertainty in understanding a system as complex as the world's climate. However there is now strong evidence that significant global warming is occurring[1]. The evidence comes from direct measurements of rising surface air temperatures and subsurface ocean temperatures and from phenomena such as increases in average global sea levels, retreating glaciers, and changes to many physical and biological systems. It is likely that most of the warming in recent decades can be attributed to human activities (IPCC 2001[2]). This warming has already led to changes in the Earth's climate.

The existence of greenhouse gases in the atmosphere is vital to life on Earth – in their absence average temperatures would be about 30 centigrade degrees lower than they are today. But human activities are now causing atmospheric concentrations of greenhouse gases – including carbon dioxide, methane, tropospheric ozone, and nitrous oxide – to rise well above pre-industrial levels. Carbon dioxide levels have increased from 280 ppm in 1750 to over 375 ppm today – higher than any previous levels that can be reliably measured (i.e. in the last 420,000 years). Increasing greenhouse gases are causing temperatures to rise; the Earth's surface warmed by approximately 0.6 centigrade degrees over the twentieth century. The Intergovernmental Panel on Climate Change (IPCC) projected that the average global surface temperatures will continue to increase to between 1.4 centigrade degrees and 5.8 centigrade degrees above 1990 levels, by 2100.

Reduce the causes of climate change

The scientific understanding of climate change is now sufficiently clear to justify nations taking prompt action. It is vital that all nations identify cost-effective steps that they can take now, to contribute to substantial and long-term reduction in net global greenhouse gas emissions.

Action taken now to reduce significantly the build-up of greenhouse gases in the atmosphere will lessen the magnitude and rate of climate change. As the United Nations Framework Convention on Climate Change (UNFCCC) recognises, a lack of full scientific certainty about some aspects of climate change is not a reason for delaying an immediate response that will, at a reasonable cost, prevent dangerous anthropogenic interference with the climate system.

As nations and economies develop over the next 25 years, world primary energy demand is estimated to increase by almost 60%. Fossil fuels, which are responsible for the majority of carbon dioxide emissions produced by human activities, provide valuable resources for many nations and are projected to provide 85% of this demand (IEA 2004[3].) Minimising the amount of this carbon dioxide reaching the atmosphere presents a huge challenge. There are many potentially cost-effective technological options that could contribute to stabilising greenhouse gas concentrations. These are at various stages of research and development. However barriers to their broad deployment still need to be overcome.

Carbon dioxide can remain in the atmosphere for many decades. Even with possible lowered emission rates we will be experiencing the impacts of climate change throughout the 21st century and beyond. Failure to implement significant reductions in net greenhouse gas emissions now, will make the job much harder in the future.

Prepare for the consequences of climate change

Major parts of the climate system respond slowly to changes in greenhouse gas concentrations. Even if greenhouse gas emissions were stabilised instantly at today's levels, the climate would still continue to change as it adapts to the increased emission of recent decades. Further changes in climate are therefore unavoidable. Nations must prepare for them.

The projected changes in climate will have both beneficial and adverse effects at the regional level, for example on water resources, agriculture, natural ecosystems and human health. The larger and faster the changes in climate, the more likely it is that adverse effects will dominate. Increasing temperatures are likely to increase the frequency and severity of weather events such as heat waves and heavy rainfall. Increasing temperatures could lead to large-scale effects such as melting of large ice sheets (with major impacts on low-lying regions throughout the world). The IPCC estimates that the combined effects of ice melting and sea water expansion from ocean warming are projected to cause the global mean sea-level to rise by between 0.1 and 0.9 metres between 1990 and 2100. In Bangladesh alone, a 0.5 metre sea-level rise would place about 6 million people at risk from flooding.

Developing nations that lack the infrastructure or resources to respond to the impacts of climate change will be particularly affected. It is clear that many of the world's poorest people are likely to suffer the most from climate change. Long-term global efforts to create a more healthy, prosperous and sustainable world may be severely hindered by changes in the climate.

The task of devising and implementing strategies to adapt to the consequences of climate change will require worldwide collaborative inputs from a wide range of experts, including physical and natural scientists, engineers, social scientists, medical scientists, those in the humanities, business leaders and economists.

Conclusion

We urge all nations, in the line with the UNFCCC principles[4], to take prompt action to reduce the causes of climate change, adapt to its impacts and ensure that the issue is included in all relevant national and international strategies. As national science academies, we commit to working with governments to help develop and implement the national and international response to the challenge of climate change.

G8 nations have been responsible for much of the past greenhouse gas emissions. As parties to the UNFCCC, G8 nations are committed to showing leadership in addressing climate change and assisting developing nations to meet the challenges of adaptation and mitigation.

We call on world leaders, including those meeting at the Gleneagles G8 Summit in July 2005, to:

- Acknowledge that the threat of climate change is clear and increasing.
- Launch an international study[5] to explore scientifically-informed targets for atmospheric greenhouse gas concentrations, and their associated emissions scenarios, that will enable nations to avoid impacts deemed unacceptable.
- Identify cost-effective steps that can be taken now to contribute to substantial and long-term reduction in net global greenhouse gas emissions. Recognise that delayed action will increase the risk of adverse environmental effects and will likely incur a greater cost.
- Work with developing nations to build a scientific and technological capacity best suited to their circumstances, enabling them to develop innovative solutions to mitigate and adapt to the adverse effects of climate change, while explicitly recognising their legitimate development rights.
- Show leadership in developing and deploying clean energy technologies and approaches to energy efficiency, and share this knowledge with all other nations.
- Mobilise the science and technology community to enhance research and development efforts, which can better inform climate change decisions.

Notes and references

1 This statement concentrates on climate change associated with global warming. We use the UNFCCC definition of climate change,which is 'a change of climate which is attributed directly or indirectly to human activity that alters the composition of the global atmosphere and which is in addition to natural climate variability observed over comparable time periods'.

2 IPCC (2001). Third Assessment Report. We recognise the international scientific consensus of the Intergovernmental Panel on Climate Change (IPCC).

3 IEA (2004). World Energy Outlook 4. Although long-term projections of future world energy demand and supply are highly uncertain, the World Energy Outlook produced by the International Energy Agency (IEA) is a useful source of information about possible future energy scenarios.

4 With special emphasis on the first principle of the UNFCCC, which states: 'The Parties should protect the climate system for the benefit of present and future generations of humankind, on the basis of equity and in accordance with their common but differentiated responsibilities and respective capabilities. Accordingly, the developed country Parties should take the lead in combating climate change and the adverse effects thereof'.

5 Recognising and building on the IPCC's ongoing work on emission scenarios.

Academia Brasiliera de Ciências
Brazil

Royal Society of Canada,
Canada

Chinese Academy of Sciences,
China

Academié des Sciences,
France

Deutsche Akademie der Naturforscher
Leopoldina, Germany

Indian National Science Academy,
India

Accademia dei Lincei,
Italy

Science Council of Japan,
Japan

Russian Academy of Sciences,
Russia

Royal Society,
United Kingdom

National Academy of Sciences,
United States of America

teams are called Coordinating Lead Authors. Scientists selected by the Lead Authors to provide brief specialized material are called Contributing Authors. In the most recent (2007) IPCC assessment report, AR4, three drafts were written and reviewed.

The task of editing and revising takes more than two years in order to provide the governments with a text that has the approval of the scientific community. As an example, more than 1,000 scientists participated in the preparation in the third report (the TAR) of WG I, of whom 122 were Lead Authors, selected from a large number of candidates nominated by their governments, 515 were Contributing Authors, 420 were Reviewers, and 21 were Review Editors. The review comments provided by various different sources (the scientific community, government authorities, and also non-governmental organizations) were taken into account by the Lead Authors, and the chapter texts were revised as a result. The last step before publication is that of acceptance and approval by the governments. The Summaries for Policymakers are discussed line by line and subsequently modified and approved during formal plenary meetings at which representatives of non-governmental organizations attend as observers. The contents of the extended summaries are also submitted for approval, and substantial attention is devoted to ensuring consistency between the different stages of the report. Finally, a complete set of the reports is published. The time elapsing between successive assessment reports to date has typically been five or six years. The IPCC assessment reports are available in print versions (see References), and the more recent ones plus many other IPCC reports are also available as free downloads from the IPCC website: http://www.ipcc.ch.

The IPCC is sometimes described as producing a consensus report, and it might reasonably be asked whether a consensus goal risks suppressing minority views. However, IPCC authors have been keenly aware of this problem and have made strenuous efforts to avoid such a risk. Where the science has been settled, and experts agree, and the research community has moved on to work on other issues, the IPCC states the resulting consensus. But where there are still unanswered questions, or where there are differing views among experts, and where further research thus remains to be done, the IPCC states that too. Thus, to take an example from the Fourth Assessment Report, if the question is whether Arctic sea ice has decreased in recent years, the IPCC report says that the answer is yes. If the question is, will parts of the Greenland ice sheet destabilize in the near future and cause a large increase in sea level, the IPCC says that topic is an area of active research, and the answer is not yet known.

Thus, by carefully assessing the peer-reviewed research literature, the Intergovernmental Panel on Climate Change (IPCC) has acquired a reputation for objectivity and trustworthiness and has effectively become the voice of the mainstream climate-change scientific community as the world seeks to understand the findings of climate science and their relevance to public policy.

Public understanding of how the IPCC works is important in gaining widespread public acceptance of the IPCC assessment conclusions, and yet the process may sometimes seem complex and mysterious. For example, one common misconception is that the Summary for Policymakers is written by and for the government delegations, and that scientifically substantive changes could be made by governments for political purposes before and during the plenary session. Thus, the question may arise as to whether the Summary for Policymakers accurately reflects the scientific findings of the full report itself. In fact, however, the Summary for Policymakers is written by a subgroup of the same scientists who also wrote the underlying chapters. The purpose of the subsequent plenary session is to make clarifications in the draft in order to more succinctly and accessibly communicate the science to the policymakers. The scientist Lead Authors are present throughout the plenary process to ensure scientific accuracy and consistency with the underlying report. The IPCC Fourth Assessment Report is currently the most comprehensive assessment of the scientific literature on climate change, and many careful steps were taken to ensure that it effectively and accurately communicates to policymakers and the public the state of scientific understanding.

10.7 An example of model evolution in the IPCC context: cloud processes

We turn now to an examination of selected scientific topics illustrating both the IPCC process and the way in which models are used in climate research. Our discussion is largely condensed from a historical overview of climate-change science published as Chapter One of the IPCC Fourth Assessment Report (Le Treut *et al.*, 2007), for which the author of this chapter was a Coordinating Lead Author and which contains many references. Clouds have long been an aspect of the climate system that is far from fully understood. On the key issue of uncertainties in model predictions of equilibrium climate change, here is the FAR's assessment:

> "One of the largest sources of uncertainty in the simulation of equilibrium climate change lies in the prediction of clouds. It has been shown that clouds can produce either a positive or negative feedback, depending on the model and parameterization of cloud used, giving an uncertainty of a factor of two or more in the equilibrium warming. Earlier schemes base cloud cover on relative humidity and prescribed radiative properties; later models use schemes which explicitly represent cloud water and allow cloud radiative properties to vary. The latter are more detailed but not necessarily more accurate as more parameters have to be specified."

Clouds and cloud feedbacks are still a topic of active research in which many unknowns remain. The modeling of cloud processes and feedbacks provides a striking example of the unequal pace of progress in climate science. On the one hand, cloud representation in atmospheric models has become much more comprehensive and physically based since the simple algorithms of early GCMs, in order to take into account key physical processes. On the other hand, clouds are still considered as a major source of uncertainty in the simulation of climate changes.

In the early 1980s, most GCMs were still using cloud amounts prescribed as functions of location and altitude, as well as prescribed cloud radiative properties, to compute atmospheric radiation. The cloud amounts were very often derived from a zonally averaged observationally derived climatology. Subsequent generations of models first used relative humidity and other simple predictors to diagnose cloudiness, thus providing some measure of increased realism for the models, A more explicit representation of clouds was progressively introduced into climate models, beginning in the late 1980s. Models first used simplified representations of cloud microphysics, but more recent generations of models generally incorporate a much more comprehensive and detailed representation of clouds, based on consistent physical principles. The results, when compared to extensive observations, generally yield a more realistic depiction of clouds.

In spite of this undeniable progress, uncertainty in cloud feedbacks remains today as one of the key factors explaining the spread in GCM simulations of future climate. This cannot be regarded as a surprise: the sensitivity of the Earth's climate to changing atmospheric greenhouse gas concentrations can indeed depend strongly on cloud feedbacks. This truism can be illustrated on the simplest theoretical grounds, using data which have been available for many years. Satellite measurements have provided meaningful estimates of the Earth radiation budget since the early seventies. Clouds, which cover about 60% of the Earth's surface, are responsible for up to two-thirds of the planetary albedo, or reflectivity for solar radiation, a quantity which is observed to be about 30%. An albedo change of only 1%, changing the Earth's albedo from 30% to either 31% or 29%, would then cause a change in the black-body radiative equilibrium temperature of about 1°C. This temperature change is a highly significant value, roughly equivalent to the direct radiative effect of a CO_2 doubling. Simultaneously, in addition to their influence on the albedo, clouds also contribute importantly to the planetary greenhouse effect. In addition, changes in cloud cover constitute only one of the many parameters that affect cloud radiative interactions: cloud optical thickness, cloud height, and cloud microphysical properties can also be modified by atmospheric temperature changes, through a bewildering variety of physical processes adding greatly to the complexity of feedbacks.

The importance of simulated cloud feedbacks had been highlighted by the detailed analysis of model results prior to the FAR. One especially insightful GCM experiment by Senior and Mitchell (1993) produced global average surface temperature changes (due to doubled carbon dioxide) ranging from 1.9 to 5.4 °C, depending solely on how cloud radiative properties were treated in the model. Thus, the most fundamental results of a complex state-of-the-art climate model could be drastically altered merely by substituting one reasonable cloud parameterization for another.

10.8 Coupled models

The Charney *et al.* (1979) estimate of a range of global-mean surface temperature increases due to doubled atmospheric carbon dioxide, from 1.5°C to 4.5 °C, remained part of conventional wisdom as recently as the TAR. The early GCM climate projections, as well as most of those presented in the FAR, were the results of atmospheric models coupled with simple slab ocean models, i.e. models omitting explicit ocean dynamics, as noted earlier.

The actual climate system, of course, involves many components other than the atmosphere; GCMs have gradually introduced these components, beginning with the ocean. Our discussion of the history of coupled ocean–atmosphere models is based on that of Le Treut *et al.* (2007) in AR4. The first attempts at coupling atmospheric and oceanic models were carried out during the late 1960s and early 1970s. Replacing GCMs containing slab ocean components by fully coupled atmosphere–ocean models (discussed in Chapter 7) may arguably have constituted one of the most significant advances in the history of climate modeling, although both the atmospheric and oceanic components of coupled models have also undergone highly significant improvements over time. However, the development of coupled atmosphere–ocean GCMs (AOGCMs) quickly led to significant modifications in the simulated patterns of climate change, particularly in oceanic regions. It also opened up the possibility of exploring transient climate scenarios, rather than the equilibrium integrations featured in the FAR, and it constituted a critical step toward the development of comprehensive Earth system models that include explicit representations of chemical and biogeochemical cycles.

Throughout their short history, coupled models have faced difficulties which have considerably impeded their development. One simple obstacle is that the initial state of the ocean is not precisely known. A second problem is that a surface flux imbalance (in one or more of the key quantities: energy, momentum, and fresh water) that is much smaller than observational accuracy can easily be large enough to cause a drifting of coupled GCM simulations into unrealistic states. A third issue is that there is no direct stabilizing feedback that can compensate for errors in the simulated salinity. The strong emphasis attached to having a realistic simulation of the observed climate state provided a rationale for introducing flux adjustments or flux corrections in early simulations. These were essentially empirical corrections that could not be justified on physical principles, and which consisted of arbitrary additions of surface fluxes of heat and salinity in order to prevent the slow drift of the simulated climate away from a realistic state. Models eventually evolved to the point where non-flux-corrected coupled simulations were possible, thus enabling simulations of climate change over the twenty-first century, for example. Some of the first such coupled model runs without flux corrections nevertheless displayed a persistent climate drift that still affected their simulations. Both the FAR and the SAR pointed out the apparent need for flux adjustments in coupled models as a problematic feature of climate modeling.

By the time of the TAR, however, the situation had evolved, and about half the coupled GCMs assessed in the TAR did not employ flux adjustments. That report noted that

> "some non-flux-adjusted models are now able to maintain stable climatologies of comparable quality to flux-adjusted models."

Since that time, evolution away from flux correction (or flux adjustment) has continued at many modeling centers, although some models continue to rely on it. Chapter 7 describes the effect of removing flux adjustments on coupled models between TAR and AR4. The design of coupled model simulations is also strongly linked with the methods chosen for model initialization. In flux-adjusted models, the initial ocean state is typically obtained from preliminary and long simulations, so as to bring the ocean model into equilibrium. Non-flux-adjusted models often employ a procedure based on ocean observations, although some spin-up phase may be required even then.

These considerable advances in model design have not diminished the existence of a range of model results. This is not a surprise, however, because it is known that climate predictions are intrinsically affected by uncertainty. It is often useful to distinguish between two fundamental kinds of idealized prediction problems. The first kind is defined as the prediction of the actual time-dependent properties of the climate system in response to a given initial state. Predictions of the first kind are initial-value problems. Because of the nonlinearity and instability of the governing equations, such systems generally are not predictable indefinitely far into the future.

Predictions of the second kind deal with the determination of the response of the climate system to changes in the external forcings. These predictions are not concerned directly with the chronological evolution of the climate state, but rather with the long-term average of the statistical properties of climate. Originally it was thought that predictions of the second kind do not at all depend on initial conditions. Instead, they are intended to determine how the statistical properties of the climate system change as some external forcing parameter, such as CO_2 content, is altered. Estimates of future climate scenarios as a function of the concentration of atmospheric greenhouse gases are typical examples of predictions of the second kind. However, ensemble simulations show that the projections tend to form clusters around a number of attractors as a function of their initial state.

Uncertainties in climate predictions of the second kind arise mainly from model uncertainties and errors. To assess and disentangle the factors affecting the accuracy of the simulations, the scientific community has organized a series of systematic comparisons of the different existing models. A number of these ambitious and comprehensive "model intercomparison projects" (MIPs) were set up in the 1990s under the auspices of WCRP to undertake controlled conditions for model evaluation.

One of the first was AMIP (where "A" denotes atmospheric), which studied atmospheric GCMs. The development of coupled models has induced the development of CMIP (where "C" denotes coupled), which studied coupled atmosphere–ocean global climate models (AOGCMs), and their response to idealized forcings, such as 1% annual increase in the atmospheric CO_2 concentration. It proved important in carrying out the various MIPs to standardize the model forcing parameters and the model output so that file formats, variable names, units, etc., are easily recognized by data users. The fact that the model results were stored separately and independently of the modeling centers, and that the analysis of the model output was performed mainly by research groups independent of the modelers, has added to confidence in the results.

AMIP and CMIP opened a new era for climate modeling, setting standards of quality control, providing organizational continuity, and ensuring that results are generally reproducible. Results from AMIP have provided a number of insights into climate model behavior, and quantified improved agreement between simulated and observed atmospheric properties as new versions of models are developed. In general, results of the MIPs suggest that the most problematic areas of coupled model simulations involve cloud–radiation processes, the cryosphere, the deep ocean, and ocean–atmosphere interactions.

Comparing different models is not sufficient, however. Using multiple simulations from a single model (the so-called Monte Carlo, or ensemble, approach) has proved to be a necessary and complementary approach to assess the stochastic nature of the climate system. Computational constraints limited early ensembles to a relatively small number of samples (typically fewer than ten). These ensemble simulations clearly indicated that even with a single model a large spread in the climate projections can be obtained.

Intercomparison of existing models and ensemble model studies, i.e. those involving many integrations of the same model, are still undergoing rapid development. Carrying out ensemble AOGCM integrations was impractical until recent advances in computer power occurred, as these systematic comprehensive climate model studies are exceptionally demanding on computer resources.

10.9 The Fourth Assessment Report (AR4) and the future of the IPCC

The most recent IPCC assessment report was published in 2007, so it is very recent history indeed. Although it is far too early to pass final judgment on it, the WGI portion of AR4, like its predecessor in the IPCC Third Assessment Report (TAR), has rapidly become recognized as an authoritative summary of the state of the science. As discussed above (Box 10.1), the TAR had been evaluated and endorsed by the National Academy of Sciences in the United States and by its counterparts in many other countries, as well as by leading scientific professional societies. In its turn, AR4 has also already been endorsed by many scientific societies. For example, in October 2009, eighteen such organizations in the United States cited AR4 in an open letter to the U.S. Senate (Box 10.2).

AR4 contains two summary statements, and it is instructive to compare them with the summary statements of the SAR and TAR:

SAR: *"The balance of evidence suggests a discernible human influence on global climate."*

(IPCC, 1996)

TAR: *"There is new and stronger evidence that most of the warming observed over the last 50 years is attributable to human activities."*

(IPCC, 2001)

AR4: *"Warming of the climate system is unequivocal, as is now evident from observations of increases in global average air and ocean temperatures, widespread melting of snow and ice, and rising global average sea level."*

(IPCC, 2007)

AR4: *"Most of the observed increase in globally averaged temperatures since the mid-20th century is very likely due to the observed increase in anthropogenic greenhouse gas concentrations."*

(IPCC, 2007)

Here '"very likely" is calibrated IPCC uncertainty terminology denoting a confidence estimate of 90% or more. Clearly, these summary statements from three successive IPCC assessment reports reflect the evolution of climate-change science itself. In little more than a decade, a tentative attribution couched in cautious language has evolved into a robust scientific finding. In its two statements, AR4 points out that there are many observations confirming that the climate is warming, and it expresses the sense of strong scientific confidence in attributing most of the warming in recent decades to a human-caused augmentation of the natural greenhouse effect.

GCMs played a major role in reaching these findings. Recently, for example, several climate modeling groups have attempted to simulate the evolution of climate over the twentieth century with different GCMs, using various combinations of natural and anthropogenic forcings as inputs to the models. The simplest measure of climate for such numerical experiments is the global average surface temperature. A clear conclusion of this line of research is that models driven with only natural forcings (solar variability and volcanic aerosols) do not accurately simulate the observed changes in climate over the twentieth century, nor do models driven with

Box 10.2 **Joint statement addressed to the U.S. Senate, issued in October, 2009 by eighteen scientific societies in the United States, stating the "consensus scientific view" and reaffirming conclusions that "reflect the scientific consensus represented by, for example, the Intergovernmental Panel on Climate Change and U.S. Global Change Research Program. Many scientific societies have endorsed these findings in their own statements. . ." Available at http://www.aaas.org/news/releases/2009/1021climate_letter.shtml. Reprinted with permission from the American Association for the Advancement of Science, Washington, DC, USA.**

October 21, 2009

American Association for the Advancement of Science
American Chemical Society
American Geophysical Union
American Institute of Biological Sciences
American Meteorological Society
American Society of Agronomy
American Society of Plant Biologists
American Statistical Association
Association of Ecosystem Research Centers
Botanical Society of America
Crop Science Society of America
Ecological Society of America
Natural Science Collections Alliance
Organization of Biological Field Stations
Society for Industrial and Applied Mathematics
Society of Systematic Biologists
Soil Science Society of America
University Corporation for Atmospheric Research

Dear Senator:

As you consider climate change legislation, we, as leaders of scientific organizations, write to state the consensus scientific view.

Observations throughout the world make it clear that climate change is occurring, and rigorous scientific research demonstrates that the greenhouse gases emitted by human activities are the primary driver. These conclusions are based on multiple independent lines of evidence, and contrary assertions are inconsistent with an objective assessment of the vast body of peer-reviewed science. Moreover, there is strong evidence that ongoing climate change will have broad impacts on society, including the global economy and on the environment. For the United States, climate change impacts include sea level rise for coastal states, greater threats of extreme weather events, and increased risk of regional water scarcity, urban heat waves, western wildfires, and the disturbance of biological systems throughout the country. The severity of climate change impacts is expected to increase substantially in the coming decades.[1]

If we are to avoid the most severe impacts of climate change, emissions of greenhouse gases must be dramatically reduced. In addition, adaptation will be necessary to address those impacts that are already unavoidable. Adaptation efforts include improved infrastructure design, more sustainable management of water and other natural resources, modified agricultural practices, and improved emergency responses to storms, floods, fires and heat waves.

We in the scientific community offer our assistance to inform your deliberations as you seek to address the impacts of climate change.

---

[1] The conclusions in this paragraph reflect the scientific consensus represented by, for example, the Intergovernmental Panel on Climate Change and U.S. Global Change Research Program. Many scientific societies have endorsed these findings in their own statements, including the American Association for the Advancement of Science, American Chemical Society, American Geophysical Union, American Meteorological Society, and American Statistical Association.

American Association for the Advancement of Science
1200 NewYork Avenue, NW, Washington, DC 20005 USA
Tel: 202 326 6600 Fax: 202 289 4950 www.aaas.org

Alan I. Leshner
Executive Director
American Association for the Advancement of Science

Thomas Lane
President
American Chemical Society

Timothy L. Grove
President
American Geophysical Union

May R. Berenbaum
President
American Institute of Biological Sciences

Keith Seitter
Executive Director
American Meteorological Society

Mark Alley
President
American Society of Agronomy

Tuan-hua David Ho
President
American Society of Plant Biologists

Sally C Morton
President
American Statistical Association

Lucinda Johnson
President
Association of Ecosystem Research Centers

Kent E. Holsinger
President
Botanical Society of America

Kenneth Quesenberry President Crop Science Society of America	Mary Power President Ecological Society of America
William Y. Brown President Natural Science Collections Alliance	Brian D. Kloeppel President Organization of Biological Field Stations
Douglas N. Arnold President Society for Industrial and Applied Mathematics	John Huelsenbeck President Society of Systematic Biologists
Paul Bertsch President Soil Science Society of America	Richard A. Anthes President University Corporation for Atmospheric Research

only anthropogenic forcings (both human-caused greenhouse gases and aerosols). However, when the natural and the anthropogenic forcings are used together as inputs to the models, the most realistic twentieth century climate simulations are obtained (Figure 10.3). This figure stands as a key summary result illustrating the power of a branch of modern climate-change science called detection and attribution. This research is devoted to determining whether any given observed phenomena represent a significant departure from natural variability, and if so, to identifying the cause of such a departure.

General circulation models have also played a key role in another prominent feature of the IPCC reports, especially AR4, namely projecting future climate change. Figure 10.4 illustrates such projections. Here, IPCC provided modeling groups with a set of hypothetical scenarios distinguished by different greenhouse

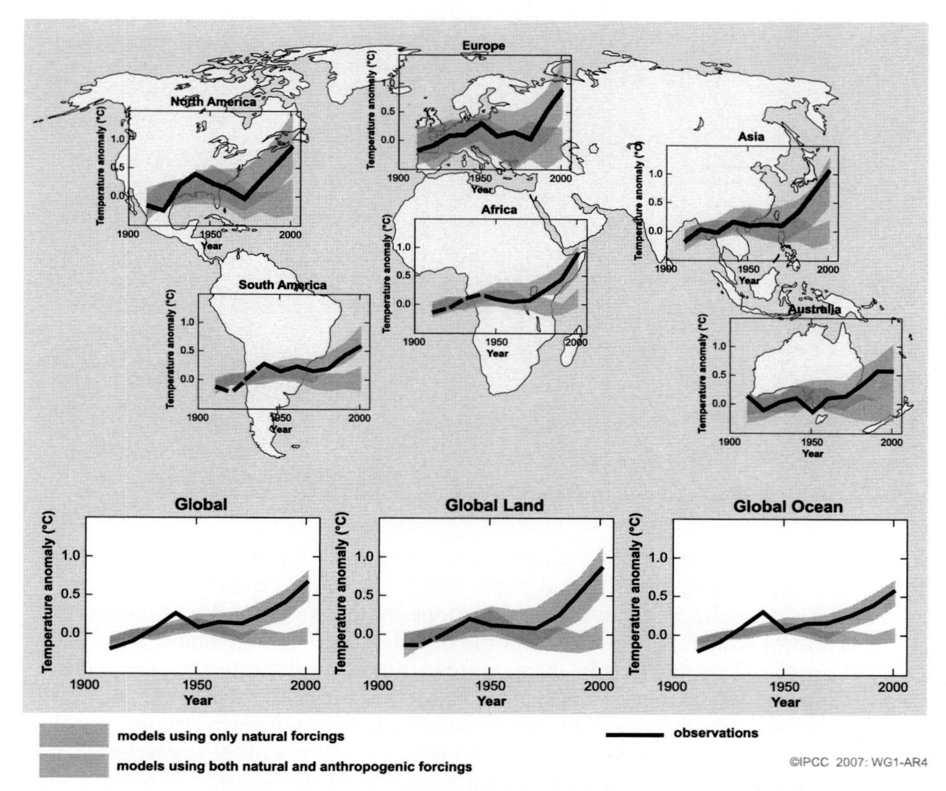

Figure 10.3 Temperature changes relative to the corresponding average for 1901–1950 (°C) from decade to decade from 1906 to 2005 over the Earth's continents, as well as the entire globe, global land area, and the global ocean (lower graphs). The black line indicates observed temperature change, while the shaded bands show the combined range covered by 90% of recent model simulations. The upper shaded band indicates simulations that include natural and human factors, while the lower shaded band indicates simulations that include only natural factors. Dashed black lines indicate decades and continental regions for which there are substantially fewer observations. This is FAQ 9.2, Figure 1 in Hegerl *et al.* (2007). See also color plate.

gas emissions. Using these inputs, GCM research groups then produce projections of future climates.

Figures 10.3 and 10.4 exemplify the ability of GCMs and other accomplishments of modern climate-change science to provide valuable scientific information to policymakers that is policy-relevant without being policy-prescriptive. The IPCC has clearly been a success at its mandated mission. However, many climate scientists have concerns about its future. One is that the level of effort required of participating scientists is a heavy burden on the research community. Another is

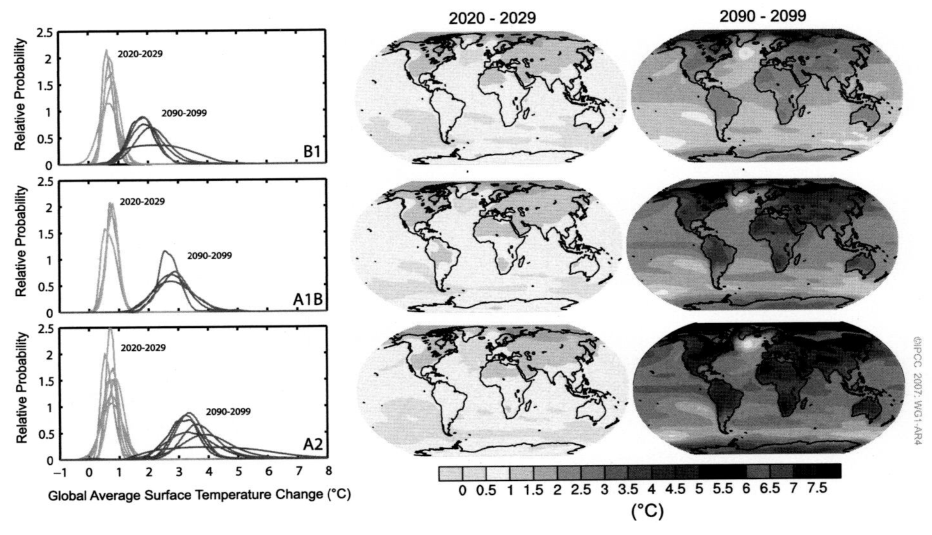

Figure 10.4 Projected surface temperature changes for the early and late twenty-first century relative to the period 1980–1999. The central and right panels show the AOGCM multi-model average projections for the B1 (top), A1B (middle) and A2 (bottom) scenarios averaged over the decades 2020–2029 (centre) and 2090–2099 (right). The left panels show corresponding uncertainties as the relative probabilities of estimated global average warming from several different AOGCM and Earth System Model of Intermediate Complexity studies for the same periods. Some studies present results only for a subset of the scenarios, or for various model versions. Therefore the difference in the number of curves shown in the left-hand panels is due only to differences in the availability of results. This is adapted from Figures 10.8 and 10.28 of Meehl *et al.* (2007) and is Figure TS.28 from the Technical Summary of IPCC (2007). See also color plate.

that IPCC, which in 1990 simply assessed the peer-reviewed literature, now clearly influences the research agenda. For example, even before AR4 was completed, many major GCM research groups world-wide had begun planning and seeking resources, such as supercomputer time, in order to carry out the GCM scenario runs needed for the next IPCC assessment report, AR5, which is scheduled to appear in 2013. A third concern is that IPCC is not especially nimble and cannot respond quickly to new scientific developments. Its painstaking and thorough process of reviewing and revising several drafts of its lengthy assessment reports is time consuming. For example, AR4 generated more than 30,000 review comments on preliminary drafts, and the authors responded in writing to each of them. A fourth is that IPCC is written by professional scientists, not professional science communicators, so that its reports do not make easy reading for non-scientists, and its procedures may sometimes seem mysterious to the wider world.

Today, IPCC has become a unique resource, surely the leading source for authoritative scientific information about climate change. The wide participation of the scientific community, the scientific accuracy, and the absence of any policy prescription in the assessment reports are the characteristics that render these reports so powerful. The IPCC serves a unique role in informing policymakers, as well as others such as industry, media, and the broad public.

This chapter began with a quotation from C. E. P. Brooks in 1951, illustrating the prevailing scientific opinion in the mid twentieth century that human-made carbon dioxide emissions were not important causes of climate change. In little more than half a century, rapid scientific progress has now reversed that view. Global climate models have appeared and have proven to be essential tools in understanding the climate system and climate change. The IPCC has also come into being, and its reports have become definitive summaries of scientific findings.

Today, a prevailing concern among many climate scientists is that society will not take these findings sufficiently to heart and that policymakers will be unable to muster the political will to act successfully, in the words of the UNFCCC,

"*to stabilize the concentrations of greenhouse gases in the atmosphere at a level that will prevent dangerous anthropogenic interference with the climate system.*"

For these concerned scientists, the words of the Nobel laureate F. Sherwood Rowland in 1984, quoted by Somerville and Jouzel (2008), are a somber warning:

After all, what's the use of having developed a science well enough to make predictions, if in the end all we're willing to do is stand around and wait for them to come true!

References

Bolin, B. (2007). *A History of the Science and Politics of Climate Change: The Role of the Intergovernmental Panel on Climate Change*, United Kingdom and New York, NY, USA: Cambridge University Press.

Brooks, C. E. P. (1951). Geological and Historical Aspects of Climate Change. In *Compendium of Meteorology*, ed. Malone, T. F., Boston, MA: American Meteorological Society, pp. 1004–1018.

Callendar, G. S. (1939). The composition of the atmosphere through the ages. *Meteorology Magazine*, 74, 33–39.

Charney, J. G., *et al.* (1979). *Carbon Dioxide and Climate: A Scientific Assessment*. Washington, DC: National Academy of Sciences.

Hegerl, G. C., Zwiers, F. W., Braconnot, P. *et al.* (2007). Understanding and Attributing Climate Change. In *Climate Change 2007: The Physical Science Basis. Contribution of Working Group I to the Fourth Assessment Report of the Intergovernmental Panel on Climate Change*, eds. Solomon, S., Qin, D., Manning, M., Chen, Z., Marquis, M., Averyt, K. B., Tignor, M., and Miller, H. L. Cambridge, United Kingdom and New York, NY, USA: Cambridge University Press.

IPCC (1990). Climate Change: The IPCC Scientific Assessment. In *The IPCC First Assessment Report*, eds. Houghton, J. T., Jenkins, G. J., and Ephraums, J. J.. Cambridge, United Kingdom and New York, NY, USA: Cambridge University Press.

IPCC (1996). Climate Change 1995: The Science of Climate Change. In *The IPCC Second Assessment Report,or SAR*, eds. Houghton, J. T., *et al.* Cambridge, United Kingdom and New York, NY, USA: Cambridge University Press.

IPCC (2001). Climate Change 2001: The Scientific Basis. Contribution of Working Group I to the Third Assessment Report of the Intergovernmental Panel on Climate Change. In *The IPCC Third Assessment Report*, eds. Houghton, J. T., *et al.* Cambridge, United Kingdom and New York, NY, USA: Cambridge University Press.

IPCC (2007). Climate Change 2007: The Physical Science Basis. Contribution of Working Group I to the Fourth Assessment Report of the Intergovernmental Panel on Climate Change. In *The IPCC Fourth Assessment Report*, eds. Solomon, S., Qin, D., Manning, M., Chen, Z., Marquis, M., Averyt, K. B., Tignor, M. and Miller, H. L. Cambridge, United Kingdom and New York, NY, USA: Cambridge University Press.

Le Treut, H., Somerville, R., Cubasch, U. *et al.* (2007). Historical Overview of Climate Change. In *Climate Change 2007: The Physical Science Basis. Contribution of Working Group I to the Fourth Assessment Report of the Intergovernmental Panel on Climate Change*, eds. Solomon, S., Qin, D., Manning, M., Chen, Z., Marquis, M., Averyt, K. B., Tignor, M., and Miller, H. L. Cambridge, United Kingdom and New York, NY, USA: Cambridge University Press.

Meehl, G. A., Stocker, T. F., Collins, W. D. *et al.* (2007). Global Climate Projections. In *Climate Change 2007: The Physical Science Basis. Contribution of Working Group I to the Fourth Assessment Report of the Intergovernmental Panel on Climate Change*, eds. Solomon, S., Qin, D., Manning, M., Chen, Z., Marquis, M., Averyt, K. B., Tignor, M., and Miller, H. L. Cambridge, United Kingdom and New York, NY, USA: Cambridge University Press.

Phillips, N. A. (1956). The general circulation of the atmosphere: a numerical experiment. *Quarterly Journal of the Royal Meteorological Society*, 82, 123–164.

Senior, C. A. and Mitchell, J. F. B. (1993). Carbon dioxide and climate: the impact of cloud parameterization. *Journal of Climate*, 6, 393–418.

Somerville, R. C. J. (2008). Bert Bolin 1925–2007. *Bulletin of the American Meteorological Society*, 89, 1046–1048.

Somerville, R. C. J. and Jouzel, J. (2008). The Global Consensus and the Intergovernmental Panel on Climate Change. In *Facing Climate Change Together*, eds. Gautier, C. and Fellous, J. -L., Cambridge, United Kingdom and New York, NY, USA: Cambridge University Press, (the 1984 quotation from F. Sherwood Rowland originally appears in an article by Paul Brodeur, *The New Yorker*, June 9, 1986, p. 81).

Index

Printed in the United States
By Bookmasters